Low-dimensional and Symplectic Topology

Proceedings of Symposia in
PURE MATHEMATICS

Volume 82

Low-dimensional and Symplectic Topology

Michael Usher
Editor

American Mathematical Society
Providence, Rhode Island

2009 GEORGIA INTERNATIONAL TOPOLOGY CONFERENCE

with support from the National Science Foundation,
grant DMS-0852505

2010 *Mathematics Subject Classification.* Primary 57–06, 20F36, 53D12, 53D35, 55P48, 57M25, 57M27, 57R17, 57R91.

Library of Congress Cataloging-in-Publication Data

Georgia International Topology Conference (2009 : University of Georgia)
Low-dimensional and symplectic topology : Georgia International Topology Conference, May 18–29, 2009, University of Georgia, Athens, Georgia / Michael Usher, editor.
p. cm. — (Proceedings of symposia in pure mathematics ; v. 82)
Includes bibliographical references.
ISBN 978-0-8218-5235-4 (alk. paper)
1. Low-dimensional topology—Congresses. 2. Manifolds (Mathematics)—Congresses. 3. Simplexes (Mathematics)—Congresses. I. Usher, Michael, 1978– II. Title.

QA612.14.G46 2009
514′.22—dc23

2011025453

♾ The paper used in this book is acid-free and falls within the guidelines
established to ensure permanence and durability.
Visit the AMS home page at `http://www.ams.org/`

10 9 8 7 6 5 4 3 2 1 16 15 14 13 12 11

To Clint McCrory
on the occasion of his retirement.

Contents

Preface

The 2009 Georgia International Topology Conference was held at the University of Georgia in Athens, Georgia, from May 18-29, 2009. This event, attracting 222 participants from around the world, continued a longstanding tradition of major international topology conferences held in Athens every eight years since 1961.

The two main goals of the conference were to give wide exposure to new and important results, and to encourage interaction among graduate students and researchers in different stages of their careers. The conference featured 39 plenary talks aimed at a general audience of topologists by distinguished speakers from around the world, touching on breakthroughs in such topics as hyperbolic geometry, geometric group theory, symplectic and contact topology, Heegaard Floer theory, and knot theory, among others. There was also a session of informal presentations by graduate students during the weekend, as well as six evening introductory lectures by leading experts, aimed at graduate students, on a variety of topics in low-dimensional, contact, and symplectic topology. Slides for most of the talks remain available on the internet, at http://math.uga.edu/~topology/2009/schedule.htm.

A problem session was also held near the end of the conference, and a report on it is included in these proceedings. The other articles in the proceedings represent an array of survey and original research articles related to the topics discussed in the conference. I am grateful to both the authors of these articles and to the referees for the efforts that they have contributed toward the publication of the volume.

The conference was organized by Michael Ching, William Kazez, Gordana Matić, Clint McCrory and myself. The speakers were selected with the assistance of our Scientific Advisory Committee, consisting of Simon Donaldson, Yakov Eliashberg, David Gabai, Rob Kirby, Bruce Kleiner, Dusa McDuff, Dennis Sullivan, Cliff Taubes and Karen Vogtmann. The conference also benefited greatly from logistical support provided by Julie McEver, Connie Poore, Gail Suggs, Laura Ackerley, and Christy McDonald. Finally, the organizers are very grateful to the National Science Foundation (grant DMS-0852505) and to the University of Georgia for support which made the conference possible.

M.U.
June 2011

Proceedings of Symposia in Pure Mathematics
Volume **82**, 2011

Algebra, Topology and Algebraic Topology of 3D Ideal Fluids

Dennis Sullivan

Abstract. There is a remarkable and canonical problem in 3D geometry and topology: To understand existing models of 3D fluid motion or to create new ones that may be useful. We discuss from an algebraic viewpoint the PDE called Euler's equation for incompressible frictionless fluid motion. In part I we define a "finite dimensional 3D fluid algebra," write its Euler equation and derive properties related to energy, helicity, transport of vorticity and linking that characterize this equation. This is directly motivated by the infinite dimensional fluid algebra associated to a closed riemannian three manifold whose Euler equation as defined above is the Euler PDE of fluid motion. The classical infinite dimensional fluid algebra satisfies an additional identity related to the Jacobi identity for the lie bracket of vector fields. In part II we discuss informally how this Jacobi identity can be reestablished in finite dimensional approximations as a Lie infinity algebra. The main point of a developed version of this theory would be a coherence between various levels of approximation. It is hoped that a better understanding of the meaning of the Euler equation in terms of such infinity structures would yield algorithms of computation that work well for conceptual reasons.

1. Algebra and Topology of Ideal Fluids

A finite dimensional $3D$-fluid algebra is a finite dimensional vector space V provided with three structures:

1) an alternating trilinear form $\{\ ,\ ,\ \}$ on V, called the triple intersection form.
2) a symmetric nondegenerate bilinear form $\langle\ ,\ \rangle$ on V, called the vorticity linking form.
3) a positive definite inner product $(\ ,\ )$ on V, called the metric.

If M is a $3D$ closed oriented Riemannian manifold there is a classical example of a fluid algebra which is infinite dimensional and which is constructed inside the differential forms:

- V consists of the coexact one forms, the image of two forms by the operator $*d*$ where $*$ is the Hodge star operator of the metric. Under the correspondence between one forms and vector fields given by the metric,

2010 *Mathematics Subject Classification.* Primary 35Q35, 55P48.
Key words and phrases. Euler equations, finite-dimensional models, infinity structures.

elements in V correspond to volume preserving vector fields which have flux or net flow zero across any closed hypersurface.

- The vorticity linking form on V is defined by setting

$$\langle a, b\rangle = \int_M a \wedge db.$$

This is equal to $\int_M (da) \wedge b$, depends only on da and db, and may be construed as a linking number of the two one-dimensional transversally measured foliations defined by the kernels of da and db and transversally measured by the two forms da and db (see Arnold-Khesin[**AK**] and Sullivan[**S76**]). Here da and db determine the vorticities of the vector fields corresponding to a and to b. da and db may be approximated, in the sense of integrating against a smooth test form, by weighted sums of closed curves which bound weighted sums of surfaces approximating a and b. Recall that linking numbers are defined by intersecting one set of curves with surfaces bounding a second set of curves. Thus the integral approximately computes the total linking of the weighted family of curves approximating da with those approximating db.

One can see from the Hodge decomposition that the vorticity linking form is nondegenerate on coexact one-forms.

- The triple intersection form $\{a, b, c\}$ is the integral of a wedge b wedge c over M, which may be construed as a triple intersection of the surfaces with boundary approximating a, b, and c.

The Euler evolution in V describing ideal incompressible frictionless fluid motion in M is an ODE whose solutions may be described in words as follows: an isotopy from the identity is a fluid motion iff the path in V corresponding to the instantaneous velocity of the motion satisfies: the vorticity at time t is the two form transported by the motion from the vorticity two form at time zero.

We can write out this Euler ODE just using the elements of a fluid algebra.

Namely, the right hand side of the evolution ODE for $X(t)$ is described by its inner product with any vector Z as follows: $\left(\frac{dX}{dt}, Z\right) = \{X, DX, Z\}$ where D is the operator on V defined by $(DX, Y) = \langle X, Y\rangle$. D is called the curl operator on the elements of a fluid algebra. Note that $(DX, Y) = (X, DY)$ since $\langle\ ,\ \rangle$ is symmetric by assumption.

Proposition 1.1. *The "energy" $= (X, X)$ and the "helicity" $= (X, DX)$ are each preserved by the evolution in a fluid algebra.*

Proof. 1) $\frac{d}{dt}(\text{energy}) = 2\left(\frac{dX}{dt}, X\right) = 2\{X, DX, X\} = 0$ by the alternating property of $\{\ ,\ ,\ \}$.

2) $\frac{d}{dt}(\text{helicity}) = 2\left(\frac{dX}{dt}, DX\right) = 2\{X, DX, DX\} = 0$ again by the alternating property of $\{\ ,\ ,\ \}$.

□

Corollary 1.2. *For a finite dimensional fluid algebra, there is no finite time blowup and the flow stays on the intersection of the energy sphere with the helicity level set.*

Notice in this classical three manifold case the alternating trilinear form on all one-forms and the bilinear form on all one-forms are purely topological and don't depend on the metric. The metric is needed to define the subspace of coexact forms inside all one-forms to which one restricts the trilinear form, the bilinear form and the inner product to obtain the infinite dimensional fluid algebra.

Also notice that even though the Proposition follows for this classical Euler case by the same short proof, the Corollary of course does not follow because of the non compactness of the infinite dimensional sphere. The truth or falsity of infinite time existence of smooth solutions of the Euler evolution for arbitrary smooth initial conditions is a celebrated open problem.

Here is another family of examples:

EXAMPLE 1.3. Given a finite dimensional Lie algebra $[\, ,\,]$ with a nondegenerate invariant symmetric pairing $\langle\, ,\, \rangle$ define a fluid algebra by:

- for the linking form take $\langle\, ,\, \rangle$.
- for the intersection form take $\{X, Y, Z\} = \langle [X, Y], Z\rangle$ and
- for the metric take any positive definite inner product.

Euler flows for these examples were described differently by Arnold and Khesin [**AK**]. The interested reader may verify that their evolution equation (see [**AK**, Theorem I.4.9]) is related to ours by a linear coordinate change.

Notice:

1) For a general fluid algebra one may reverse the above formula to define a bracket by the formula $\langle [X, Y], Z\rangle = \{X, Y, Z\}$. If the Jacobi identity were satisfied for this bracket the fluid algebra would arise from the Lie algebra example.
2) In the infinite dimensional example related to $3D$ manifolds and the Euler ODE this Jacobi identity is satisfied and corresponds to the Lie algebra of volume preserving flux zero vector fields.
3) In order to have models which satisfy the invariance of energy, helicity, and more generally circulation (see below) and the transport property of vorticity (see below) this Jacobi relation is not required. The significance of the Jacobi relation needs exploring (see Part II). One interpretation of Jacobi is the following: if we extend the transport (defined below) to higher tensors by the derivation property then Jacobi is equivalent to the statement that transport fixes the tensor defining transport itself.

DEFINITION 1.4. (Velocity and Vorticity) We think of elements in V as analogous to velocity and vorticity fields of the ideal fluid. The vorticity of a velocity is obtained by applying the curl operator D to the velocity.

PROPOSITION 1.5. *If $X(t)$ satisfies the Euler equation of the fluid algebra*

$$\left(\frac{dX(t)}{dt}, Y\right) = \{X, DX, Y\},$$

then the vorticity $DX(t) = Y(t)$ satisfies

$$\left(\frac{dY(t)}{dt}, Z\right) = \{D'Y, Y, DZ\}$$

where D' is the inverse of D.

PROOF. $\frac{d}{dt}DX(t) = D\left(\frac{d}{dt}X(t)\right)$. So

$$\left(\frac{d}{dt}Y(t), Z\right) = \left(D\left(\frac{d}{dt}X(t)\right), Z\right) = \left(\frac{d}{dt}X(t), DZ\right)$$
$$= \{X, DX, DZ\} = \{X, Y, DZ\} = \{D'Y, Y, DZ\}.$$

□

DEFINITION 1.6. (Transport) Define the infinitesimal transport of Z by X, $T(X,Z)$, by the condition that its inner product with arbitrary W is given by $(T(X,Z),W) = \{X, Z, DW\}$.

Note $T(X,Z) = -T(Z,X)$, but Jacobi is not assured for $T(X,Y)$. The transport is meant to model for volume preserving vector fields the Lie derivative action, which by the way is canonically equivalent to the the Lie derivative action on closed two forms.

COROLLARY 1.7. *If $X(t)$ satisfies the Euler equation of a fluid algebra then the vorticity $Y(t) = curlX(t) = DX(t)$ satisfies $\frac{dY(t)}{dt} = T(X(t), Y(t))$, namely the vorticity is transported from one time to another by the motion.*

PROOF. Since $Y = DX$, by the proposition and the definition of transport the inner product of each with Z equals $\{X, DX, DZ\}$. □

Note: This definition of transport may be derived in the classical case from the expression of the Lie bracket of two volume preserving vector fields $[V, V']$ in terms of the corresponding one-forms by the formula $[V, V'] = *d*(V \wedge V')$ where $*d*$ is the adjoint of d.

COROLLARY 1.8. *(Invariance of circulation) If the vorticity $DZ(t)$ of a time dependent field $Z(t)$ satisfies $\frac{d}{dt}(DZ(t)) = T(X(t), DZ(t))$, then the linking number of $DZ(t)$ and $DX(t)$ stays constant.*

Here, consistently with the situation discussed in the classical example, the linking number of $DZ(t)$ and $DX(t)$ refers to the value $\langle X, Z\rangle$ of the vorticity linking form on the fields X and Z.

PROOF. The linking number of DZ and DX is equal to (X, DZ). So

$$\frac{d}{dt}(X, DZ) = \left(\frac{d}{dt}X(t), DZ(t)\right) + \left(X, \frac{d}{dt}DZ(t)\right)$$
$$= \{X, DX, DZ\} + (T(X, DZ), X)$$
$$= \{X, DX, DZ\} + \{X, DZ, DX\}$$

which is zero by the alternating property. □

CONCLUSION 1.9. *The Euler ODE for velocity associated to a fluid algebra is the only ODE whose evolution keeps constant the linking of the transported vorticity of a general field with the evolving vorticity of the velocity solution of the ODE (as expressed in Corollary 1.8).*

PROOF. The calculation of the proof of Corollary 1.8 shows cancellation takes place iff the evolution of the velocity $X(t)$ satisfies $\left(\frac{d}{dt}X(t), DZ\right) = \{X, DX, DZ\}$. Since D is invertible this is equivalent to the definition of the Euler ODE. □

2. Algebraic Topology of Ideal Fluids

In the classical example of ideal fluid motion on a metric three manifold M without boundary, the fluid algebra is embedded in the differential algebra of all differential forms. So we are adding to the fluid algebra non constant functions, non constant volume forms, more one-forms and more two forms. New operators and operations appear: exterior derivative, the wedge product, the integral of three forms over the manifold, the Hodge star operator and the Hodge decomposition with the associated projections: all forms = exact forms + coexact forms + harmonic forms. (exact means image d and coexact means image of $*d*$). The fluid algebra in these terms is made from the subspace of coexact forms in degree one, the curl operator obtained by doing d then star, the alternating form and metric as described above using wedge, star and the integral. Alas, all of this is infinite dimensional so compactness and the easy proof above of long time existence disappears.

However there are many finite dimensional models of the algebraic topology residing in the above structure. Two types are:

I (grid type). Divide the manifold into cells and average the forms over oriented faces of the correct dimension. By Stokes' theorem one gets a map of chain complexes I:(all smooth forms with d) $\to$ (cellular cochains with d) which induces an isomorphism on cohomology. There are maps of chain complexes in the opposite direction using the heat flow and dual cells which are inverse to I up to chain homotopy and invariant under star (from class lectures and an unpublished manuscript).

II (eigenvalue type). Consider the eigendecomposition of all p forms for the laplacian operator $(-1)^{p-1}(*d*d - d*d*)$ and project onto finite dimensional pieces by putting bounds on the eigenvalues. These are invariant under d and star as well and induce isomorphisms on cohomology. For the flat torus these are used in numerical calculation for fluids.

A third type which is due to Whitney, is very elegant and useful for algebraic topology (see [**S77**]) but it is not invariant under star for a fundamental reason which to my knowledge has never been corrected. So we do not consider Whitney forms here.

The first two finite dimensional approximations have everything they need for defining the fluid algebra approximating the classical fluid algebra for the classical Euler evolution EXCEPT the wedge product. This lack can be corrected by forming the wedge product on the finite dimensional image inside forms and then integrating or projecting back to the finite dimensional model. This is how we multiply numbers on a computer using finite parts of the decimal expansion and how numerical computations for fluids are performed. In each case the multiplication obtained is appropriately commutative but NOT associative.

Now nontrivial ideas of algebraic topology enter. First recall that a chain mapping of chain complexes of vector spaces (or free $\mathbb{Z}$-modules) inducing a homology isomorphism has an inverse up to chain homotopy. Such inverses and the chain homotopies can be used to transport algebraic structures (up to homotopy) between chain complexes that are very different as vector spaces, for example the infinite dimensional deRham complex and the finite dimensional models. The associator of this finite dimensional wedge product $((a \wedge b) \wedge c)$ - $(a \wedge (b \wedge c))$ is a three to one operation which is a mapping of chain complexes which commutes with the

the natural differentials. Moreover because of the homology isomorphism above this associator is the commutator with d of a correcting three to one operation of degree one less which we denote $\wedge_1$. Now we continue using ideas of Stasheff's Princeton Thesis (1959). Consider the five ways to associate four entities. These arrange naturally at the vertices of a pentagon. We sum the corrections $\wedge_1$ on each edge combined with $\wedge$ to build a four to one operation of degree -1 which again is a cycle in that it commutes with d extended to multivariables by the Leibniz rule. By the homology isomorphism above there is a degree -2 four to one operation $\wedge_2$ which fills in the cycle corresponding to the boundary of the pentagon, i.e. one whose commutator with d is the four to one operation of degree -1 obtained by going around the pentagon. This process continues indefinitely producing n to 1 correction operations of degree $-n+2$ whose commutator with d is an appropriate more and more complex looking formula in the inductively constructed corrections.

There are two versions of this: the original Stasheff one is very beautiful. The complex looking formulae are nothing more than the combinatorics of the moduli spaces of complete hyperbolic surfaces with geodesic boundary boundedly related to the two disk with three or more punctures on the boundary, then naturally compactified by geometric limits. This is the model also controlling genus zero part of open string theories. The Stasheff polyhedra can also be described as a moduli space of planar rooted trees.

The second way remembers the commutativity and seems more appropriate here. It may be modeled on the moduli spaces of rooted trees (in space) whose leaves are labeled and whose interior edges are painted black or white. This model is a picture of what is called the bar cobar construction for algebras over an operad (see Vallette[**V**] and Wilson[**W**]).

Interestingly enough there is a specific procedure for computing these corrections by placing the chain homotopies mentioned above on the interior edges of the trees and using the wedge at the vertices. This procedure is identical to the tree part of the Feynman diagram algorithm in perturbative Chern-Simons quantum field theory where the "propagator" there is the chain homotopy here.

After doing this work we obtain the derived or "infinity" version of the graded commutative wedge product compressed onto the finite dimensional approximating models.

The same ideas may be applied to the bracket or transport discussed above which satisfies Jacobi in the infinite dimensional model. The Jacobi identity may be encoded in the the algebraic statement that on forms the wedge product and the adjoint of $d = (-1)^p * d*$ satisfy the following identity in words: the deviation of $*d*$ from being a derivation of $\wedge$ is, as a two variable operator, itself a derivation in each variable. Adding the equation that $*d * *d* = 0$ yields formally that the two variable operation satisfies Jacobi. This Lie bracket on forms becomes via the metric isomorphism the usual Lie bracket of vector fields extended by Leibniz to all multivector fields. (This is referred to as the Schouten-Nijenhuis bracket.) This relationship between $*d*$ and $\wedge$ means we have a so called Batalin-Vilkovisky algebra or briefly a BV algebra. The BV formalism in perturbative quantum field theory due to Batalin-Vilkovisky is perhaps the most natural for mathematicians (see Costello [**C**]). The BV structure $(*d*, \wedge)$ may also be compressed to a new structure of BV algebra up to homotopy into our finite dimensional models. The bracket so obtained is called a Lie infinity structure.

So now we have a commutative infinity structure related to d and a Lie infinity structure related to $*d*$ whose leading term is the transport above.

Before making models that might be useful for fluid simulation these ideas need I think to be completed in the following way:

1) describe the complete structure used above d,star, $\wedge$ and the integral as an algebraic structure. Then develop the diagrammatic compression algorithms for this structure which yield the derived or infinity version of the structure on the finite dimensional approximations. To my knowledge the current abstract homotopical algebra just falls short of this task. General multilinear operations with outputs can be treated as in [**S09**], but pairings are not treated there or anywhere else to my knowledge (although Kevin Costello's work on renormalization and perturbative quantum field theory may come close [**C**]).
2) understand the Euler evolution as a functorial construction on the derived or infinity version of this algebraic structure. The efforts of Arnold and Khesin and others are a beginning but to my knowledge this goal is not yet achieved (although describing the flow by an action principle may lead to such a functorial principle).

If these two tasks are completed, we will have conceptually natural effective theories of fluid motion at every scale that fit together in an appropriate sense. There will be, by definition almost, natural algorithms for fluid computation based on the corrections that emerge from the compression of the algebraic structure into finite dimensions. These may play a role in proving long term existence of the classical ODE if that long term existence is true. If it is not true, this may also be revealed in these models which work at every scale. In any case we will have natural algorithms for computations of real fluids at every scale which of course do have long time existence and which are potentially observable at extremely small scales.

References

[AK] V. Arnold and B. Khesin, *Topological methods in hydrodynamics.* Applied Mathematical Sciences, **125**. Springer-Verlag, New York, 1998.

[C] K. Costello, *Renormalization and effective field theory.* Math. Surveys Monogr. **170**, AMS, Providence, 2011.

[S76] D. Sullivan, *Cycles for the dynamical study of foliated manifolds and complex manifolds.* Invent. Math. **36** (1976), 225–255.

[S77] D. Sullivan, *Infinitesimal computations in topology.* Publ. Math. Inst. Hautes Études Sci. **47** (1977), 269–331.

[S09] D. Sullivan. *Homotopy theory of the master equation package applied to algebra and geometry: a sketch of two interlocking programs.* In *Algebraic topology—old and new*, 297–305, Banach Center Publ., **85**, Polish Acad. Sci. Inst. Math., Warsaw, 2009.

[V] B. Vallette, *Koszul duality for PROPs.* C. R. Math. Acad. Sci. Paris **338** (2004), no. 12, 909–914.

[W] S. Wilson, *Free Frobenius algebra on the differential forms of a manifold.* Preprint, 2007, available at arXiv:math/0710.3550.

CUNY Graduate Center, 365 Fifth Avenue, Room 4208 New York, NY 10016-4309

SUNY, Stony Brook New York 11794
E-mail address: dsullivan@gc.cuny.edu

Proceedings of Symposia in Pure Mathematics
Volume **82**, 2011

Subgroups and quotients of automorphism groups of RAAGs

Ruth Charney and Karen Vogtmann

ABSTRACT. We study subgroups and quotients of outer automorphsim groups of right-angled Artin groups (RAAGs). We prove that for all RAAGs, the outer automorphism group is residually finite and, for a large class of RAAGs, it satisfies the Tits alternative. We also investigate which of these automorphism groups contain non-abelian solvable subgroups.

1. Introduction

A right-angled Artin group, or RAAG, is a finitely-generated group determined completely by the relations that some of the generators commute. A RAAG is often described by giving a simplicial graph Γ with one vertex for each generator and one edge for each pair of commuting generators. RAAGs include free groups (none of the generators commute) and free abelian groups (all of the generators commute). Subgroups of free groups and free abelian groups are easily classified and understood, but subgroups of right-angled Artin groups lying between these two extremes have proved to be a rich source of examples and counterexamples in geometric group theory. For details of this history, we refer to the article [**Ch07**].

Automorphism groups and outer automorphism groups of RAAGs have received less attention than the groups themselves, with the notable exception of the two extreme examples, i.e. the groups $Out(F_n)$ of outer automorphism groups of a free group and the general linear group $GL(n,\mathbb{Z})$. The group $Out(F_n)$ has been shown to share a large number of properties with $GL(n,\mathbb{Z})$, including several kinds of finiteness properties and the Tits alternative for subgroups. These groups have also been shown to differ in significant ways, including the classification of solvable subgroups. In a series of recent papers [**CCV07, CV08, BCV09**], we have begun to address the question of which properties shared by $Out(F_n)$ and $GL(n,\mathbb{Z})$ are in fact shared by the entire class of outer automorphism groups of right-angled Artin groups. We are also interested in the question of determining properties which depend on the shape of Γ and in determining exactly how they depend on it.

In our previous work, an important role was played by certain restriction and projection homomorphisms, which allow one to reduce questions about the full

2010 *Mathematics Subject Classification.* Primary 20F36.
Key words and phrases. right-angled Artin Groups, automorphisms, Tits alternative.
R. Charney was partially supported by NSF grant DMS 0705396.
K. Vogtmann was partially supported by NSF grant DMS 0705960.

outer automorphism group of a RAAG to questions about the outer automorphism groups of smaller subgroups. In the first section of this paper we recall these tools and develop them further. In the next section we apply them to prove

Theorem 10. *For any defining graph* Γ, *the group* $Out(A_\Gamma)$ *is residually finite.*

This result was obtained independently by A. Minasyan [**Mi09**], by different methods. We next prove the Tits' alternative for a certain class of *homogeneous* RAAGs (see section 5).

Theorem 17. *If* Γ *is homogeneous, then* $Out(A_\Gamma)$ *satisfies the Tits' alternative.*

In the last section, we investigate solvable subgroups of $Out(A_\Gamma)$. We provide examples of non-abelian solvable subgroups and we determine an upper bound on the virtual derived length of solvable subgroups when A_Γ is homogeneous. Finally, by studying translation lengths of infinite order elements, we find conditions under which all solvable subgroups of $Out(A_\Gamma)$ are abelian. We show that excluding "adjacent transvections" from the generating set of $Out(A_\Gamma)$ gives rise to a subgroup $\widetilde{O}ut(A_\Gamma)$ satisfying a strong version of the Tits alternative.

Corollary 30. *If* Γ *is homogeneous of dimension* n, *then every subgroup of* $\widetilde{O}ut(A_\Gamma)$ *is either virtually abelian or contains a non-abelan free group.*

Thus for graphs which do not admit adjacent transvections, the whole group $Out(A_\Gamma)$ satisfies this property. One case which is simple to state is the following.

Corollary 31. *If* Γ *is connected with no triangles and no leaves, then all solvable subgroups of* $Out(A_\Gamma)$ *are virtually abelian.*

Charney would like to thank the Forschungsinstitut für Mathematik in Zurich and Vogtmann the Hausdorff Institute for Mathematics in Bonn for their hospitality during the writing of this paper. Both authors would like to thank Talia Fernós for helpful conversations and Richard Wade for his comments on the first version of this paper.

2. Some combinatorics of simplicial graphs

Certain combinatorial features of the defining graphs Γ for our right-angled Artin groups will be important for studying their automorphisms. In this section we establish notation and recall some basic properties of these features.

DEFINITION 1. Let v be a vertex of Γ. The *link of* v, denoted $lk(v)$, is the full subgraph spanned by all vertices adjacent to v. The *star of* v, denoted $st(v)$, is the full subgraph spanned by v and $lk(v)$.

DEFINITION 2. Let Θ be a subgraph of Γ. The *link* of Θ, denoted $lk(\Theta)$, is the intersection of the links of all vertices in Θ. The *star* of Θ, denoted $st(\Theta)$ is the full subgraph spanned by $lk(\Theta)$ and Θ. The *perp* of Θ, denoted $\Theta^\perp$, is the intersection of the stars of all vertices in Θ. (See Figure 1.)

These can be expressed in terms of distance in the graph as follows:

- $v \in lk(\Theta)$ iff $d(v,w) = 1$ for all $w \in \Theta$
- $v \in \Theta^\perp$ iff $d(v,w) \leq 1$ for all $w \in \Theta$
- $v \in st(\Theta)$ iff $v \in lk(\Theta) \cup \Theta$

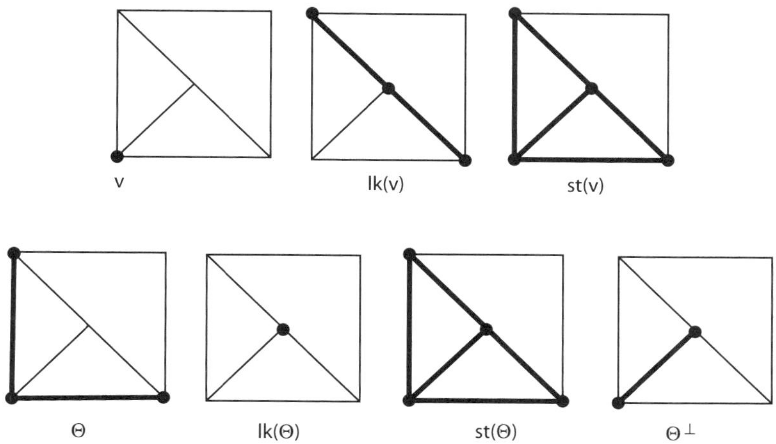

FIGURE 1. Links, stars and perps

Recall that a complete subgraph of Γ is called a *clique*. (In this paper, cliques need not be maximal.) If Δ is a clique, then $st(\Delta) = \Delta^\perp$; otherwise $st(\Delta)$ strictly contains $\Delta^\perp$.

LEMMA 3. *If Δ is a clique, then $st(\Delta)^\perp$ is also a clique and $st(\Delta) \supseteq st(\Delta)^\perp \supseteq \Delta$.*

PROOF. Since Δ is a clique, $v \in st(\Delta)$ implies $st(v) \supseteq \Delta$. Therefore

$$st(\Delta)^\perp = \cap_{v \in st(\Delta)} st(v) \supseteq \Delta.$$

If $x \in st(\Delta)^\perp$, then $d(x, v) \leq 1$ for all vertices $v \in st(\Delta)$, including all $v \in \Delta$, i.e. $x \in st(\Delta)$. If y is another vertex in $st(\Delta)^\perp$, then similarly $d(y, v) \leq 1$ for all vertices $v \in st(\Delta)$, so in particular $d(y, x) = 1$. Since any two vertices of $st(\Delta)^\perp$ are adjacent, $st(\Delta)^\perp$ is a clique.

□

We define $v \leq w$ to mean $lk(v) \subseteq st(w)$. This relation is transitive and induces a partial ordering on equivalence classes of vertices $[v]$, where $w \in [v]$ if and only if $v \leq w$ and $w \leq v$ ([**CV08**], Lemma 2.2). The links $lk[v]$ and stars $st[v]$ of equivalence classes of maximal vertices v will be of particular interest to us.

REMARK 4. In the authors' previous paper [**CV08**], the notation $J_{[v]}$ was used to denote the star of an equivalence class $[v]$. This notation was chosen to emphasize that $st[v]$ has the structure of the "join" of two smaller graphs, $[v]$ and $lk[v]$. In the current, more general setting, we find the notation $st(\Theta)$ to be more intuitive.

For a full subgraph $\Theta \subset \Gamma$, the right-angled Artin group A_Θ embeds into A_Γ in the natural way. The image is called a *special subgroup* of A_Γ, and we use the same notation A_Θ for it. An important observation is that the centralizer of A_Θ is equal to $A_{\Theta^\perp}$ (see, e.g., [**CCV07**], Proposition 2.2).

We remark that if v is a vertex in $\Theta \subset \Gamma$, then it is possible for v to be maximal in Θ but not in Γ. Unless otherwise stated, the term "maximal vertex" will always mean maximal with respect to the original graph Γ.

The subgraph spanned by $[v]$ is either a clique, or it is disconnected and discrete ([**CV08**], Lemma 2.3). In the first case the subgroup $A_{[v]}$ is abelian and we call v an *abelian vertex*; in the second, $A_{[v]}$ is a non-abelian free group, and we call v a *non-abelian vertex*. Note that for any vertex v, $st[v]$ is the union of the stars of the vertices $w \in [v]$.

A *leaf* of Γ is a vertex which is an endpoint of only one edge. A *leaf-like* vertex is a vertex v whose link contains a unique maximal vertex w, and $[v] \leq [w]$. In particular, a leaf is leaf-like. If Γ has no triangles, then every leaf-like vertex is in fact a leaf.

3. Key tools

Generators for $Out(A_\Gamma)$ were determined by M. Laurence [**Lau95**], extending work of H. Servatius [**Ser89**]. They consist of

- graph automorphisms
- inversions of a single generator v
- transvections $v \mapsto vw$ for generators $v \leq w$
- partial conjugations by a generator v on one component of $\Gamma - st(v)$

As in [**CV08**], we consider the finite-index subgroup $Out^0(A_\Gamma)$ of $Out(A_\Gamma)$ generated by inversions, transvections and partial conjugations. This is a normal subgroup, called the *pure outer automorphism group*

If Γ is connected and v is a maximal vertex, then any pure outer automorphism ϕ of A_Γ has a representative f_v which preserves both $A_{[v]}$ and $A_{st[v]}$ ([**CV08**], Prop. 3.2). This allows us to define several maps from $Out^0(A_\Gamma)$ to the outer automorphism groups of various special subgroups, as follows.

(1) Restricting f_v to $A_{st[v]}$ gives a *restriction map*
$$R_v \colon Out^0(A_\Gamma) \to Out^0(A_{st[v]}).$$

(2) The map $A_\Gamma \to A_{\Gamma-[v]}$ which sends each generator in $[v]$ to the identity induces an *exclusion map*
$$E_v \colon Out^0(A_\Gamma) \to Out^0(A_{\Gamma-[v]}).$$

(3) Since v is maximal with respect to the graph $st[v]$ and $lk[v] = st[v] - [v]$, we can compose the restriction map on A_Γ with the exclusion map on $A_{st[v]}$ to get a *projection map*
$$P_v \colon Out^0(A_\Gamma) \to Out^0(A_{lk[v]}).$$

If Γ is the star of a single vertex v, then $[v]$ is the unique maximal equivalence class, and R_v is the identity. If Γ is a complete graph, then $\Gamma = [v]$ and $lk[v]$ is empty, in which case we define $P_v = E_v$ to be the trivial map.

The reader can verify that these maps are well-defined homomorpisms. For the restriction map this follows from the fact that $A_{st[v]}$ is its own normalizer. For the exclusion map it follows from the fact that the normal subgroup generated by a maximal equivalence class $[v]$ is charactersitic. (See [**CV08**] for details).

3.1. The amalgamated restriction homomorphism R. Let Γ be a connected graph. We can put all of the restriction maps R_v together to obtain an amalgamaged restriction map
$$R = \prod R_v \colon Out^0(A_\Gamma) \to \prod Out^0(A_{st[v]}),$$

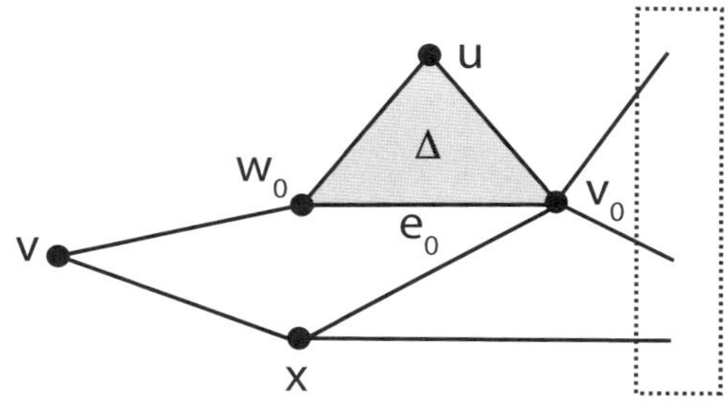

FIGURE 2. Notation for proof of Lemma 6

where the product is over all maximal equivalence classes $[v]$. It was proved in [**CV08**] that the kernel K_R of R is a finitely-generated free abelian group, generated by partial conjugations. If Γ has no triangles, we also found a set of generators for K_R [**CCV07**]. We will need this information for general Γ in what follows, so we will now present another (and simpler) proof that K_R is free abelian which also identifies a set of generators for K_R. The proof will use the following fact due to Laurence.

THEOREM 5 ([**Lau95**], Thm 2.2). *An automorphism of A_Γ which takes every vertex to a conjugate of itself is a product of partial conjugations.*

By definition, any automorphism representing an element of K_R acts on the star of each maximal equivalence class of vertices as conjugation by some element of A_Γ. We begin by showing that the same is true for every equivalence class:

LEMMA 6. *Let f be an automorphism representing an element of K_R. Then for every vertex $v \in \Gamma$, f acts on $st[v]$ as conjugation by some $g \in A_\Gamma$.*

PROOF. This is by definition of the kernel if v is maximal. Since every vertex of Γ is in the star of some maximal vertex, f sends every vertex to a conjugate of itself. By Theorem 5, this implies that f is a product of partial conjugations.

If v is not maximal, then choose a maximal vertex v_0 with $v < v_0$. After adjusting by an inner automorphism if necessary, we may assume f is the identity on $st[v_0]$. If v is adjacent to v_0, then $st[v] \subset st[v_0]$ and we are done.

If v is not adjacent to v_0, choose a maximal vertex $w_0 \in lk(v) \cap lk(v_0)$ (note that one always exists). Then f acts as conjugation by some g on $st[w_0]$. Let e_0 be the edge from v_0 to w_0. Since $st(e_0) \subset st(w_0)$, f acts as conjugation by g on all of $st(e_0)$. Since $st(e_0) \subset st(v_0)$, g centralizes $st(e_0)$, i.e. g is in the subgroup generated by $st(e_0)^\perp$. By Lemma 3, $st(e_0)^\perp = \Delta$ is a clique containing e_0, so the subgroup A_Δ is abelian. (See Figure 2.)

Since A_Δ is abelian, we can write $g = g_2 g_1$ where g_1 is a product of generators in $lk[v]$ and g_2 a product of generators not in $lk[v]$. We claim that f acts as conjugation by g_2 on all of $st[v]$. Since $[v] \subset st[w_0]$, f acts as conjugation by g on $[v]$, and since g_1 commutes with $[v]$, this is the same as conjugation by g_2. The action of f on $lk[v]$ is trivial, since $lk[v] \subset st[v_0]$, so it suffices to show that g_2 commutes with $lk[v]$. For suppose $u \in \Delta$ does not lie in $lk[v]$, and $x \in lk[v]$. Then either x lies in $st(u)$, hence commutes with u, or x and v lie in the same component of $\Gamma - st(u)$. In the latter case, since f is a product of partial conjugations, the

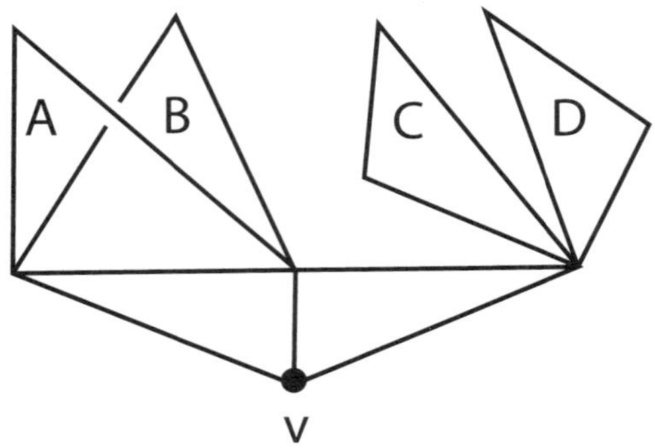

FIGURE 3. The $\hat{v}$-components are $A \cup B$ and $C \cup D$

total exponent of u in the conjugating element must be the same at v and at x; but $f(x) = x$, so this total exponent must be 0. That is, u can appear as a factor in g_2 only if it commutes with all of $lk[v]$. □

Next, we describe some automorphisms contained in the kernel K_R. If Γ is a connected graph and v is a vertex of Γ, say vertices x and y are in the same *$\hat{v}$-component* of Γ if x and y can be connected by an edge-path which contains no edges of $st(v)$ (though it may contain vertices of $lk(v)$). A $\hat{v}$-component lying entirely inside $st(v)$ is called a *trivial* $\hat{v}$-component, and any other $\hat{v}$-component is *non-trivial*. In Figure 3, there are two non-trivial $\hat{v}$ components, one consisting of $A \cup B$, and one consisting of $C \cup D$. If $st(v)$ has no triangles, a non-trivial $\hat{v}$-component is the same thing as a non-leaf component of $\Gamma - v$. In general, each component of $\Gamma - st(v)$ is contained in a single $\hat{v}$-component, but a single $\hat{v}$-component may contain several components of $\Gamma - st(v)$.

DEFINITION 7. A *$\hat{v}$-component conjugation* is an automorphism of A_Γ which conjugates all vertices in a single nontrivial $\hat{v}$-component of Γ by v.

By the remarks above, a $\hat{v}$-component conjugation is in general a product of partial conjugations by v on components of $\Gamma - st(v)$. To see that such conjugations lie in K_R, note that for any w, all of the vertices of $st[w]$ which do not lie in $st(v)$ lie in the same $\hat{v}$-component as w. Hence any $\hat{v}$-component conjugation acts as an inner automorphsim on $st[w]$.

Let $\hat{c}(v)$ be the number of non-trivial $\hat{v}$-components in Γ.

THEOREM 8. *The kernel K_R of the restriction map is free abelian, generated by non-trivial $\hat{v}$-component conjugations for all $v \in \Gamma$. The rank of the kernel is $\sum_{v\in\Gamma}(\hat{c}(v) - 1)$.*

PROOF. Let $\widehat{\mathcal{PC}}$ denote the set of all non-trivial $\hat{v}$-component conjugations for all $v \in \Gamma$. We first prove that $\widehat{\mathcal{PC}}$ generates K_R.

Let $\phi \in K_R$. For each representative f of ϕ, let V_f be the set of vertices v such that $st[v]$ is pointwise fixed. Now fix a representative f such that V_f is of maximal size. We proceed by induction on the number of vertices in $\Gamma - V_f$.

If $V_f = \Gamma$, then f is the trivial automorphism and there is nothing to prove. If not, choose a vertex w at distance 1 from V_f, so w is connected by an edge e to

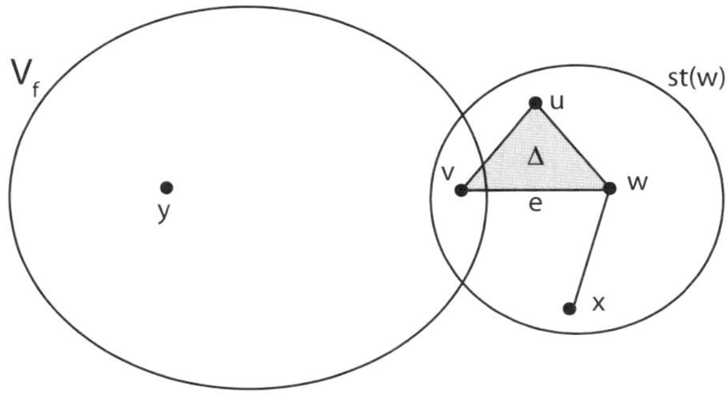

FIGURE 4. Notation for proof of Theorem 8

some $v \in V_f$. Then f acts non-trivially on $st[w]$ as conjugation by some $g \in A_\Gamma$. Since f acts trivially on $st[v]$, g fixes $st(e)$. The centralizer of $A_{st(e)}$ is equal to $A_{st(e)^\perp}$. By Lemma 3, $st(e)^\perp = \Delta$ for some clique Δ containing e, so g is in the abelian subgroup A_Δ and we can write $g = u_1^{\epsilon_1} \dots u_k^{\epsilon_k}$ for distinct vertices $u_i \in \Delta$.

If $st(w) \subseteq st(u_i)$ (e.g. if $u_i = w$), conjugation by u_i is trivial on $st(w)$ and we may assume $\epsilon_i = 0$, i.e. u_i does not appear in the expression for g. If $V_f \subset st(u_i)$, replace f by f composed with the inner automorphism by $u_i^{-\epsilon_i}$; the new V_f contains (so is equal to) the old one. We may now assume neither $st(w)$ nor V_f are contained in the star of any u_i.

Fix u_i and $x \in st(w) - st(u_i)$ and $y \in V_f - st(u_i)$. We claim that x and y are in different connected components of $\Gamma - st(u_i)$. To see this, suppose x and y are in the same connected component of $\Gamma - st(u_i)$. Since f sends each vertex to a conjugate of itself, Theorem 5 implies that f is a product of partial conjugations, hence x and y must be conjugated by the same total power of u_i. For y this power is zero, since $y \in V_f$, and so ϵ_i must also be zero, i.e. u_i does not occur in the expression for g.

We claim further that x and y must be in different $\hat{u}_i$-components of Γ. Suppose they were in the same $\hat{u}_i$-component. Let γ be an edge-path joining y to x which avoids edges of $st(u_i)$, with vertices $y = x_0, x_1, x_2, \dots, x_k = x$. We know that y is fixed by f and x is conjugated by a non-trivial power of u_i. Therefore there is some x_j in $lk(u_i)$ with the property that x_{j-1} is not conjugated by u_i but x_{j+1} is conjugated by a non-trivial power of u_i. Since γ does not use edges of $st(u_i)$, neither x_{j-1} nor x_{j+1} is in $lk(u_i)$, i.e. neither commutes with u_i. Thus f does not act as conjugation by the same total power of u_i on all of $st[x_j]$, contradicting Lemma 6.

The vertices of $st(w) - st(u_i)$ lie in a single, non-trivial $\hat{u}_i$-component (the component containing w) and by the discussion above, this $\hat{u}_i$-component contains no vertices of $V_f - st(u_i)$. Thus, there is a non-trivial $\hat{u}_i$-component conjugation f_i which affects vertices of $st(w)$ but not V_f. The automorphism $f' = f_k^{-\epsilon_k} \circ \cdots \circ f_1^{-\epsilon_1} \circ f$ has a strictly larger $V_{f'}$, which includes w as well as V_f. By induction, f' is a product of elements of $\widehat{\mathcal{PC}}$, hence so is f.

It remains to check that any two elements of $\widehat{\mathcal{PC}}$ commute in $Out(A_\Gamma)$. Let f_v be $\hat{v}$-component conjugation, and f_w a $\hat{w}$-component conjugation. If v and w

are adjacent, these commute. If $d(v,w) > 1$, then $st(w)$ is contained in a single $\hat{v}$-component D_v, and $st(v)$ is contained in a single $\hat{w}$-component D_w. It follows that D_v contains every $\hat{w}$-component except D_w, and D_w contains every $\hat{v}$-component except D_v. It is now easy to check that for any $\hat{v}$-component C_v and $\hat{w}$-component C_w, one of the following holds: C_w and C_v are disjoint, $C_v \subset C_w$, $C_w \subset C_v$, or $\Gamma - C_v$ and $\Gamma - C_w$ are disjoint. In any of these cases, the corresponding partial conjugations f_v and f_w commute in $Out(A_\Gamma)$.

The only other relation among the generators of $\widehat{\mathcal{PC}}$ is that for a fixed v, the product of all non-trivial $\hat{v}$-component conjugations is an inner automorphism. The last statement of the theorem follows. □

3.2. The amalgamated projection homomorphism P. We can combine the projection homomorphisms P_v for maximal equivalence classes $[v]$ in the same way we combined the restriction homomorphisms, to obtain an amalgamated projection homomorphism

$$P = \prod P_v \colon Out^0(A_\Gamma) \to \prod Out^0(A_{lk[v]}).$$

Recall that a vertex v is called leaf-like if there is a unique maximal vertex w in $lk(v)$ and this vertex satisfies $[v] \leq [w]$. The transvection $v \mapsto vw$ is called a *leaf transvection*. It is proved in [**CV08**] that if Γ is connected and is not the star of a single vertex (and hence has more than one maximal equivalence class), then the kernel K_P of P is a free abelian group generated by K_R and the set of all leaf transvections.

4. Residual finiteness

It is easy to see using congruence subgroups that $GL(n,\mathbb{Z})$ is residually finite, and E. Grossman proved that $Out(F_n)$ is also residually finite ([**Gr74**]). In this section we use these facts together with our restriction and exclusion homomorphisms to show that in fact $Out(A_\Gamma)$ is residually finite for every defining graph Γ. The same result has been obtained by A. Minasyan [**Mi09**] by different methods. Both proofs use a fundamental result of Minasyan and Osin which takes care of the case when the defining graph is disconnected:

THEOREM 9. [**MiOs09**] *If G is a finitely generated, residually finite group with infinitely many ends, then $Out(G)$ is residually finite.*

THEOREM 10. *For any right-angled Artin group A_Γ, $Out(A_\Gamma)$ is residually finite.*

PROOF. Every right angled Artin group A_Γ is finitely generated and residually finite (it's linear), and A_Γ has infinitely many ends if and only if Γ is disconnected. Therefore, by Theorem 9, we may assume that Γ is connected.

We proceed by induction on the number of vertices in Γ.

Consider first the case in which $\Gamma = st[v]$ for a single equivalence class $[v]$. If $[v]$ is abelian, we know by Proposition 4.4 of [**CV08**] that

$$Out(A_\Gamma) = Tr \rtimes (GL(A_{[v]}) \times Out(A_{lk[v]}))$$

where Tr is the free abelian group generated by the leaf transvections. Since $[v]$ is abelian, $GL(A_{[v]}) = GL(k,\mathbb{Z})$, which is residually finite, and $Out(A_{lk[v]})$ is residually finite by induction. The result now follows because semi-direct products of finitely generated residually finite groups are residually finite [**Mi71**]. If $[v]$ is non-abelian, then $Out(A_\Gamma) = Out(A_{[v]}) \times Out(A_{lk[v]})$ (or possibly a $\mathbb{Z}/2\mathbb{Z}$-extenion of this). Since $A_{[v]}$ is a free group, $Out(A_{[v]})$ is residually finite and $Out(A_{lk[v]})$ is residually finite by induction, so this case also follows.

Now suppose that Γ is not the star of a single equivalence class. Since $Out^0(A_\Gamma)$ has finite index in $Out(A_\Gamma)$, it suffices to prove that $Out^0(A_\Gamma)$ is residually finite. For any maximal equivalence class $[v]$, $Out^0(st[v])$ is residually finite by induction, so any element of $Out^0(A_\Gamma)$ which maps non-trivially under R is detectable by a finite group. It remains to show that the same is true for elements in the kernel K_R of R.

Let ϕ be an element of K_R. It follows from Lemma 8 and the fact that K_R is abelian that ϕ can be factored as

$$\phi = \phi_1 \circ \cdots \circ \phi_k$$

where ϕ_i is a product of $\hat{v}_i$-component conjugations, and the classes $[v_1], \ldots, [v_k]$ are distinct. Let $[w]$ be a maximal vertex adjacent to $[v_1]$. Consider the image of ϕ under the exclusion homomorphism $E_w : Out^0(A_\Gamma) \to Out^0(A_{\Gamma-[w]})$. By induction, the target group $Out^0(A_{\Gamma-[w]})$ is residually finite, so it suffices to show that this image, $\bar{\phi}$, is non-trivial.

Write $\bar{\phi} = \bar{\phi}_1 \circ \cdots \circ \bar{\phi}_k$. Note that $\bar{\phi}_1$ is still a nontrivial partial conjugation on $A_{\Gamma-[w]}$ since the vertices which were removed commuted with all elements of $[v_1]$. Moreover, for $i > 1$, the partial conjugations in $\bar{\phi}_i$ are either trivial, or are partial conjugations by elements distinct from $[v_1]$. It follows that $\bar{\phi}$ acts non-trivially on $A_{\Gamma-[w]}$ as required. □

5. Homogeneous graphs and the Tits alternative

Recall that the Tits alternative for a group G states that every subgroup of G is either virtually solvable or contains a non-abelian free group. Both $GL(n,\mathbb{Z})$ and $Out(F_n)$ are known to satisfy the Tits alternative [**Ti72**], [**BFH00, BFH05**]. We will show that $Out(A_\Gamma)$ satisfies the Tits alternative for a large class of graphs Γ.

DEFINITION 11. Let Γ be a finite simplicial graph. We say Γ is *homogeneous of dimension* 0 if it is empty, and *homogeneous of dimension* 1 if it is non-empty and discrete (no edges). For $n > 1$, we say Γ is *homogeneous of dimension* n if it is connected and the link of every vertex is homogeneous of dimension $n-1$.

If Δ is a k-clique in Γ and v is a vertex in Δ, then the link of Δ in Γ is equal to the link of $\Delta - v$ in $lk(v)$. A simple inductive argument now shows that if Γ is homogeneous of dimension n, then the link of any k-clique is homogeneous of dimension $n-k$. In particular, every maximal clique in Γ is an n-clique (hence the terminology "homogeneous").

LEMMA 12. *If Γ is homogeneous of dimension $n > 1$, then any two n-cliques α and β are connected by a sequence of n-cliques $\alpha = \sigma_1, \sigma_2, \ldots, \sigma_k = \beta$ such that $\sigma_{i-1} \cap \sigma_i$ is an $(n-1)$-clique.*

PROOF. We proceed by induction on n. For $n = 2$, this is simply the statement that Γ is connected. For $n > 2$, since Γ is connected and every edge is contained in an n-clique, we can find a sequence of n-cliques from α to β such that consecutive n-cliques share at least a vertex. Thus is suffices to consider the case where α and

β share a vertex v. In this case, there are $(n-1)$-cliques α' and β' in $lk(v)$ that together with v span α and β. By induction, α' and β' can be joined by a sequence of $(n-1)$-cliques in $lk(v)$ that intersect consecutively in $(n-2)$-cliques. Taking the join of these with v gives the desired sequence. □

We can also express this lemma in topological terms. If K_Γ is the flag complex associated to Γ (that is, the simplicial complex whose k-simplices correspond to the k-cliques of Γ), then the lemma states that for Γ homogeneous, K_Γ is a chamber complex.

EXAMPLES 13. (1) For $n=2$, a graph Γ is homogeneous if and only if Γ is connected and triangle-free. These are precisely the RAAGs studied in [**CCV07**]. (2) The join of two homogeneous graphs is again homogeneous so, for example, the join of two connected, triangle-free graphs is homogeneous of degree 4. (3) If Γ is the 1-skeleton of a connected triangulated n-manifold, then Γ is homogeneous of dimension n.

Our main concern is to be able to do inductive arguments on links of vertices; in particular, we will need such links to be connected or discrete at all stages of the induction. It may appear that homogeneity is a stronger condition than necessary. This is not the case.

LEMMA 14. *Γ is homogeneous of dimension $n>1$ if and only if Γ is connected and the link of every (non-maximal) clique is either discrete or connected.*

PROOF. If Γ is homogeneous, then so is the link of every k-clique, $k<n$, so by definition it is either discrete or connected.

Conversely, assume that Γ is connected and the link of every non-maximal k-clique is either discrete or connected. We proceed by induction on the maximal size m of a clique in Γ. If $m=2$, then the link of every vertex (1-clique) in Γ is discrete and non-empty, so by definition, Γ is homogeneous of dimension 2.

For $m>2$, we claim first that the link of every vertex is connected. For if Γ contains some vertex with a discrete link, then there exists an adjacent pair of vertices v,w such that the link of v is discrete while the link of w is not. In this case, v lies in $lk(w)$ but v is not adjacent to any other vertex in $lk(w)$. This contradicts the assumption that the link of w is connected.

If Δ is a k-clique in $lk(v)$, then $\Delta * v$ is a $(k+1)$-clique in Γ. Since the link of Δ in $lk(v)$ is equal to the link of $\Delta * v$ in Γ, it is either discrete or connected. Thus, by induction, $lk(v)$ is homogeneous. Moreover, every link must be homogeneous of the same dimension, for if v,w are adjacent vertices, then the homogeneous dimension of $lk(v)$ and $lk(w)$ are both equal to $r-1$ where r is the size of the maximal clique containing v and w. □

The next lemma contains some other elementary facts about homogeneous graphs.

LEMMA 15. *Let Γ be homogeneous of dimension n and assume that Γ is not the star of a single vertex. Let $[v]$ be a maximal equivalence class in Γ.*

1. *If $[v]$ is abelian, then $[v]$ is a singleton.*
2. *For any maximal $[v]$, $lk[v]$ is homogeneous of dimension $n-1$ and is not the star of a single vertex.*

PROOF. (1) Suppose $[v]$ is abelian and contains k vertices. Then $[v]$ it spans a k-clique and $st[v] = st(v)$. By hypothesis, there is some n-clique σ not contained in $st[v]$ and by Lemma 12, we can choose σ so that $\sigma \cap st[v]$ is an $(n-1)$-clique. It follows that if $k > 1$, then σ contains some vertex of $[v]$ and hence every vertex of $[v]$ (since they are all equivalent), contradicting our assumption that σ does not lie in $st[v]$. We conclude that $k = 1$, or in other words, $[v]$ is a single point. This proves (1).

For (2), let $[v]$ be any maximal equivalence class. Then either $[v]$ is free, or a singleton and in either case, $lk[v] = lk(v)$, so it is homogeneous of dimension $n-1$. If $lk[v]$ is contained in the star of a single vertex $w \in lk[v]$, then $[v] < [w]$. But this is impossible since $[v]$ is maximal. □

REMARK 16. Suppose Γ is the star of a single vertex v. Then $[v]$ is necessarily abelian so it spans a k-clique, and $lk[v]$ is homogeneous of dimension $n-k$. By Proposition 4.4 of [**CV08**], we have

$$Out(A_\Gamma) = Tr \rtimes (GL(k, \mathbb{Z}) \times Out(A_{lk[v]}))$$

where Tr is free abelian. Moreover, $lk[v]$ cannot be the star of a single vertex. To see this, note that if $lk[v] = st(w)$, then w is also adjacent to every vertex of Γ, hence it is equivalent to v, contradicting the assumption that $w \in lk[v]$.

We can now prove the main theorem of this section.

THEOREM 17. *If Γ is homogeneous of dimension n, then $Out(A_\Gamma)$ satisfies the Tits alternative, that is, every subgroup of $Out(A_\Gamma)$ is either virtually solvable or contains a non-abelian free group.*

PROOF. For Γ a complete graph, $Out(A_\Gamma) = GL(n, \mathbb{Z})$ so this follows from Tits' original theorem. So assume this is not the case. It suffices to prove the Tits Alternative for the finite index subgroup $Out^0(A_\Gamma)$. We proceed by induction on n. For $n = 1$, A_Γ is a free group and the theorem follows from [**BFH00, BFH05**].

If $n > 1$, then for every maximal $[v]$, $lk[v]$ is homogeneous of lower dimension, so by induction, $Out^0(A_{lk[v]})$ satisfies the Tits alternative. It is straightforward to verify that the Tits alternative is preserved under direct products, subgroups, and abelian extensions, so if Γ is not a star, the theorem follows from the exact sequence

$$1 \to K_P \to Out^0(A_\Gamma) \to \prod Out^0(A_{lk[v]}).$$

If Γ is a star, it follows from Remark 16. □

6. Solvable subgroups

6.1. Virtual derived length.

DEFINITION 18. Let G be a solvable group and $G^{(i)}$ its derived series. The *derived length* of G is the least n such that $G^{(n)} = \{1\}$. The *virtual derived length* of G, which we denote by $vdl(G)$, is the minimum of the derived lengths of finite index subgroups of G.

For an arbitrary group H, define

$$\mu(H) = \max\{vdl(G) \mid G \text{ is a solvable subgroup of } H\}.$$

Note that if H is itself solvable, then $\mu(H) = vdl(H)$.

The following properties of $\mu(H)$ are easy exercises.

LEMMA 19. (1) *If* $H = \prod H_i$, *then* $\mu(H) = \max\{\mu(H_i)\}$.

(2) *If* N *is a subgroup of* H, *then* $\mu(N) \leq \mu(H)$. *If* $[H : N] < \infty$, *then* $\mu(N) = \mu(H)$.

(3) *If* $N \lhd H$ *is a solvable normal subgroup of derived length* k, *then* $\mu(H) \leq \mu(H/N) + k$.

A group G has $vdl(G) = 1$ if and only if G is virtually abelian, and hence $\mu(H) = 1$ if and only if every solvable subgroup of H is virtually abelian. By [**BFH05**], $\mu(Out(F_n)) = 1$ for any free group F_n.

The situation for $GL(n, \mathbb{Z})$ is more complicated. Let U_n denote the unitriangular matrices in $GL(n, \mathbb{Z})$, that is, the (lower) triangular matrices with 1's on the diagonal.

PROPOSITION 20. $\mu(U_n) = \lfloor \log_2(n-1) \rfloor + 1$, *and* $\mu(U_n) \leq \mu(GL(n, \mathbb{Z})) \leq \mu(U_n) + 1$.

PROOF. It is easy to verify that $U_n(R)$ is solvable with derived length less than $log_2(n) + 1$ for any ring R. Let $e^a_{i,j}$ denote the elementary matrix with a in the (i,j)-th entry. For any finite index subgroup G of U_n, there exists $m \in \mathbb{Z}$ such that G contains all of the elementary matrices $e^m_{i,j}$ with $i > j$. The relation

$$[e^m_{i,k}, e^m_{k,j}] = e^{m^2}_{i,j}$$

then implies that the kth commutator subgroup $G^{(k)}$ contains all of the elementary matrices of the form $e^a_{i,j}$ with $i \geq j + 2^k$ and $a = m^{(2^k)}$. In particular, $G^{(k)}$ is non-trivial if $2^k < n$. Thus the derived length of G satisfies $\log_2(n) \leq dl(G) \leq dl(U_n) < log_2(n) + 1$, which translates to the first statement of the proposition.

The first inequality of the second statement follows from Lemma 19(2). For the second inequality, we use a theorem of Mal'cev [**Ma56**], which implies that every solvable subgroup $H \subset GL(n, \mathbb{Z})$ is virtually isomorphic to a subgroup of $T_n(\mathcal{O})$, the lower triangular matrices over the ring of integers $\mathcal{O}$ in some number field. The first commutator subgroup of $T_n(\mathcal{O})$ lies in $U_n(\mathcal{O})$, so $vdl(H) \leq dl(T_n(\mathcal{O})) \leq dl(U_n(\mathcal{O})) + 1 = \mu(U_n) + 1$. □

REMARK 21. The exact relation between $\mu(U_n)$ and $\mu(GL(n, \mathbb{Z}))$ is not completely clear. Dan Segal has shown us examples demonstrating that $\mu(GL(n, \mathbb{Z})) = \mu(U_n) + 1$ for $n = 1 + 3 \cdot 2^t$, while for $n = 1 + 2^t$ he shows $\mu(U_n) = \mu(GL(n, \mathbb{Z}))$ [**Se09**].

6.2. Maximum derived length for homogeneous graphs. In the case of a homogeneous graph Γ, it is easy to obtain an upper bound on the virtual derived length of solvable subgroups of $Out(A_\Gamma)$:

THEOREM 22. *If* Γ *is homogeneous of dimension* n, *then* $\mu(Out(A_\Gamma)) \leq n$.

PROOF. First assume Γ is not the star of a single vertex. Since $Out^0(A_\Gamma)$ has finite index in $Out(A_\Gamma)$, their maximal derived lengths μ agree. We proceed by induction on n. For $n = 1$, Γ is discrete so $\mu(Out(A_\Gamma)) = 1$ by [**BFH05**]. For $n > 1$, we apply Lemma 19 to the abelian extension,

$$1 \to K_P \to Out^0(A_\Gamma) \to \prod Out^0(A_{lk[v]})$$

to conclude that $\mu(Out^0(A_\Gamma)) \leq 1 + \max\{\mu(Out^0(A_{lk[v]})\}$. By Lemma 15, $lk[v]$ is homogeneous of dimension $n-1$ and not a star, so the theorem follows by induction.

Now suppose $\Gamma = st(v)$. If Γ is a complete graph, $Out(A_\Gamma) \cong GL(n,\mathbb{Z})$, which has virtual derived length $\mu(GL(n,\mathbb{Z})) < log_2(n) + 2 \le n+1$. If it is not a complete graph, then by Remark 16, $Out(A_\Gamma)$ is an abelian extension of $GL(k,\mathbb{Z}) \times Out(A_{lk[v]})$, where $0 < k < n$ and $lk[v]$ is homogeneous of dim $n-k$ and not a star. Hence by induction,

$$\mu(Out(A_\Gamma)) \le 1 + \max\{\mu(GL(k,\mathbb{Z})), \mu(Out(A_{lk[v]}))\} \le 1 + \max\{k, n-k\} \le n.$$

□

Here is a stronger formulation of the previous theorem. If Γ is homogeneous of dimension n and Δ is an $(n-1)$-clique, then $lk(\Delta)$ is discrete, hence generates a free group $F(lk(\Delta))$.

THEOREM 23. *Let Γ be homogeneous of dimension $n \ge 2$ and assume Γ is not the star of a single vertex. Then there is a homomorphism*

$$Q : Out^0(A_\Gamma) \to \prod Out(F(lk(\Delta))),$$

where the product is taken over some collection of $(n-1)$-cliques, such that the kernel of Q is a solvable group of derived length at most $n-1$.

PROOF. Induction on n. For $n = 2$, take $Q = P$. The kernel K_P is abelian.

Suppose $n > 2$. Then P maps $Out^0(A_\Gamma)$ to a product of groups $Out^0(A_{lk[v]})$, where $lk[v]$ is homogeneous of dimension $n-1$ and not the star of a single vertex. By induction, there exists a homomorphism Q_v from $Out^0(A_{lk[v]})$ to a product of groups $Out(F(lk_v(\Delta)))$ where Δ is an $n-2$ clique in $lk(v)$ and $lk_v(\Delta)$ is its link. The kernel H_v of Q_v is solvable of derived length at most $n-2$.

Let $\Delta' = \Delta * v$. Then Δ' is an $(n-1)$-clique in Γ whose link $lk(\Delta')$ is exactly $lk_v(\Delta)$. Thus the composite $Q = (\prod Q_v) \circ P$ gives the desired homomorphism. The kernel of Q fits in an exact sequence

$$1 \to K_P \to ker\ Q \to \prod H_v.$$

It follows that $ker\ Q$ is solvable of derived length at most $n-1$. □

6.3. Examples of solvable subgroups. We now investigate lower bounds on the virtual derived length of $Out(A_\Gamma)$. If $[v]$ is an abelian equivalence class with k elements, then $GL(k,\mathbb{Z})$ embeds as a subgroup of $Out(A_\Gamma)$; in particular, $Out(A_\Gamma)$ contains solvable subgroups of virtual derived length at least $log_2(k)$. In homogeneous graphs, abelian equivalence classes have only one element, so one cannot construct non-abelian solvable subgroups in this way. However, non-abelian solvable subgroups do exist, and we show two ways of constructing them in this section. More examples may be found in [**Da09**].

PROPOSITION 24. *Let Γ be any finite simplicial graph. Suppose Γ contains k distinct vertices, $v_1, \dots, v_k$ satisfying*

(1) *$v_2, \dots, v_k$ span a $(k-1)$-clique*
(2) *$[v_1] \le [v_2] \le \dots \le [v_k]$*

Then $Out(A_\Gamma)$ contains a subgroup isomorphic to the unitriangular group U_k. In particular, $\mu(Out(A_\Gamma)) \ge log_2(k)$.

PROOF. Let α_i denote the transvection $v_i \mapsto v_i v_{i+1}$. Let H denote the subgroup of $Out(A_\Gamma)$ generated by α_i, $1 \leq i \leq k-1$. Since H preserves the subgraph Γ' spanned by the v_i's, it restricts to a subgroup of $Out(A_{\Gamma'})$. It is easy to see that this restriction maps H isomorphically onto its image, so without loss of generality, we may assume that $\Gamma = \Gamma'$.

Abelianizing A_Γ gives a map ρ from H to $GL(k, \mathbb{Z})$. The image of α_i under ρ is the elementary matrix $e^1_{i+1,i}$. It follows that the image of H in $GL(k, \mathbb{Z})$ is precisely U_k. Thus, it suffices to verify that the kernel of ρ is trivial. Let Δ be the clique spanned by $v_2, \dots, v_k$. Note that an element of H takes each v_i to $v_i w_i$ for some $w_i \in A_\Delta$. Since A_Δ is already abelian, a non-trivial w_i cannot be killed by abelianizing A_Γ. Thus, ρ is injective. □

PROPOSITION 25. *Let Γ be any finite simplicial graph. Suppose Γ contains $k-1$ distinct vertices, $v_1, \dots, v_{k-1}$ satisfying*

(1) *$v_1, \dots, v_{k-1}$ span a $(k-1)$-clique*
(2) *$[v_1] \leq \cdots \leq [v_{k-1}]$*
(3) *$\Gamma - st(v_1)$ has at least two distinct components that are not contained in $st(v_{k-1})$*

Then $Out(A_\Gamma)$ contains a subgroup isomorphic to U_k. In particular, $\mu(Out(A_\Gamma)) \geq log_2(k)$.

PROOF. For $i = 1, \dots k-2$, take α_i, to be the transvection $v_i \mapsto v_i v_{i+1}$. Let C be a component of $\Gamma - st(v_1)$ which is not contained in $st(v_{k-1})$. For $i = 1, \dots, k-1$, take β_i to be the partial conjugation of C by v_i. Note that β_i is non-trivial in $Out(A_\Gamma)$ since condition (3) guarantees that $\Gamma - st(v_i)$ contains at least two components. Let H denote the subgroup of $Out(A_\Gamma)$ generated by the α_i's and β_i's. We claim that H is isomorphic to U_k.

Since α_i acts only on vertices in $st(v_1)$ while β_i acts only on vertices not in $st(v_1)$, the subgroups H_α and H_β generated by the α_i's and β_i's respectively, are disjoint and H_β is easily seen to be normal in H. Hence H is the semi-direct product, $H = H_\beta \ltimes H_\alpha$. The subgroup H_α is isomorphic to U_{k-1}, as shown in the proof of the previous proposition, while H_β is isomorphic to the free abelian group A_Δ generated by the clique Δ spanned by the v_i's. It is now easy to verify that H is isomorphic to U_k as claimed. □

EXAMPLE 26. Suppose A_Γ is homogeneous of dimension 2. Then by Theorem 22, $\mu(Out(A_\Gamma)) \leq 2$. In the next section (Corollary 31), we will show that if Γ has no leaves, then $\mu(Out(A_\Gamma)) = 1$. If Γ does have leaves, and for some leaf v, $st(v)$ separates Γ, then $\mu(Out(A_\Gamma)) = 2$. For if the components of $\Gamma - st(v)$ are not leaves, then Proposition 25 implies that $\log_2(3) \leq \mu(Out(A_\Gamma))$, and if some component is a leaf v' attached at the same base w, then Proposition 24 applied to $[v] \leq [v'] \leq [w]$ gives the same result.

6.4. Translation lengths and solvable subgroups. In constructing the non-abelian solvable subgroups above, a key role was played by transvections $v \mapsto vw$ between adjacent (i.e. commuting) vertices. We call these *adjacent transvections.* In this section, we will show that without adjacent transvections, no such subgroups can exist for homogenous graphs.

Recall that $Out(A_\Gamma)$ is generated by the finite set S consisting of graph symmetries, inversions, partial conjugations and transvections. Define the following subsets of S,

$$\tilde{S} = S - \{\text{adjacent transvections}\}$$
$$\tilde{S}^0 = \tilde{S} - \{\text{graph symmetries}\}$$

and the subgroups of $Out(A_\Gamma)$ generated by them,

$$\widetilde{Out}(A_\Gamma) = \langle \tilde{S} \rangle$$
$$\widetilde{Out}^0(A_\Gamma) = \langle \tilde{S}^0 \rangle.$$

We will prove that for Γ homogeneous, all solvable subgroups of $\widetilde{Out}(A_\Gamma)$ are virtually abelian.

The proof proceeds by studying the translation lengths of infinite-order elements. The connection between solvable subgroups and translation lengths was first pointed out by Gromov [**Gr87**].

DEFINITION 27. Let G be a group with finite generating set S, and let $\|g\|$ denote the word length of g in S. The *translation length* $\tau(g) = \tau_{G,S}(g)$ is the limit

$$\lim_{k\to\infty} \frac{\|g^k\|}{k}.$$

Elementary properties of translation lengths include the following (see [**GS91**], Lemma 6.2):

- $\tau(g^k) = k\tau(g)$.
- If S' is a different finite generating set, then $\tau_{G,S}(g)$ is positive if and only if $\tau_{G,S'}(g)$ is positive.
- If $H \le G$ is a finitely-generated subgroup, and the generating set for G includes the generating set for H, then $\tau_H(h) \ge \tau_G(h)$.

Note that the kernel K_R of the amalgamated restriction homomorphism lies in $\widetilde{Out}^0(A_\Gamma)$ since it is generated by products of partial conjugations.

PROPOSITION 28. *Assume Γ is connected. Then every element of the kernel K_R of the amalgamated restriction homomorphism has positive translation length in $\widetilde{Out}^0(A_\Gamma)$.*

PROOF. Fix $\phi \in K_R$. We will find a function $\lambda = \lambda_\phi \colon \widetilde{Out}^0(A_\Gamma) \to \mathbb{R}_{\ge 0}$ satisfying

(1) $\lambda(\phi^k) \ge \frac{k}{2}$ and
(2) $\lambda(\gamma_1 \dots \gamma_k) \le 2k$ for $\gamma_i \in \tilde{S}^0$

The proof is then finished by the following argument. Let $m_k = \|\phi^k\|$, and write $\phi^k = \gamma_1 \dots \gamma_{m_k}$ with $\gamma_i \in \tilde{S}^0$. Then

$$\frac{k}{2} \le \lambda(\phi^k) = \lambda(\gamma_1 \dots \gamma_{m_k}) \le 2m_k$$

so

$$\frac{m_k}{k} \ge \frac{1}{4} > 0$$

for all k, so

$$\tau(\phi) = \lim_{k\to\infty} \frac{m_k}{k} \ge \frac{1}{4} > 0.$$

To define λ recall that by Theorem 8, the kernel K_R is free abelian, generated by $\hat{w}$-component conjugations. We write

$$\phi = \phi_{w_1}\phi_{w_2}\dots\phi_{w_k}$$

where ϕ_{w_i} is a nontrivial product of conjugations by w_i and the w_i are distinct.

First observe that the only transvections onto w_i are adjacent transvections. For if u is not adjacent to w_i and $w_i \leq u$, then there is only one non-trivial $\hat{w}_i$-component (the component of u), hence the unique $\hat{w}_i$-component conjugation is an inner automorphism. It follows that every element of $\widetilde{O}ut^0(A_\Gamma)$ fixes w_i up to conjugacy and inversion.

Set $w = w_1$. For an arbitrary element $x \in A_\Gamma$, define $p(x) = p_w(x)$ to be the absolute value of the largest power of w which can occur in a minimal-length word representing x. For example, if u or v does not commute with w, then $p(wuvw^{-2}) = 2$. If a minimal word representing x does not contain any powers of w, then $p(x) = 0$.

In [**HM95**], Hermiller and Meier describe a "left greedy" normal form for words in A_Γ, obtained by shuffling letters as far left as possible using the commuting relations and canceling inverse pairs whenever they occur. In particular, any reduced word can be put in normal form just by shuffling. It follows that the highest power of w that can occur in a minimal word for x is equal to the highest power of w appearing in the normal form for x.

For any automorphism $f \in Aut(A_\Gamma)$, define $p(f)$ to be the maximum over all vertices $v \neq w$ of $p(f(v))$. For an outer automorphism $\phi \in \widetilde{O}ut^0(A_\Gamma)$, define $\lambda(\phi)$ to be the minimum value of $p(f)$ as f ranges over automorphisms f representing ϕ. We must show that λ satisfies properties (1) and (2) above.

(1) Let f_w be a $\hat{w}$-component conjugation on the $\hat{w}$ component C, and let $v \neq w$ be a vertex of Γ. If $v \in C - st(w)$, then $p(f_w^k(v)) = k$, and $p(f_w^k(v)) = 0$ otherwise. An inner automorphism can reduce the power of w by shifting it to vertices in the complement of C, but cannot reduce the maximum power of w over all vertices by more than $[k/2]$. Since $\phi_w = \phi_{w_1}$ is non-trivial on at least one $\hat{w}$-component, this implies $\lambda(\phi_w^k) \geq k/2$. Since the partial conjugations ϕ_i for $i > 1$ do not change the power of w occuring at any vertex, we conclude that $\lambda(\phi^k) \geq k/2$.

(2) To prove property (2) we need to first establish some properties of the power function p. By abuse of notation, we will view $\tilde{S}^0$ as a subset of $Aut(A_\Gamma)$ in the obvious way.

CLAIM. Let $x \in A_\Gamma$. If $p(x) = 0$ and $f \in \tilde{S}^0$ then $p(f(x)) \leq 1$.

PROOF. If f is a transvection or partial conjugation by some $u \neq w$, then $p(f(x)) = 0$. Likewise for inversions. So the only cases we have to consider are when f is either a non-adjacent transvection of w onto v or partial conjugation of $C \subset \Gamma$ by w.

Suppose f is a (non-adjacent) transvection $f\colon v \mapsto vw$ or $f\colon v \mapsto wv$. Then $f(x)$ has the property that any two copies of w are separated by v and any two copies of w^{-1} are separated by v^{-1}. "Shuffling left" can never switch the order of v and w, so this must also be true in the normal form for $f(x)$.

If f is a partial conjugation by w, then the w's in $f(x)$ alternate, i.e.

$$f(x) = a_1wa_2w^{-1}a_3w\dots$$

where the a_i are words which do not use w or w^{-1}, so shuffling left can only cancel w-pairs, never increase the power to more than 1. ☐

A minimal word representing $x \in A_\Gamma$ can be put in the form

$$a_0 w^{k_1} a_1 w^{k_2} \dots w^{k_n} a_n$$

where

- a_i contains no w and
- w does not commute with a_i for $1 \le i \le n-1$.

so that $p(x) = max\{k_i\}$.

CLAIM. For any $f \in \tilde{S}^0$ and $x \in A_\Gamma$, $p(f(x)) \le p(x) + 2$.

PROOF. First assume that $f(w) = w$. This holds for all generators in $\tilde{S}^0$ with the exception of a partial conjugation by u of a component C containing w. Write $x = a_0 w^{k_1} a_1 w^{k_2} \dots, w^{k_n} a_n$ as above. Let b_i be the normal form for $f(a_i)$. Then $f(x) = b_0 w^{k_1} b_1 w^{k_2} \dots w^{k_n} b_n$, where b_i does not commute with w.

Case 1: f is a partial conjugation or transvection by w. Then no w can shuffle across an entire b_i, so we need only consider the highest power appearing in $b_{i-1} w^{k_i} b_i$. Now by the previous claim, b_i is of the form

$$b_i = c_1 w^{\pm 1} c_2 w^{\pm 1} \dots w^{\pm 1} c_k$$

where w does not commute with $c_2, \dots, c_{k-1}$ and similarly for b_{i-1}. It follows that left shuffling of $b_{i-1} w^{k_i} b_i$ can at worst combine w^{k_i} with the last w in b_{i-1} and the first w in b_i, producing a power of at most $|k_i| + 2$.

Case 2: f is a partial conjugation (on a component not containing w) or transvection by some $u \neq w$. Then no new w's appear and no w can shuffle across an entire b_i, so the maximum power of w does not change, i.e., $p(f(x)) = p(x)$.

It remains to consider the case where f is a partial conjugation by u with $f(w) = uwu^{-1}$. Then f can be written as the composite of an inner automorphism by u followed by a product of partial conjugations fixing w. By case 2 above, we have $p(f(x)) = p(uxu^{-1})$. Since u does not commute with w, conjugating by u changes only the factors a_0 and a_n in the normal form for x. Thus $p(uxu^{-1}) = p(x)$. □

If $f \in Aut(A_\Gamma)$ can be written as a product of m elements of $\tilde{S}^0$, then the above claim shows that $p(f(v)) \le 2m$, for any vertex $v \neq w$. If $\phi \in \widetilde{Out}^0(A_\Gamma)$ can be written as a product $\phi = \phi_1 \dots \phi_m$, with ϕ_i represented by $f_i \in \tilde{S}^0$, then $\lambda(\phi) \le p(f) \le 2m$. This completes the proof of the proposition. □

COROLLARY 29. *If Γ is homogeneous of dimension n, then the translation length of every infinite-order element of $\widetilde{Out}^0(A_\Gamma)$ is positive.*

PROOF. If $\Gamma = st(v)$, then by Remark 16, $Out(A_\Gamma) = Tr \rtimes (GL(k,\mathbb{Z}) \times Out(A_{lk[v]}))$. In this case, $\widetilde{Out}^0(A_\Gamma) = (\mathbb{Z}/2)^k \times \widetilde{Out}^0(A_{lk[v]})$ where the $(\mathbb{Z}/2)^k$ factor is generated by inversions of elements in $[v]$. Thus every infinite order element of $\widetilde{Out}^0(A_\Gamma)$ projects to an infinite order element of $\widetilde{Out}^0(A_{lk[v]}))$, so it suffices to prove the corollary in the case where Γ is not a star.

We proceed by induction on n. For $n = 1$, A_Γ is a free group F_k and $\widetilde{Out}^0(A_\Gamma) = Out(F_k)$. Alibegovic proved that infinite order elements of $Out(F_k)$ have positive translation length [**Al02**].

For $n > 1$, we will make use of the amalgamated projection homomorphism

$$P = \prod P_v \colon Out^0(A_\Gamma) \to \prod_{v\ maximal} Out(A_{lk[v]}).$$

Note that each projection P_v maps generators in $\tilde{S}^0$ to either the trivial map or to a generator of the same form in $Out(A_{lk[v]})$. Thus the image of $\widetilde{O}ut^0(A_\Gamma)$ lies in the product of the subgroups $\widetilde{O}ut^0(A_{lk[v]})$. Moreover, the kernel of P restricted to $\widetilde{O}ut^0(A_\Gamma)$ is just K_R. This follows from the fact that K_P is generated by K_R and leaf-transvections, which by definition, are adjacent transvections.

By [**CV08**], all of the groups we are considering are virtually torsion-free. Let $G \leq \widetilde{O}ut^0(A_\Gamma)$ be the inverse image of a torsion-free finite-index subgroup of $\prod \widetilde{O}ut^0(A_{lk[v]})$. If $\phi \in \widetilde{O}ut^0(A_\Gamma)$ has infinite order, then some power of ϕ is a non-trivial (infinite-order) element in G, so we need only prove that elements of G have positive translation length in $\widetilde{O}ut^0(A_\Gamma)$.

If the image of $\phi \in G$ is non-trivial in some $\widetilde{O}ut^0(A_{lk[v]})$, then it has positive translation length by induction. If the image is trivial, then ϕ lies in K_R so we are done by Proposition 28. □

COROLLARY 30. *If Γ is homogeneous of dimension n, then $\widetilde{O}ut(A_\Gamma)$ satisfies the strong Tits alternative, that is, every subgroup of $\widetilde{O}ut(A_\Gamma)$ is either virtually abelian or contains a non-abelan free group.*

PROOF. By Theorem 17, every subgroup not containing a free group is virtually solvable. So it remains to show that every solvable subgroup is virtually abelian. Since $\widetilde{O}ut^0(A_\Gamma)$ has finite index in $\widetilde{O}ut(A_\Gamma)$, it suffices to prove the same statement for $\widetilde{O}ut^0(A_\Gamma)$.

Bestvina [**Be99**], citing arguments from Conner [**Co00**] and Gersten and Short [**GS91**], shows that if a finitely-generated group G satisfies (i) G is virtually torsion-free, (ii) every abelian subgroup of G is finitely generated, and (iii) the translation length of every infinite order element is positive, then solvable subgroups of G are virtually abelian. $\widetilde{O}ut^0(A_\Gamma)$ satisfies (iii) by Corollary 29 and (i) by [**CV08**]. Since Γ is homogeneous, the fact that abelian subgroups of $Out^0(A_\Gamma)$ are finitely generated follows by a simple induction from the same fact for $Out(F_n)$ and $GL(n, \mathbb{Z})$ using the projection homomorphisms. □

In dimension 2, the only adjacent transvections are leaf transvections, so if Γ has no leaves, then $\widetilde{O}ut(A_\Gamma) = Out(A_\Gamma)$. Thus the following is a special case of Corallary 30.

COROLLARY 31. *If Γ is connected with no triangles and no leaves, then $Out(A_\Gamma)$ satisfies the strong Tits alternative.*

7. Questions

Since the projection homomorphism $P\colon Out^0(A_\Gamma) \to \prod Out^0(A_{lk[v]})$ is defined only for connected graphs Γ, inductive arguments using P break down if the links of maximal vertices are not connected, unless the desired result is known by some other argument for outer automorphism groups of free products. For homogeneous graphs, the links are always connected so this is not an issue, but several of the questions answered in this paper remain open for non-homogeneous graphs. Specifically, we can ask

(1) Is the maximal virtual derived length of a solvable subgroup of $Out(A_\Gamma)$ bounded by the dimension of A_Γ?
(2) Does $Out(A_\Gamma)$ satisfy the Tits alternative?

References

[Al02] E. Alibegovich, *Translation lengths in $Out(F_n)$*, Geom. Dedicata **92** (2002) 87–93.

[Be99] M. Bestvina, *Non-positively curved aspects of Artin groups of finite type*, Geom. Topol. **3** (1999) 269–302.

[BFH00] M. Bestvina, M. Feighn and M. Handel, *The Tits alternative for* $\mathrm{Out}(F_n)$*. I: Dynamics of exponentially-growing automorphisms*, Ann. of Math. (2) **151** (2000), 517–623.

[BFH05] M. Bestvina, Mark Feighn and Michael Handel, *The Tits alternative for* $\mathrm{Out}(F_n)$*. II. A Kolchin type theorem*, Ann. of Math. (2) **161** (2005), 1–59.

[BCV09] K.-U. Bux, R. Charney, and K. Vogtmann, *Automorphisms of two-dimensional RAAGs and partially symmetric automorphisms of free groups*, Groups Geom. Dyn. **3** (2009) no. 4, 541–554.

[Ch07] R. Charney, *An introduction to right-angled Artin groups*, Geom. Dedicata **125** (2007) 141–158.

[CCV07] R. Charney, J. Crisp and K. Vogtmann, *Automorphisms of 2-dimensional right-angled Artin groups*, Geom. Topol. **11** (2007), 2227–2264.

[CV08] R. Charney and K. Vogtmann, *Finiteness properties of automorphism groups of right-angled Artin groups*, Bull. Lond. Math. Soc. **41** (2009), no. 1, 94–102.

[Co00] G. Connor, *Discreteness properties of translation numbers in solvable groups*, J. Group Theory **3** (2000), no. 1, 77–94.

[Da09] M. Day, *On solvable automorphism groups of right-angled Artin groups*, to appear in proceedings of the 2009 International Conference on Geometric & Combinatorial Methods in Group Theory & Semigroup Theory, arXiv:0910.4789.

[GS91] S. Gersten and H. Short, *Rational subgroups of biautomatic groups*, Ann. of Math. (2) **134 (1)** (1991), 125–158.

[Gr74] E. Grossman, *On the residual finiteness of certain mapping class groups*, J. Lond. Math. Soc. (2), **9** (1974), 160–164.

[Gr87] M. Gromov, *Hyperbolic groups*, in *Essays on Group Theory*, MSRI series, vol. 8, edited by S. Gersten, Springer-Verlag, 1987.

[GL07] V. Guirardel and G. Levitt, *The Outer space of a free product*, Proc. Lond. Math. Soc. (3) **94** (2007), 695–714.

[HM95] S. Hermiller and J. Meier, *Algorithms and geometry for graph products of groups*, J. Algebra **171** (1995), no. 1, 230–257.

[Lau95] M. R. Laurence, *A generating set for the automorphism group of a graph group*, J. Lond. Math. Soc. (2) 52 (1995) 318–334.

[Ma56] A. I. Mal'cev, *On certain classes of innite soluble groups*, Mat. Sbornik 28 (1951) 567-588 (Russian); Amer. Math. Soc. Translations (2) 2 (1956) 1–21.

[Mi71] C. F. Miller, *On Group-Theoretic Decision Problems and their Classification*, Princeton Univ. Press, Princeton, NJ, 1971.

[Mi09] A. Minasyan, *Hereditary conjugacy separability of right angled Artin groups and its applications*, to appear in Groups, Geometry and Dynamics, arXiv:0905.1282.

[MiOs09] A. Minasyan, D. Osin, *Normal automorphisms of relatively hyperbolic groups*, Trans. Amer. Math. Soc. **362** (2010), 6079–6103.

[Se09] D. Segal, private communication.

[Ser89] H. Servatius, *Automorphisms of graph groups*, J. Algebra **126** (1989) 34–60.

[Ti72] J. Tits, *Free subgroups in linear groups*, J. Algebra **20** (1972), 250–270.

Department of Mathematics, Brandeis University, Waltham, Massachusetts 02453
E-mail address: `charney@brandeis.edu`

Department of Mathematics, Cornell University, Ithaca, New York 14853
E-mail address: `vogtmann@math.cornell.edu`

Proceedings of Symposia in Pure Mathematics
Volume **82**, 2011

Abelian ρ–invariants of iterated torus knots

Maciej Borodzik

ABSTRACT. We compute the ρ–invariant for iterated torus knots K for the standard representation $\pi_1(S^3 \setminus K) \to \mathbb{Z}$ given by abelianisation. For algebraic knots, this invariant turns out to be very closely related to an invariant of a plane curve singularity, coming from algebraic geometry.

1. Introduction

A von Neumann ρ–invariant (also called L^2–signature, or L^2–eta invariant) of a real closed 3–manifold M is a real number $\rho_\phi(M)$ associated to every representation $\phi : \pi_1(M) \to \Gamma$, where Γ is any countable discrete group. As a special case, if K is a knot in a 3–sphere, and we consider representations of the fundamental group of the manifold $S^3_0(K)$ (i.e. a zero framed surgery along K), then we can talk about the ρ–invariants of knots. In particular, the representation $ab : \pi_1(S^3 \setminus K) \to \mathbb{Z}$, given by abelianization, gives rise to the representation $\widetilde{ab} : \pi_1(S^3_0(K)) \to \mathbb{Z}$ and the corresponding invariant, $\rho_0(K)$, turns out to be the integral over normalised unit circle of the Tristram–Levine signature of a knot.

The ρ–invariants for knots have been introduced first in [**ChG**]. They were then deeply studied in [**COT1**]. In their seminal paper, the authors observed that they are a very subtle obstruction for some knots to be slice. Namely, let us be given a knot K bounding a disk D in the ball B^4. Let $Y = \partial(B^4 \setminus \nu(D))$, where ν denotes the tubular neighbourhood. Then Y is canonically isomorphic to $S^3_0(K)$, and, for any representation $\phi : \pi_1(Y) \to \Gamma$ that can be extended to $\tilde{\phi} : \pi_1(B^4 \setminus \nu(D)) \to \Gamma$, the corresponding ρ–invariant must vanish. This allows to construct examples of non-slice knots, indistinguishable from slice knots by previously known methods as the Tristram–Levine signature or the Casson–Gordon invariants.

The difficulty of computability of ρ–invariants is the cost of their subtlety. Only in the first nontrivial case of the representation given by abelianisation of π_1, there is a general method of computing this invariant, namely integrating the Tristram–Levine signature. In papers [**COT2**], [**Ha**], and others, these invariants

2010 *Mathematics Subject Classification.* Primary 57M25; Secondary 14H20.

Key words and phrases. ρ–invariant, L^2–signature, Tristram–Levine signature, torus knot, algebraic knot, plane curve singularity.

The author is supported by Polish MNiSzW Grant No N N201 397937 and also by the Foundation for Polish Science FNP.

were computed also for some other representations of the knot group. But there, the choice of knots is very specific.

In this paper we focus on ρ_0–invariant and compute it for all iterated torus knots. The computation consists of integrating the Tristram–Levine signature, which is not a completely trivial task. In fact, we do even more: we compute the Fourier transform of the Tristram–Levine signature function of iterated torus knot. This transform can be expressed by a surprisingly simple formula. In particular, this method can be used to detect knots, which are connected sums of iterated torus knots and which have identical Tristram–Levine signature.

What we find most interesting and striking about ρ_0 of algebraic knots, is its relation with deep algebro-geometrical invariants of the plane curve singularity. We state this relation, in terms of a uniform bound (see Proposition 4.6) but, honestly speaking, we are far from understanding it. Moreover, this relation is not that clear for algebraic links, as we show on an example.

The structure of the paper is the following. In Section 2 we recall, how to compute the Tristram–Levine signature for iterated torus knots and formulate Theorem 2.8. Then we deduce some of its corollaries. In Section 3 we prove Theorem 2.8. In Section 4 we recall definitions of some invariants of plane curve singularities and compare them to ρ_0 for algebraic knots. We end this section by computing the ρ_0 for a (d,d) torus link, i.e., the link of singularity $x^d - y^d = 0$.

We apologise the reader for not giving a definition of the ρ–invariant. A precise definition from scratch, including necessary definitions of twisted signature of a 4–manifold, would make this paper at least twice as long. Instead we refer to [**COT1**, Section 5] or to papers [**Cha, ChW, Ha**].

We end this introduction by remarking that the ρ–invariants were also studied in the context of mixed Hodge structures of hypersurface singularities. The η invariant, defined, for instance, in [**Ne1**, Section I], is closely related to the ρ_0 invariant in the case of plane curve singularities. We refer to [**Ne3, Ne4**] for the detailed study of this invariant.

2. Tristram–Levine signature of torus knots

We begin this section with some definitions, which we give also to fix the notation used in the article.

DEFINITION 2.1. A knot is called an *iterated torus knot* if it arises from an unknot by finitely many cabling operations. An iterated torus knot is of type $(p_1, q_1, \dots, p_n, q_n)$ if it is a (p_1, q_1) cable of (p_2, q_2) cable of ... of (p_n, q_n) cable of an unknot. Fore example, a torus knot $T_{p,q}$ is an iterated torus knot of type (p,q).

DEFINITION 2.2. Let K be a knot, S its Seifert matrix. Let $\zeta \in \mathbb{C}$, $|\zeta| = 1$. The *Tristram–Levine signature*, $\sigma_K(\zeta)$ is the signature of the hermitian form given by

$$(1 - \zeta)S + (1 - \bar{\zeta})S^T. \tag{2.1}$$

It is well-known that the form (2.1) is degenerate (i.e. has non-trivial kernel) if and only if ζ is a root of the Alexander polynomial Δ_K of K. The function $\zeta \to \sigma_K(\zeta)$ is piecewise constant with possible jumps only at the roots of the Alexander polynomial $\Delta_K(\zeta)$. The value of σ_K at such root can *a priori* be different than left or right limit of σ_K at that point. However, there are only finitely many such values and they do not influence the integral. As we do not want to take care of this values, we introduce a very handy notion.

DEFINITION 2.3. We shall say that two piecewise-constant functions from a unit circle (or a unit interval) to real numbers are *almost equal* if they are equal at all but finitely many points.

We would like to compute ρ_0 for an iterated torus knot. We will use Proposition 5.1 from **[COT2]**, which we can formulate as follows.

PROPOSITION 2.4. *For any knot $K \subset S^3$ we have*

$$\rho_0(K) = \int_0^1 \sigma_K(e^{2\pi i x})dx.$$

Therefore, what we have to do, is to compute the integral of the Tristram–Levine signature for an iterated torus knot. We begin with recalling results from **[Li]**, where the function σ_K is computed for iterated torus knots.

Let p, q be coprime positive integers. Let x be in the interval $[0, 1]$. Consider the set

$$\Sigma = \Sigma_{p,q} = \left\{ \frac{k}{p} + \frac{l}{q} : 1 \leq k < p,\ 1 \leq l < q \right\} \subset [0, 2] \cap \mathbb{Q}.$$

The function $s_{p,q}(x)$ is defined as

$$s_{p,q}(x) = -2|\Sigma \cap (x, x+1)| + |\Sigma|,$$

where $|\cdot|$ denotes the cardinality of the corresponding set.

LEMMA 2.5 (**[Li]**). *If $\zeta = e^{2\pi i x}$ is not a root of the polynomial $(t^{pq} - 1)(t - 1)/(t^p - 1)(t^q - 1)$, then the Tristram–Levine signature of the torus knot $T_{p,q}$ at ζ is equal to $s_{p,q}(x)$.*

Therefore, computing the ρ–invariant of a torus knot boils down to computing the integral of the function $s_{p,q}(x)$. Before we do this, let us show, how one can compute the Tristram–Levine signatures of an iterated torus knot. We shall need another lemma from **[Li]**.

LEMMA 2.6. *Let K be a knot and $K_{p,q}$ be the (p, q)–cable on K. Then for any $\zeta \in \mathbb{C}$, $|\zeta| = 1$, we have*

$$\sigma_{K_{p,q}}(\zeta) = \sigma_K(\zeta^q) + \sigma_{T_{p,q}}(\zeta).$$

This allows a recursive computation for an iterated torus knot. Namely, let for $r > 1$

$$s_{p,q;r}(x) = s_{p,q}(rx - \lfloor rx \rfloor).$$

Here $\lfloor \alpha \rfloor = \max\{n \in \mathbb{Z}, n \leq \alpha\}$.

COROLLARY 2.7. *Let K be an iterated torus knot of type $(p_1, q_1, \ldots, p_n, q_n)$. Let $x \in [0, 1]$ be such that $e^{2\pi i x}$ is not a root of the Alexander polynomial of K. Denote by $r_k = q_1 \ldots q_{k-1}$. Then*

$$\sigma_K(e^{2\pi i x}) = \sum_{k=1}^{n} s_{p_k, q_k, r_k}(x).$$

The core of this section is

THEOREM 2.8. *For any $\beta \in \mathbb{C}$ which is not an integer divisible by r we have*

$$\int_0^1 e^{\pi i \beta x} s_{p,q,r}(x)\,dx = \frac{2e^{\pi i\beta/2}\sin\frac{\pi\beta}{2}}{\pi\beta} n_{p,q;r}(\frac{\pi\beta}{2}), \tag{2.2}$$

where

$$n_{p,q;r}(t) = \cot\frac{t}{pqr}\cot\frac{t}{r} - \cot\frac{t}{pr}\cot\frac{t}{qr}.$$

In particular, by passing to a limit $\beta \to 0$ we get

$$\int_0^1 s_{p,q,r} = -\frac{1}{3}(p - \frac{1}{p})(q - \frac{1}{q}).$$

REMARK 2.9. The function $n_{p,q;r}(t)$ will be called *normalised Fourier transform* of the signature function.

We prove Theorem 2.8 in Section 3. Now we pass to corollaries.

COROLLARY 2.10. *An iterated torus knot of type $(p_1, q_1, \dots, p_n, q_n)$ has ρ_0 invariant equal to*

$$-\frac{1}{3}\sum_{k=1}^{n}(p_k - \frac{1}{p_k})(q_k - \frac{1}{q_k}).$$

REMARK 2.11. This result appears first in the paper of Kirby and Melvin [**KM**, Remark 3.9] and in [**Ne2**, Example 4.3]. I owe this information to Andrew Ranicki and András Némethi.

Apart of this corollary, Theorem 2.8 has its interest of its own. In fact, it might help to study possible cobordism relations between iterated torus knot. For example, Litherland showed in [**Li**], that the connected sum of knots $T_{2,3}$, $T_{3,5}$ and a $(2,5)$-cable on $T_{2,3}$ has the same Tristram–Levine signature as a $T_{6,5}$. It might be possible that normalised Fourier transforms can help studying similar phenomena. This could be done as follows.

LEMMA 2.12. *Let us be given two finite sets I and J of triples of integers $\{p, q, r\}$. Then the difference*

$$\Delta_{IJ}(x) := \sum_{i \in I} s_{p_i,q_i;r_i}(x) - \sum_{j \in J} s_{p_j,q_j;r_j}(x) \tag{2.3}$$

is almost equal to zero for $x \in [0, 1]$, if and only if the difference

$$\widehat{\Delta}_{IJ}(t) := \sum_{i \in I} n_{p_i,q_i;r_i}(t) - \sum_{j \in J} n_{p_j,q_j;r_j}(t) \tag{2.4}$$

is equal to zero on some open subset in $\mathbb{C}$.

SKETCH OF PROOF. The 'only if' part is trivial. To prove the 'if' part we observe that $\widehat{\Delta}_{IJ}(t) \cdot \frac{t}{e^t \sin t}$ is, up to a multiplicative constant, and up to rescaling of the parameter t, the Fourier transform of $\Delta_{IJ}(x)$, when we extend $\Delta_{IJ}(x)$ by 0 to the whole real line. On the other hand, vanishing of $\widehat{\Delta}_{IJ}(t)$ on some open subset of $\mathbb{C}$ implies that it is everywhere 0. □

PROPOSITION 2.13. *The property that $\Delta_{IJ}(x)$ is almost equal to zero is equivalent to the fact, that two following conditions are satisfied at once*

(a) $\sum_{i \in I}(p_i - \frac{1}{p_i})(q_i - \frac{1}{q_i}) = \sum_{j \in J}(p_j - \frac{1}{p_j})(q_j - \frac{1}{q_j})$.

(b) *For any $t_0 \in \mathbb{C}$ such that there exists $k \in I \cup J$ such that $\pi r_k t_0 \in \mathbb{Z}$, the residuum at t_0 of $\widehat{\Delta}_{IJ}(t)$ is zero.*

REMARK 2.14. If T is the least common multiplier of $p_k q_k r_k$ for $k \in I \cup J$, then $T\pi$ is the period of $\widehat{\Delta}_{IJ}(t)$. It follows that the condition (b) involves only finitely many equations.

PROOF OF PROPOSITION 2.13. Vanishing of $\widehat{\Delta}_{IJ}(t)$ clearly implies (b). The equality in (a) is equivalent to $3\widehat{\Delta}_{IJ}(0) = 0$. We shall prove that (b) implies that $\widehat{\Delta}_{IJ}(t)$ is bounded on $\mathbb{C}$. This is done as follows.

Observe that, in general, $\widehat{\Delta}_{IJ}(t)$ can have poles only at such t_0's, that $\pi r_k t_0 \in \mathbb{Z}$, for some $k \in I \cup J$. Moreover, these poles are at most of order 1: in fact, it is a matter of simple computation, that $n_{p,q;r}$ does not have a pole of order 2. Therefore, condition (b) implies that the $\widehat{\Delta}_{IJ}(t)$ extends holomorphically across points $\frac{n}{\pi r_k}$, where $k \in I \cup J$ and $n \in \mathbb{Z}$. As this function is periodic with real period, for any $\delta > 0$ it is bounded on the strip $|\operatorname{Im} t| \leq \delta$ by some constant, which depends, of course, on δ.

A uniform bound on $\widehat{\Delta}_{IJ}(t)$ for $|\operatorname{Im} t| \geq \delta$ results from the standard estimate $|\cot t|^2 \leq 1 + \frac{1}{(\operatorname{Im} t)^2}$. Hence, if (b) holds, then the function $\widehat{\Delta}_{IJ}(t)$ is a bounded holomorphic function, by Liouville's theorem it is then constant. The condition (a) implies then that it vanishes at 0, so it is zero everywhere. □

3. Proof of Theorem 2.8

To make computations at least a bit more transparent, let us first assume that $r = 1$. The function $s_{p,q}$ can be expressed as the sum

$$s_{p,q}(x) = 2 \sum_{\substack{\alpha<1/2 \\ \alpha\in\Sigma_{p,q}}} \chi_{(\alpha,1-\alpha)}(x) - 2 \sum_{\substack{\alpha\in(1/2,1) \\ \alpha\in\Sigma_{p,q}}} \chi_{(1-\alpha,\alpha)}(x),$$

where $\chi_{(a,b)}$ is the characteristic function of the interval (a, b). Therefore

$$\int_0^1 s_{p,q}(x) e^{\pi i \beta x} dx = -\frac{2}{\pi i \beta} \sum_{\substack{\alpha<1 \\ \alpha\in\Sigma_{p,q}}} e^{\pi i \alpha\beta} - e^{\pi i \beta(1-\alpha)}. \tag{3.1}$$

We have

$$\sum_{\substack{\alpha<1 \\ \alpha\in\Sigma_{p,q}}} e^{\pi i \alpha \beta} = \sum_{k=1}^{p-1} \sum_{\substack{l=1 \\ l<q(1-k/p)}}^{q-1} e^{\pi i \beta(k/p+l/q)}.$$

The internal sum on the right hand side is the sum of geometric series (here we use the assumption that β is not an integer) and can be expressed as

$$\frac{1}{1-e^{\pi i \beta/q}} (e^{\pi i \beta k/p} - e^{\pi i \beta(k/p+l_k/q)}),$$

where l_k satisfies

$$k/p + l_k/q > 1 > k/p + (l_k - 1)/q.$$

So we have

$$\sum_{\substack{\alpha<1\\ \alpha\in\Sigma_{p,q}}} e^{\pi i\alpha\beta} = \frac{\sum\limits_{k=1}^{p-1} e^{\pi i\beta k/p} - \sum\limits_{k=1}^{p-1} e^{\pi i\beta(k/p+l_k/q)}}{1-e^{\pi i\beta/q}}. \tag{3.2}$$

The first sum in the numerator is again geometric series. As to the second one, let us denote

$$\gamma_k = k/p + l_k/q.$$

Then γ_k's have the following obvious properties

(a) γ_k's are all different;
(b) $1+\frac{1}{pq} \le \gamma_k \le 1+\frac{p-1}{pq}$;
(c) each γ_k is of the form $1+a_k/pq$ with a_k an integer.

By the Dirichlet principle the set $\{\gamma_1,\dots,\gamma_{p-1}\}$ is the same as the set

$$\left\{1+\frac{1}{pq},\dots,1+\frac{p-1}{pq}\right\}.$$

Therefore, the second sum in the numerator (3.2), upon reordering, can be expressed as

$$\sum_{m=1}^{p-1} e^{\pi i\beta(1+m/pq)},$$

which again is geometric series. Putting all of this together we get

$$\sum_{\substack{\alpha<1\\ \alpha\in\Sigma_{p,q}}} e^{\pi i\alpha\beta} = \frac{1}{1-e^{\pi i\beta/q}}\left(\frac{e^{\pi i\beta/p}-e^{\pi i\beta}}{1-e^{\pi i\beta/p}} - \frac{e^{\pi i\beta(1+1/pq)}-e^{\pi i\beta(1+1/q)}}{1-e^{\pi i\beta/pq}}\right). \tag{3.3}$$

On the other hand, we have

$$\sum_{\substack{\alpha<1\\ \alpha\in\Sigma_{p,q}}} e^{\pi i(1-\alpha)\beta} = e^{\pi i\beta}\sum_{\substack{\alpha<1\\ \alpha\in\Sigma_{p,q}}} e^{\pi i\alpha(-\beta)},$$

and the sum on the right hand side is just (3.3) with $-\beta$ substituted in place of β. Substituting this into (3.1), and applying the formula

$$e^{\pi ia}-e^{\pi ib} = 2ie^{\pi i(a+b)/2}\sin\frac{\pi(a-b)}{2}$$

several times, we arrive finally at

$$\int_0^1 s_{p,q}(x)e^{\pi i\beta}dx = \frac{2e^{\pi i\beta/2}\sin\frac{\pi\beta}{2}}{\pi\beta}\left(\cot\frac{\pi\beta}{2pq}\cot\frac{\pi\beta}{2} - \cot\frac{\pi\beta}{2p}\cot\frac{\pi\beta}{2q}\right).$$

To conclude the proof in the case $r>1$ we observe that

$$\begin{aligned} s_{p,q;r}(x) =& 2\sum_{\substack{\alpha<1/2\\ \alpha\in\Sigma_{p,q}}}\sum_{k=0}^{r-1}\chi_{\left(\frac{\alpha+k}{r},\frac{1-\alpha+k}{r}\right)}(x)+\\ &-2\sum_{\substack{\alpha\in(1/2,1)\\ \alpha\in\Sigma_{p,q}}}\sum_{k=0}^{r-1}\chi_{\left(\frac{1-\alpha+k}{r},\frac{\alpha+k}{r}\right)}(x)\end{aligned}$$

Thus

$$\int_0^1 s_{p,q;r} e^{\pi i \beta x} = \frac{-2}{\pi i \beta} \sum_{\substack{\alpha<1 \\ \alpha\in\Sigma_{p,q}}} \sum_{k=0}^{r-1} e^{\pi i \beta(\alpha/r+k/r)} - e^{\pi i \beta(1-\alpha/r-k/r)}. \tag{3.4}$$

Now, for fixed α we have

$$\sum_{k=0}^{r-1} e^{\pi i \beta(\alpha/r+k/r)} = e^{\pi i \alpha(\beta/r)} \sum_{k=0}^{r-1} e^{\pi i \beta k/r} = e^{\pi i \alpha(\beta/r)} \frac{1-e^{\pi i \beta}}{1-e^{\pi i \beta/r}}.$$

Therefore, returning to (3.4) we get

$$\sum_{\substack{\alpha<1 \\ \alpha\in\Sigma_{p,q}}} \sum_{k=0}^{r-1} e^{\pi i \beta(\alpha/r+k/r)} = \frac{1-e^{\pi i \beta}}{1-e^{\pi i \beta/r}} \sum_{\substack{\alpha<1 \\ \alpha\in\Sigma_{p,q}}} e^{\pi i \alpha(\beta/r)}.$$

We can use (3.3) again, substituting β/r in place of β. Similarly we can deal with a sum of terms $e^{\pi i \beta(1-\alpha/r-k/r)}$. Now straightforward but long computations yield the formula (2.2).

4. Relation with algebraic invariants

The setup in this section is the following. Let $(C,0) \subset \mathbb{C}^2$ be germ of a plane curve singularity with one branch. This means that there exists a local parametrisation $C = (x(t), y(t))$, with x and y analytic functions in one variable with $x(0) = y(0) = 0$. Let us assume that the Puiseux expansion of y in fractional powers of x written in the multiplicative form (see [**EN**, page 49]) is

$$y = x^{q_1/p_1}(c_1 + x^{q_2/p_1p_2}(c_2 + \ldots + x^{q_s/p_1p_2p_3\cdots p_s}(c_s + \ldots))),$$

with $q_1 > p_1$ (otherwise we switch x with y), $\gcd(q_i, p_i) = 1$ and $p_i, q_i > 0$. The pairs $(p_1, q_1), \ldots, (p_n, q_n)$ are called characteristic pairs (or Newton pairs) of the singularity. They completely determine the topological type of the singular point.

LEMMA 4.1 (see e.g. [**EN**]). *Put $a_1 = q_1$ and $a_{k+1} = p_{k+1}p_k a_k + q_{k+1}$. Then the link of the singularity $(C,0)$ is an iterated torus knot. More precisely, it is a (p_n, a_n) cable on (p_{n-1}, a_{n-1}) cable on $\ldots$ on a (p_1, a_1) torus knot.*

REMARK 4.2. The ordering of cables in [**EN**] is different than in [**Li**]. According to Definition 2.1, the link of the singularity $(C,0)$ above would be an iterated torus knot of type $(p_n, q_n, \ldots, p_2, q_2, p_1, q_1)$.

COROLLARY 4.3. *The ρ_0 invariant of an algebraic knot is equal to*

$$\rho_0 = -\frac{1}{3}\sum_{k=1}^{n}\left(a_k p_k - \frac{a_k}{p_k} - \frac{p_k}{a_k} + \frac{1}{p_k a_k}\right). \tag{4.1}$$

It is on purpose that we wrote formula (4.1) in a different shape that in Corollary 2.10.

Let us now resolve the above singularity. This means that we have a map $\pi : (X, E) \to (U, 0)$, where U is a neighbourhood of 0 in $\mathbb{C}^2$, E is the exceptional divisor and X is a complex surface. We require the strict transform C' to be smooth, $C' \cup E$ to have only normal crossings as singularities and the resolution to be minimal, so that we cannot blow-down any exceptional curve without violating one of the two above assumptions.

Put $K = K_X$ the canonical divisor on X and let $D = C' + E_{red}$. Here, the subscript 'red' means that we take a reduced divisor, i.e. coefficients with all components are equal to 1.

LEMMA 4.4 ([**OZ**]). *Using the notation from this section, we have*

$$(K+D)^2 = a_1p_1 - \left\lceil \tfrac{a_1}{p_1} \right\rceil - \left\lceil \tfrac{p_1}{a_1} \right\rceil + \sum_{k=2}^{n} \left(a_kp_k - \left\lceil \tfrac{a_k}{p_k} \right\rceil \right), \tag{4.2}$$

where $(K+D)^2$ *denotes the self-intersection of the divisor* $K+D$, *and* $\lceil x \rceil = \min(n \in \mathbb{Z}, n \geq x)$.

On the one hand $(K+D)^2$ has a very natural meaning. Namely, at least for unibranched singularities, this is the sum of the Milnor number μ and so called $\bar{M}$ number of singularity. The latter, introduced in [**Or**] and studied in [**BZ**], can be interpreted as a parametric codimension of a singular point, i.e. the number of locally independent conditions, which are imposed on a curve given in parametric form, by the appearance of the singularity of given topological type.

On the other hand there is an apparent similarity of left hand sides of formulae (4.1) and (4.2). To make it even more similar, let us take a Zariski–Fujita [**Fuj**] decomposition of the divisor $K+D$. We have then

$$K + D = H + N$$

with H nef (its intersection with any algebraic curve in X is non-negative), N effective, $N^2 < 0$, and for any divisor N' supported on $\operatorname{supp} N$, $H \cdot N' = 0$.

LEMMA 4.5 ([**OZ**]).

$$H^2 = a_1p_1 - \frac{a_1}{p_1} - \frac{p_1}{a_1} + \sum_{k=2}^{n} \left(a_kp_k - \frac{a_k}{p_k} \right). \tag{4.3}$$

In the case of unibranched singularity, the quantity H^2 is the sum of Milnor number and so called M-number (without a bar) of singular point. Its importance lies in the fact that the sum of M-numbers of all singular points of an algebraic curve in $\mathbb{C}P^2$ can be bounded from above by global topological data of the curve, as genus and first Betti number (see [**BZ**]). These bounds involve very deep Bogomolov–Miyaoka–Yau inequality from algebraic geometry.

Thus the following result seem to be a very mysterious and shows a deep link between knot theory and algebraic geometry.

PROPOSITION 4.6. *Let* ρ_0 *be the integral of the Tristram–Levine signature of an algebraic knot (see* (4.1)*) and* H^2 *be like in* (4.3). *Then*

$$0 < -3\rho_0 - H^2 < \frac{2}{9}.$$

PROOF. It easy to observe that

$$\Delta := -3\rho_0 - H^2 = \frac{1}{a_1p_1} + \sum_{k=2}^{n} \left(\frac{1}{a_kp_k} - \frac{p_k}{a_k} \right).$$

On the one hand

$$\Delta \leq \sum_{k=1}^{n} \frac{1}{a_kp_k}.$$

Recall that $a_{k+1} = a_k p_{k+1} p_k + q_{k+1}$, so $a_{k+1} p_{k+1} > a_k p_k p_{k+1}^2 \geq 4 a_k p_k$. Hence

$$\Delta \leq \frac{1}{a_1 p_1} \sum_{k=0}^{n-1} \frac{1}{4^k} < \frac{4}{3 a_1 p_1}.$$

But $a_1 p_1 \geq 6$, so one inequality is proved.

To prove the second one, let us reorganise terms of Δ as follows

$$\Delta = \sum_{k=1}^{n-1} \left(\frac{1}{a_k p_k} - \frac{p_{k+1}}{a_{k+1}} \right) + \frac{1}{a_n p_n}.$$

But

$$\frac{1}{a_k p_k} - \frac{p_{k+1}}{a_{k+1}} = \frac{1}{a_k p_k} - \frac{p_{k+1}}{a_k p_k p_{k+1} + q_{k+1}} > \frac{1}{a_k p_k} - \frac{p_{k+1}}{a_k p_k p_{k+1}} = 0.$$

□

We end up the chapter with the simplest example of multibranched singularity, i.e. with a singularity defined locally by $x^d - y^d = 0$ with $d \geq 2$. Its link at singularity is the torus link $T_{d,d}$. Let us consider a set

$$\Sigma_d = \{ \frac{i}{d} + \frac{j}{d}, \ 1 \leq i, j \leq d-1 \}.$$

Here the element k/d appears in Σ_d precisely $d - 1 - |d - 1 - k|$ times, according to possible presentations $k = i + j$, $1 \leq i, j \leq d-1$. Let $s_d(x)$ be the function counting the elements of Σ_d in $(x, x+1)$ with a '$-$' sign and the others with '$+$' sign. Then s_d is almost equal to the Tristram–Levine signature of link $T_{d,d}$. We have the formula

$$s_d = 2 \sum_{k<d/2} (k-1) \chi_{\left(\frac{k}{d}, \frac{d-k}{d}\right)} - 2(k-1) \sum_{k>d/2} \chi_{\left(\frac{d-k}{d}, \frac{k}{d}\right)} - (d-1).$$

The final term, $-(d-1)$, comes from the $d-1$ elements of the set Σ_d of type d/d. They belong to any interval $(x, x+1)$. Thus, the integral of s_d is equal to

$$\int_0^1 s_d = -2 \sum_{k=1}^{d-1} (k-1) \frac{2k-d}{d} - (d-1).$$

But an elementary calculus shows that

$$\sum_{k=1}^{d-1} (k-1)(2k-d) = \frac{d(d-1)(d-2)}{6}.$$

Hence

$$\int_0^1 s_d = -\frac{1}{3}(d-1)(d+1).$$

On the other hand, in order to resolve the singularity of C we need only one blow-up. The exceptional divisor E consists of single rational curve with $E^2 = -1$. Then $K = K_X = \alpha E$ and $C' = \beta E$ (as E spans second (co)homology of blown-up space) and $K(K+E) = -2$ by genus formula, so $K = E$ and $C' \cdot E = d$, so $C' = -d \cdot E$. Thus $K + D = K + C' + E = (2-d)E$. Moreover, this divisor is nef, so its Zariski–Fujita decomposition is trivial, $H = (2-d)E$, $N = 0$. Thus in this case

$$H^2 = (d-2)^2, \ \ H^2 + 3 \int_0^1 s_d \sim 4d.$$

This shows that, in case of general links, a trivial analogue of Proposition 4.6 does not hold.

ACKNOWLEDGEMENTS. The author is very grateful to Tim Cochran and Stefan Friedl for explaining the rudiments of ρ–invariants. He wishes also to express his thanks to Andràs Némethi for various discussions on the subject.

References

[Bo] M. Borodzik, *Morse theory for plane algebraic curves*, preprint, University of Warsaw, Warsaw 2009, available at arXiv:1101.1870.

[BZ] M. Borodzik, H. Żołądek *Complex algebraic plane curves via Poincaré–Hopf formula. III. Codimension bounds*, J. Math. Kyoto Univ. **48** (2008), no. 3, 529–570.

[Cha] J. Cha *Topological minimal genus and L^2 signatures*, Algebr. Geom. Topol. **8**(2008), 885–909.

[ChG] J. Cheeger and M. Gromov, *Bounds on the von Neumann dimension of L^2-cohomology and the Gauss–Bonnet theorem for open manifolds.* J. Differential Geom. **21**(1985), no. 1, 1–34.

[ChW] S. Chang and S. Weinberger, *On invariants of Hirzebruch and Cheeger–Gromov*, Geom. Topol. **7**(2003), 311–319.

[COT1] T. Cochran, K. Orr, and P. Teichner, *Knot concordance, Whitney towers and L^2-signatures*, Ann. of Math. (2) **157**(2003), no. 2, 433–519.

[COT2] T. Cochran, K. Orr, and P. Teichner *Structure in the classical knot concordance group*, Comment. Math. Helv. **79**(2004), no. 1, 105–123.

[EN] D. Eisenbud and W. Neumann, *Three-dimensional link theory and invariants of plane curve singularities*, Ann. of Math. Stud. **110**, Princeton University Press, Princeton, 1985.

[Fuj] T. Fujita, *On the topology of non-complete algebraic surfaces*, J. Fac. Sci. Univ. Tokyo (Ser. 1A) **29** (1982), 503–566.

[Ha] S. Harvey, *Homology cobordism invariants and the Cochran-Orr-Teichner filtration of the link concordance group*, Geom. Topol. **12**(2008), no. 1, 387–430.

[KM] R. Kirby and P. Melvin, *Dedekind sums, μ-invariants, and the signature cocycle*, Math. Ann. **29**(1994), no. 2, 231–267.

[Li] R. A. Litherland, *Signatures of iterated torus knots* in Topology of low-dimensional manifolds (Proc. Second Sussex Conf., Chelwood Gate, 1977), pp. 71–84, Lecture Notes in Math., **722**, Springer, Berlin, 1979.

[Ne1] A. Némethi, *The eta-invariant of variation structures. I.* Topology Appl., **67**(1995), 95–111.

[Ne2] A. Némethi, *On the spectrum of curve singularities*, Proceedings of the Singularity Conference, Oberwolfach, July 1996; Progress in Mathematics, vol. 162, 93-102, 1998.

[Ne3] A. Némethi, *Dedekind sums and the signature of $f(x,y)+z^N$. I.* Selecta Math. (N.S.), **4**(1998), 361–376.

[Ne4] A. Némethi, *Dedekind sums and the signature of $f(x,y)+z^N$. II.* Selecta Math. (N.S.), **5**(1999), 161–179.

[Or] S. Yu. Orevkov, *On rational cuspidal curves. I. Sharp estimate for degree via multiplicities*, Math. Ann. **324** (2002), 657–673.

[OZ] S. Yu. Orevkov, M. G. Zaidenberg, *On the number of singular points of plane curves*, arXiv:alg-geom/9507005.

INSTITUTE OF MATHEMATICS, UNIVERSITY OF WARSAW, UL. BANACHA 2, 02-097 WARSAW, POLAND

E-mail address: `mcboro@mimuw.edu.pl`

Proceedings of Symposia in Pure Mathematics
Volume **82**, 2011

A surgical perspective on quasi-alternating links

Liam Watson

ABSTRACT. We show that quasi-alternating links arise naturally when considering surgery on a strongly invertible L-space knot (that is, a knot that yields an L-space for some Dehn surgery). In particular, we show that for many known classes of L-space knots, every sufficiently large surgery may be realized as the two-fold branched cover of a quasi-alternating link. Consequently, there is considerable overlap between L-spaces obtained by surgery on S^3, and L-spaces resulting as two-fold branched covers of quasi-alternating links. By adapting this approach to certain Seifert fibred spaces, it is possible to give an iterative construction for quasi-alternating Montesinos links.

This article studies quasi-alternating links – in particular how such links arise naturally from the perspective of Dehn surgery – and constitutes an extension of certain aspects of the author's talk at the Georgia International Topology Conference. The talk discussed the application of Khovanov homology as an obstruction to exceptional surgeries on strongly invertible knots [**Wat**], and central to this work is the following: if a two-fold branched cover of S^3, branched over a link L, has finite fundamental group then the reduced Khovanov homology of the branch set L is supported on at most 2 diagonals [**Wat**, Theorem 5.2]. In establishing this result, one is led to consider Dehn surgery on certain Seifert fibred manifolds, the result of which is in turn often forced to branch cover a quasi-alternating link for topological reasons. More generally, many families of quasi-alternating links arise in this way, and it is this point that we hope to elucidate here.

1. Background and context

Quasi-alternating links were introduced by Ozsváth and Szabó as a natural extension of the class of alternating links in the context of Heegaard Floer homology for two-fold branched covers [**OS05b**].

DEFINITION 1.1. The set of quasi-alternating links $\mathcal{Q}$ is the smallest set of links containing the trivial knot, and closed under the following relation: if L admits a projection with distinguished crossing $L(\times)$ so that

$$\det(L(\times)) = \det(L(\asymp)) + \det(L()\,()))$$

2010 *Mathematics Subject Classification.* Primary 57M12, 57M25.

Key words and phrases. quasi-alternating links, two-fold branched covers, Dehn surgery, Montesinos' trick, tangles.

This work was supported by a Canada Graduate Scholarship from NSERC.

for which $L(\asymp), L()\,() \in \mathcal{Q}$, then $L = L(\times) \in \mathcal{Q}$ as well.

In particular, Ozsváth and Szabó demonstrate that a non-split alternating link is quasi-alternating [**OS05b**, Lemma 3.2], and that the two-fold branched cover of a quasi-alternating link is an L-space [**OS05b**, Proposition 3.3]. Recall that a rational homology sphere Y is an L-space if $\operatorname{rk}\widehat{\mathrm{HF}}(Y) = |H_1(Y;\mathbb{Z})|$, where $\widehat{\mathrm{HF}}$ denotes the (hat version of) Heegaard Floer homology [**OS05a**].

In a related vein, Manolescu demonstrated that the rank of the knot Floer homology of a quasi-alternating knot is given by the determinant of the knot [**Man07**], and subsequently Manolescu and Ozsváth showed that quasi-alternating links have both Khovanov [**Kho00**] and knot Floer [**OS04, Ras03**] homology groups supported in a single diagonal (or, *thin* homology) [**MO07**]. The same is true of odd-Khovanov homology [**ORS**], and as a result these homology theories contain essentially the same information as their respective underlying polynomials when applied to quasi-alternating links. While it seemed possible that the collection of thin links was the same as the collection of quasi-alternating links, the situation is in fact more complicated. Greene has demonstrated that the knot 11^n_{50} is not quasi-alternating despite having all of the aforementioned homologies supported in a single diagonal [**Gre10**].

L-spaces turn out to have interesting topological properties, and as such it would be interesting to be able to characterize such manifolds without reference to Heegaard Floer homology. Since quasi-alternating links give rise to L-spaces (by way of two-fold branched covers), attempts to better understand such branch sets may be viewed as an approach to such a characterization. In constructing the spectral sequence from the reduced Khovanov homology of a branch set, converging to the Heegaard Floer homology of the two-fold branched cover [**OS05b**], Ozsváth and Szabó show that the skein exact triangle in Khovanov homology lifts to the surgery exact sequence for Heegaard Floer homology in the cover. With this in mind, our aim is to better understand quasi-alternating knots from the perspective of Dehn surgery.

Let $S^3_r(K)$ denote r-surgery on a knot $K \hookrightarrow S^3$. If $S^3_r(K)$ is an L-space then K is referred to as an L-space knot, and it is well known that every such knot gives rise to an infinite family of L-spaces: if $S^3_{r_0}(K)$ is an L-space for $r_0 > 0$, then $S^3_r(K)$ is an L-space as well for every rational number $r \geq r_0$ [**OS11, OS05a**]. It seems natural to ask then if quasi-alternating links enjoy an analogous property.

To this end, we will be primarily interested in the knots $K \hookrightarrow S^3$ having the following property:

QA There is a positive integer N such that $S^3_r(K)$ is the two-fold branched cover of a quasi-alternating link for all rational numbers $r \geq N$.

Notice that a knot satisfying property **QA** is necessarily strongly invertible. It turns out that many L-space knots are strongly invertible, and our aim is to establish that these examples satisfy property **QA**.

THEOREM 1.2. *The following knots have property* **QA***:*

1. *Torus knots and, more generally, Berge knots (see Proposition 4.1, as well as* [**Wat**, Section 4]*).*
2. *The $(-2, 3, q)$-pretzel knot, for all positive, odd integers q (see Theorem 5.1).*

3. *All sufficiently positive cables of knots with property* **QA** *(see Theorem 6.1).*

Consequently, there is considerable overlap between known L-spaces obtained by surgery on a knot in S^3, and L-spaces obtained as the two-fold branched cover of a quasi-alternating link.

In certain instances, one may ask if property **QA** holds for knots in manifolds other than S^3 (that is, if suitably large surgeries on the knot may be realized as two-fold branched covers of quasi-alternating links). For example, it can be shown that Dehn surgery on a regular fibre of certain Seifert fibred spaces gives rise to manifolds that are two-fold branched covers of quasi-alternating links, and this in turn yields an iterative geometric construction for producing all quasi-alternating Montesinos links (see Section 7).

2. The Montesinos trick

A tangle T is a pair (B^3, τ) where $\tau \hookrightarrow B^3$ is a pair of properly embedded arcs in a three-ball (together with a possibly empty collection of closed components), meeting the boundary sphere transversally in the 4 endpoints. The two-fold branched cover of a tangle will be denoted by $\mathbf{\Sigma}(B^3, \tau)$, where τ is the branch set. Notice that $\mathbf{\Sigma}(B^3, \tau)$ is a manifold with torus boundary. Montesinos' observation is that a Dehn filling of such a manifold may be viewed as a two-fold branched cover of S^3, the branch set of which is obtained by attaching a rational tangle to T in a prescribed way [**Mon75**]. Recall that a tangle is rational if and only if the two-fold branched cover is a solid torus, and that tangles in this setting are considered up to homeomorphism of the pair (B^3, τ).

To exploit this fact, generally referred to as *the Montesinos trick*, we briefly recall the notation introduced in [**Wat**].

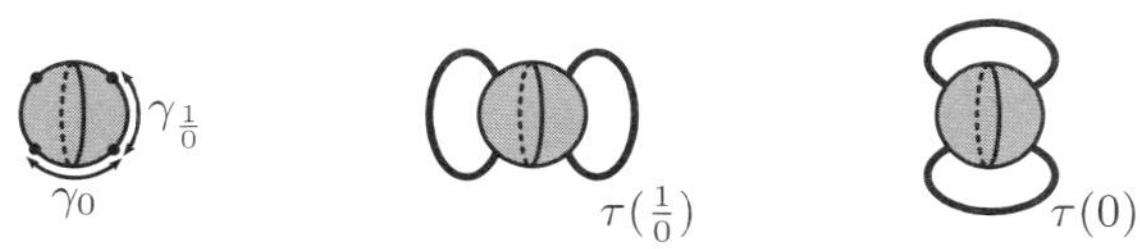

FIGURE 1. The arcs $\gamma_{\frac{1}{0}}$ and γ_0 in the boundary of a tangle (left); and the 'denominator' (centre) and 'numerator' (right) closures denoted $\tau(\frac{1}{0})$ and $\tau(0)$ respectively when applied to a given tangle with fixed framing.

Set $M = \mathbf{\Sigma}(B^3, \tau)$ and let α and β be a pair of slopes in ∂M that intersect geometrically in a single point. Then there is a choice of representative for the tangle T compatible with the pair $\{\alpha, \beta\}$ in the sense that $\widetilde{\gamma_{\frac{1}{0}}} = \alpha$ and $\widetilde{\gamma_0} = \beta$. Here, $\gamma_{\frac{1}{0}}$ and γ_0 are arcs embedded in the boundary of the tangle with endpoints on $\partial\tau$ as in Figure 1, and $\{\widetilde{\gamma_{\frac{1}{0}}}, \widetilde{\gamma_0}\}$ is the pair of slopes in $\partial\left(\mathbf{\Sigma}(B^3, \tau)\right) = \partial M$ covering $\gamma_{\frac{1}{0}}$ and γ_0.

As a result, the Dehn fillings $M(\alpha)$ and $M(\beta)$ may be obtained as the two-fold branched covers $\mathbf{\Sigma}(S^3, \tau(\frac{1}{0}))$ and $\mathbf{\Sigma}(S^3, \tau(0))$ respectively, where $\tau(\frac{1}{0})$ and $\tau(0)$ are the links resulting from the closures by the rational tangles shown in Figure 1.

In particular, $M(n\alpha + \beta) \cong \mathbf{\Sigma}(S^3, \tau(n))$ in this notation, giving the branch sets for integer surgeries relative to the basis $\{\alpha, \beta\}$ (throughout we fix an orientation of

∂M for which $\alpha \cdot \beta = +1$). The closure giving rise to the branch set $\tau(n)$ is shown in Figure 2. Notice that the half-twists in the base lift to full Dehn twists along α in the cover. More generally, we may write $M(p\alpha + q\beta) \cong \mathbf{\Sigma}(S^3, \tau(r))$ where $r = \frac{p}{q}$ is given by the continued fraction expansion $[a_1, \ldots, a_\ell]$, and $\tau(r)$ is the link obtained by adding a rational tangle specified by the continued fraction to the tangle T. A specific example is shown in Figure 2; for details see [**Wat**, Section 3.3].

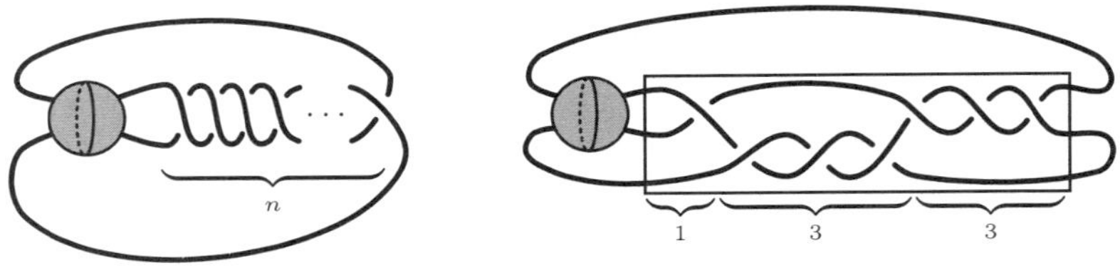

FIGURE 2. The closure $\tau(n)$ (left) giving rise to the branch set for integer surgeries (that is, Dehn fillings along slopes $n\alpha + \beta$), and the closure $\frac{13}{10} = [1,3,3]$ (right) corresponding to $13\alpha + 10\beta$ Dehn filling in the cover.

As a result, $\det(\tau(r)) = c_M \Delta(p\alpha + q\beta, \lambda_M)$ where Δ measures the minimal geometric intersection number between two slopes, λ_M is the rational longitude, and c_M is some fixed constant depending only on M [**Wat**, Lemma 3.2]. This is due to the fact that $\det(\tau(r)) = |H_1(\mathbf{\Sigma}(S^3, \tau(r)); \mathbb{Z})|$ whenever $\mathbf{\Sigma}(S^3, \tau(r))$ is a $\mathbb{Q}$-homology sphere, and $\det(\tau(r)) = 0$ otherwise. Recall that the rational longitude is characterized by the property that some finite collection of like oriented copies of λ_M in ∂M bound an essential surface in M. As a result, if $H_1(M; \mathbb{Z}) \cong H \oplus \mathbb{Z}$ for some finite abelian group H, then $c_M = \operatorname{ord}_H i_*(\lambda_M)|H|$ where i_* is the homomorphism induced by the inclusion $i\colon \partial M \hookrightarrow M$.

A particular useful class of examples is provided by strongly invertible knots. Recall that a knot K is strongly invertible if there is an involution on the complement $S^3 \smallsetminus \nu(K)$ that fixes a pair of arcs meeting the boundary torus transversally in exactly 4 points. It is an easy exercise to show that the quotient of such an involution is always a three-ball, and by keeping track of the image of the fixed point set in the quotient, τ, such a knot complement is always the two-fold branched cover of a tangle $T = (B^3, \tau)$. In particular, we have the notation $S^3_r(K) = \mathbf{\Sigma}(S^3, \tau(r))$ for r-surgery on the strongly invertible knot K, once a representative compatible with the preferred basis $\{\mu, \lambda\}$ (provided by the knot meridian μ and the Seifert longitude λ) has been fixed.

3. Quasi-alternating tangles

DEFINITION 3.1. A tangle $T = (B^3, \tau)$ is called quasi-alternating if it admits a representative (of the homeomorphism class of the pair (B^3, τ)) with the property that $\tau(\frac{1}{0})$ and $\tau(0)$ are quasi-alternating links. Such a representative, when it exists, will be referred to as a quasi-alternating framing for T.

Note that $\tau(\frac{1}{0})$ and $\tau(0)$ are quasi-alternating if and only if the mirrors, $\tau(\frac{1}{0})^\star$ and $\tau(0)^\star$, are quasi-alternating. As a result the tangle T is quasi-alternating if and only if $T^\star$ is a quasi-alternating tangle. This fact allows us to pass to the mirror and restrict to positive surgeries whenever convenient.

THEOREM 3.2. *If T is a quasi-alternating tangle then, up to taking mirrors and/or renaming $\gamma_{\frac{1}{0}}$ and γ_0, the links $\tau(r)$ are quasi-alternating for all rational numbers $r \geq 0$, where $T = (B^3, \tau)$ is a quasi-alternating framing.*

SKETCH OF PROOF. This follows from [**Wat**, Theorem 4.7 and Remark 4.5], but we sketch the argument here for the reader's convenience.

First notice that, up to taking mirrors, $\det(\tau(1)) = \det(\tau(\frac{1}{0})) + \det(\tau(0))$. This results from the observation that $\det(\tau(1)) = c_M\Delta(\alpha+\beta, \lambda_M) = c_M(\Delta(\alpha, \lambda_M) + \Delta(\beta, \lambda_M))$, where $M = \mathbf{\Sigma}(B^3, \tau)$. As a result, $\tau(1)$ must be quasi-alternating. Next notice that, up to renaming $\gamma_{\frac{1}{0}}$ and γ_0 (thereby renaming the pair of slopes α and β in the cover), we have that $\det(\tau(n)) = n\det(\tau(\frac{1}{0})) + \det(\tau(0))$ for all $n \geq 0$ by induction on n.

The choices above ensure that we may orient ∂M so that $\alpha\cdot\beta = +1$, $\alpha\cdot\lambda_M > 0$ and $\beta\cdot\lambda_M > 0$. With these choices fixed, the result follows from a second induction on ℓ, the length of the continued fraction expansion of $r = [a_1, \ldots, a_\ell]$, applying the fact that $\det(\tau(r)) = c_M\Delta(p\alpha + q\beta, \lambda_M)$ where $r = \frac{p}{q}$. In particular, resolving the final crossing of the continued fraction gives $\det(\tau(r)) = \det(\tau(r_0)) + \det(\tau(r_1))$ as required, where r_0 and r_1 correspond to the 0- and 1-resolutions specified by the continued fractions $[a_1, \ldots, a_{\ell-1}]$ and $[a_1 \ldots, a_\ell - 1]$ (see [**Wat**, Section 3.4]). □

EXAMPLE 3.3. Any alternating projection of a non-split alternating link L gives rise to a quasi-alternating tangle: it suffices to form a tangle by removing two arcs of L in such a way that the crossings adjacent to the endpoints of the tangle alternate as they are traversed in order around the diagram of the tangle. See Figure 3, for example. As a result, any alternating link gives rise to an infinite family of quasi-alternating links by applying Theorem 3.2.

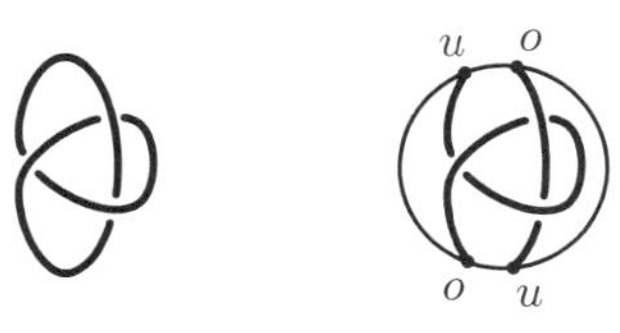

FIGURE 3. An alternating projection of the trefoil (left) gives rise to a quasi-alternating tangle (right). Notice that both $\tau(\frac{1}{0})$ and $\tau(0)$ are alternating links.

For the particular instance of this example shown in Figure 3, the resulting quasi-alternating tangle is a rational tangle. As a result, the two-fold branched cover of the tangle is a solid torus, and this gives an alternative view on why this particular tangle is quasi-alternating: since every closure of this tangle corresponds to a surgery on the trivial knot in the two-fold branched cover, it follows from work of Hodgson and Rubinstein that the resulting branch sets are always two-bridge links [**HR85**]. These are non-split alternating links – and hence quasi-alternating – for all but the zero surgery.

Figure 4 shows second quasi-alternating tangle arising in this way that is not a rational tangle.

Further examples arise naturally in the context of Dehn surgery on a strongly invertible knots more generally, and this is explored in the following sections.

FIGURE 4. Two quasi-alternating framings for a given tangle: notice that these tangles differ by a homeomorphism that does not fix the boundary. This tangle is two-fold branched covered by the complement of the right-hand trefoil (see Section 4 and, more generally, Section 7) as demonstrated in [**Wat**, Figure 5], for example.

4. Lens space surgeries and quasi-alternating links

Moser showed that the positive (p,q)-torus knot yields a lens space for $(pq-1)$-surgery [**Mos71**]. More generally, Berge gives a conjecturally complete list of knots admitting lens space surgeries that includes torus knots [**Ber**], and results due to Schreier [**Sch24**] and Osborne [**Osb81**] show that every such knot is strongly invertible by virtue of having Heegaard genus at most 2. That is, up to mirrors, there is a positive integer N, depending on K, for which $S^3_N(K)$ is a lens space whenever K is a knot in Berge's list.

As a result, the quotient tangle $T=(B^3,\tau)$ compatible with $\{\mu, N\mu+\lambda\}$ is a framed quasi-alternating tangle: $\tau(\frac{1}{0})$ is unknotted while $\tau(0)$ must be a non-split two-bridge link due to work of Hodgson and Rubinstein [**HR85**]. Therefore every Berge knot complement may be viewed as the two-fold branched cover of a quasi-alternating tangle, and in summary we have the following immediate application of Theorem 3.2 (c.f. [**Wat**, Proposition 4.8]):

PROPOSITION 4.1. *Up to mirrors, every Berge knot has property* **QA**.

5. Pretzels admitting L-space surgeries

Let P_q denote the $(-2,3,q)$-pretzel (for odd positive integers q); P_7 is illustrated in Figure 5. It is known that each P_q admits L-space surgeries [**OS05a**]. Indeed, P_1, P_3 and P_5 are the $(5,2)$-, $(3,4)$- and $(3,5)$- torus knots, respectively, and P_7 is a non-torus Berge knot (it was demonstrated by Fintushel and Stern that this hyperbolic knot admits lens space surgeries [**FS80**]). In general P_q is hyperbolic when $q>5$, however P_q does not admit finite fillings (in particular, does not admit lens space surgeries) for $q>9$ [**Mat00**].

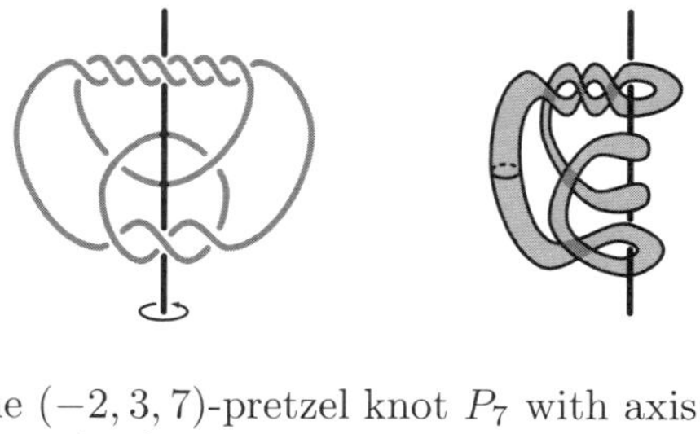

FIGURE 5. The $(-2,3,7)$-pretzel knot P_7 with axis for the strong inversion labelled (left), and the quotient of the involution on the complement of P_7 (right).

Since the knot P_q is strongly invertible, surgeries on P_q are two-fold branched covers of S^3. Moreover, we have the following:

THEOREM 5.1. *The knot P_q has property* **QA** *for every odd positive integer q.*

PROOF. Since P_1 is a torus knot, the case $q = 1$ (and, indeed, the cases $q = 3, 5, 7$) follows from Proposition 4.1. To prove the theorem, assume without loss of generality that $q > 1$. We first need to construct the associated quotient tangle for each P_q. This amounts to choosing a fundamental domain for the action of the involution on the knot complement (illustrated in Figure 5), and is done in Figure 6.

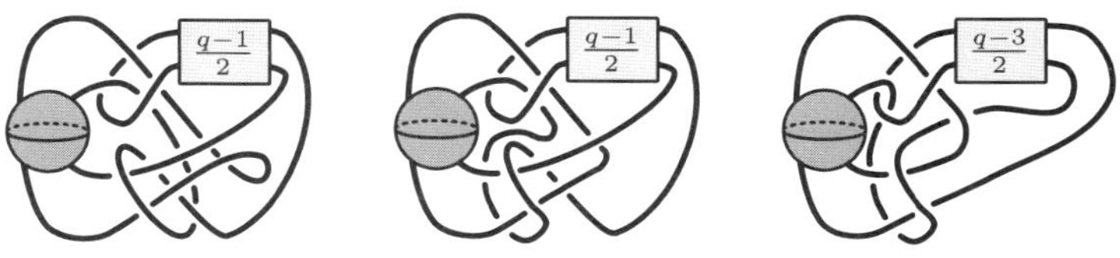

FIGURE 6. Simplifying the tangle resulting from the quotient of P_q by the strong inversion. Notice that the last step involves a homeomorphism that does not fix the boundary sphere.

The preferred framing for this tangle, which we denote by $T_q = (B^3, \tau_q)$, is shown in Figure 7. To see this, it suffices to check the image of the longitude in the quotient of P_3, then observe that the framing changes by 4 half-twists in the quotient when q is replaced by $q+2$. In particular, notice that the branch set associated to $S^3_{2q+5}(P_q) \cong \mathbf{\Sigma}(S^3, \tau_q(2q+5))$ has a fixed number of twists corresponding to the framing, so that only the twists of the second box vary in q for this family of branch sets.

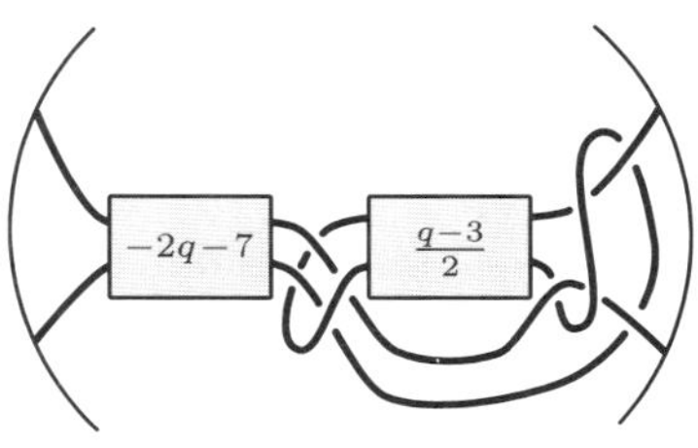

FIGURE 7. The preferred representative $T_q = (B^3, \tau_q)$ of the associated quotient tangle for the knot P_q. Note that the twists in the left-most box correspond to the framing, while adding a positive half-twist in the right-most box corresponds to exchanging the knot P_q for P_{q+2} in the cover.

We claim that the branch sets $\tau_q(2q+5)$ are quasi-alternating. To see this, consider the tangle $T' = (B^3, \tau')$ shown in Figure 8. This tangle has the property, by construction, that $\tau'(n) = \tau_{2n+3}(4n+11) = \tau_q(2q+5)$, and as such $\det(\tau'(n)) = 4n+11$.

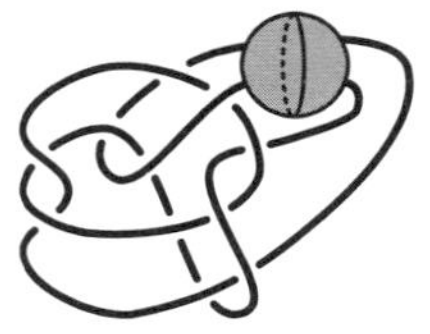

FIGURE 8. The quasi-alternating tangle $T' = (B^3, \tau')$ for which $\tau'(n) = \tau_{2n+3}(4n+11) = \tau_q(2q+5)$.

Moreover, as shown in Figure 9 the branch set $\tau'(\frac{1}{0})$ is the $(2,4)$-torus link hence $\det(\tau(\frac{1}{0})) = 4$. As a result, $\det(\tau(n)) = n\det(\tau(\frac{1}{0})) + 11$. Notice that, as an alternating link, $\tau'(\frac{1}{0})$ is quasi-alternating.

FIGURE 9. Isotopy of the link $\tau(\frac{1}{0})$ yields the $(2,4)$-torus link.

Therefore, if we can demonstrate that $\tau'(0)$ is quasi-alternating with $\det(\tau(0)) = 11$, we can conclude that $\tau'(n)$ is quasi-alternating for all $n \geq 0$ (in particular, that $T' = (B^3, \tau')$ is a framed quasi-alternating tangle). This is indeed the case: the isotopy in Figure 10 yields an alternating diagram for $\tau(0) \simeq 7_2$ and it is well known that this two-bridge knot has $\det(\tau(0)) = 11$.

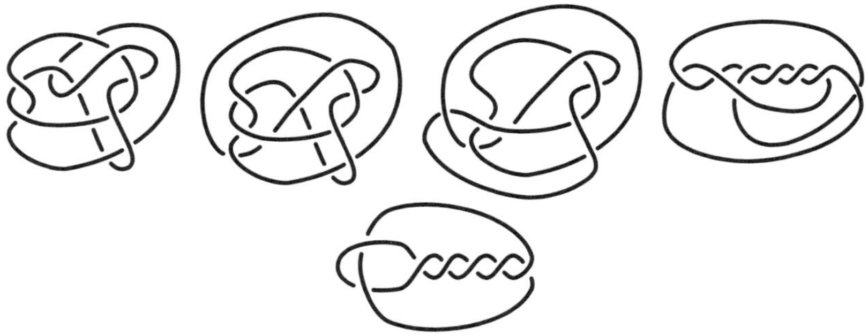

FIGURE 10. Isotopy of the knot $\tau(0)$ yields the knot 7_2.

To complete the proof notice that we have shown, for each q, that there is a positive integer surgery for which the corresponding branch set is quasi-alternating. In particular, by fixing a representative for T_q compatible with $\{\mu, (2q+5)\mu + \lambda\}$ we obtain a framed quasi-alternating tangle. Therefore, the manifold $S^3_r(P_q)$ is the two-fold branched cover of a quasi-alternating link for all rational numbers $r \geq 2q+5$ by applying Theorem 3.2. $\square$

We point out an interesting consequence of the construction used in the proof of Theorem 5.1: there are composite, quasi-alternating knots realized as quasi-alternating on a non-composite diagram. Indeed, since P_5 is the $(3,5)$-torus knot, the manifold $S^3_{15}(P_5)$ is a connect sum of lens spaces [**Mos71**]. Therefore, the associated quasi-alternating branch set $\tau_5(15) = \tau'(1)$ must be a connect sum of two-bridge knots [**HR85**]. We have demonstrated that $\tau'(n)$ is quasi-alternating for all n, and leave it as a challenge to the interested reader to demonstrate that $\tau'(1)$ is the knot $3_1\#4_1$. Notice that since both $\tau'(\frac{1}{0})$ and $\tau'(0)$ are non-trivial prime links, the resolved crossing must constitute an interaction between the connect summands of $\tau'(1)$. We point out that it is an open problem to show that $K_1\#K_2$ quasi-alternating implies that the K_i are quasi-alternating, and this example suggests that an answer to this question might be subtle.

6. On cabling

Let $\mathcal{C}_{q,p}(K)$ denote the (q,p)-cable of a knot K. By this convention, *sufficiently positive cabling* refers to taking sufficiently large p.

THEOREM 6.1. *Let K be a knot with property* **QA**. *Then all sufficiently positive cables of K have property* **QA** *as well.*

PROOF. We begin by recalling a result due to Gordon [**Gor83**]. Let $r = \frac{kpq\pm1}{k}$. Then since the cable space corresponding to $\mathcal{C}_{q,p}$ is Seifert fibred over an annulus with a single cone point of order q, r-surgery on $\mathcal{C}_{q,p}(K)$ may be obtained via surgery on the original knot K: $S^3_r(\mathcal{C}_{q,p}(K)) \cong S^3_{r/q^2}(K)$ (see [**Gor83**, Corollary 7.3]). In particular, we have that $S^3_{pq-1}(\mathcal{C}_{q,p}(K)) \cong S^3_{(pq-1)/q^2}(K)$ so that $\frac{pq-1}{q^2}$ is an increasing function in p.

Now suppose that K has property **QA**. Then for $p \gg 0$ we can be sure that $S^3_{(pq-1)/q^2}(K)$ is the two-fold branched cover of a quasi-alternating link. Therefore, $S^3_{pq-1}(\mathcal{C}_{q,p}(K))$ must be a two-fold branched cover of a quasi-alternating link as well. Now since the cable of a strongly invertible knot is strongly invertible, the representative of the associated quotient tangle to $\mathcal{C}_{q,p}(K)$ compatible with $(\mu, (pq-1)\mu+\lambda)$ is a quasi-alternating tangle for all p sufficiently large. This observation, together with an application of Theorem 3.2, proves the claim. $\square$

Notice that the proof of Theorem 6.1 depends only on properties of Dehn surgery and makes no reference to diagrams of the branch sets. Moreover, we recover a result due to Hedden that sufficiently positive cables of L-space knots provide new examples of L-space knots [**Hed09**, Theorem 1.10]. However, Hedden demonstrates that, given an L-space knot K, $\mathcal{C}_{q,p}(K)$ admits L-space surgeries whenever $p \geq q(2g(K)-1)$ where $g(K)$ denotes the Seifert genus of K. It is natural to ask if $\mathcal{C}_{q,q(2g(K)-1)}(K)$ has property **QA**, as these seem to be potential candidates for L-space knots failing property **QA**.

7. Constructing quasi-alternating Montesinos links

We conclude by observing that the interaction between Dehn surgery (in the cover) and quasi-alternating tangles (in the base) provides a recipe for constructing infinite families of quasi-alternating Montesinos links. As shown below, every Montesinos link arises in this way, and as such it would be interesting to compare this approach with the constructions of Champanerkar and Kofman [**CK09**] and

Widmer [**Wid09**]. We remark that these constructions imply that many Seifert fibred L-spaces are realized as the two-fold branched cover of a quasi-alternating link. For background on Dehn filling Seifert fibred manifolds we refer the reader to Boyer [**Boy02**].

Let M be a Seifert fibred space with base orbifold a disk with n cone points denoted $D^2(p_1,\ldots,p_n)$. Then ∂M is a torus, with a distinguished slope φ given by a regular fibre in the boundary. The following is due to Heil [**Hei74**]:

THEOREM 7.1. *Given an M as described above, $M(\varphi)$ is a connect sum of n lens spaces, and for any slope $\alpha \neq \varphi$ we have that the Dehn filling $M(\alpha)$ is Seifert fibred with base orbifold $S^2(p_1,\ldots,p_n,\Delta(\alpha,\varphi))$.*

This result generalizes Moser's results [**Mos71**] pertaining to torus knots (see Section 4), since a (p,q)-torus knot is a regular fibre in a Seifert fibration of S^3 with base orbifold $S^2(p,q)$. Note in particular that in the present setting $M(\varphi)$ necessarily a two-fold branched cover of S^3 with branch set given by a connect sum of n two-bridge links [**HR85**]. More generally, a result of Montesinos says that such manifolds may be obtained as the two-fold branched cover of a tangle $M = \mathbf{\Sigma}(B^3,\tau)$ (though in this setting it may be that the fixed point set τ includes some closed components in addition to the pair of arcs), so that $M(\alpha)\cong\mathbf{\Sigma}(S^3,L)$ where L is a Montesinos link composed of $n+1$ rational tangles whenever $\alpha\neq\varphi$ [**Mon76**].

Our aim is to identify when L is quasi-alternating, by implicitly extending property **QA** to regular fibres in certain Seifert fibred spaces.

To this end, fix a regular fibre φ in the boundary of M, and denote by λ_M the rational longitude. We will suppose that $\varphi\neq\lambda_M$; this assumption ensures that M is not the twisted I-bundle over the Klein bottle (for this case we refer the reader to [**Wat**, Section 5.2]). Given an orientation of λ_M, fix an orientation on φ so that $\varphi\cdot\lambda_M > 0$. Now for any slope μ with the property that $\mu\cdot\varphi = +1$ we have that $\alpha = \mu + k\varphi$ shares this same property for every integer k. As a result, $\alpha\cdot\lambda_M > 0$ for $k \gg 0$. Fix such a k, together with a representative for the associated quotient tangle T compatible with the resulting pair $\{\alpha,\varphi\}$. Then by Theorem 7.1 (together with the preceding discussion), the branch set for the fibre filling $\tau(0)$ is a connect sum of n two-bridge links and $\tau(\frac{1}{0})$ is a Montesinos link composed of n rational tangles.

PROPOSITION 7.1. *With the above notation, if $\tau(\frac{1}{0})$ is a quasi-alternating link, then T is a quasi-alternating tangle. In particular, $\tau(r)$ is a quasi-alternating link for all $r \geq 0$.*

PROOF. This is immediate from the set up preceding the statement of the proposition, together with an application of Theorem 3.2. □

This proposition may be used to construct infinite families of quasi-alternating links, in an obvious manner, provided the quasi-alternating requirement on $\tau(\frac{1}{0})$ may be established.

7.1. Small Seifert fibred spaces. A Seifert fibred space is called *small* if it contains exactly 3 exceptional fibres (that is, has base orbifold $S^2(p_1,p_2,p_3)$). Notice that, by applying Theorem 7.1, such manifolds may be obtained by Dehn filling of some Seifert fibred M with base orbifold $D^2(p_1,p_2)$. Moreover, there is a slope α with $\alpha\cdot\lambda_M > 0$ and $\alpha\cdot\varphi = +1$ so that $M(\alpha)$ is a lens space (with base orbifold $S^2(p_1, p_2)$). Let T be the associated quotient tangle so that $M \cong \mathbf{\Sigma}(B^3, \tau)$, with representative chosen compatible with $\{\alpha, \varphi\}$. As a particular instance of Proposition 7.1, we have demonstrated the following (compare Proposition 4.1 in the case of torus knot exteriors):

PROPOSITION 7.2. *Every M as described above is the two-fold branched cover of a quasi-alternating tangle.*

7.2. An iterative construction. Proposition 7.2 establishes that Proposition 7.1 is not vacuous. In particular, we observe that a regular fibre in a Seifert fibration of a lens space is a knot admitting a lens space surgery, and indeed provides an example of a knot satisfying a more general form of property **QA** up to mirrors.

With this as a base case, it is now clear that if Y is a Seifert fibration with base orbifold $S^2(p_1, \ldots, p_n)$, and Y is the two-fold branched cover of a quasi-alternating link, then a regular fibre in Y is a knot satisfying a more general form of property **QA**. We end with an example demonstrating how this fact may be used to generate infinite families of quasi-alternating Montesinos links, using Dehn surgery to control the construction.

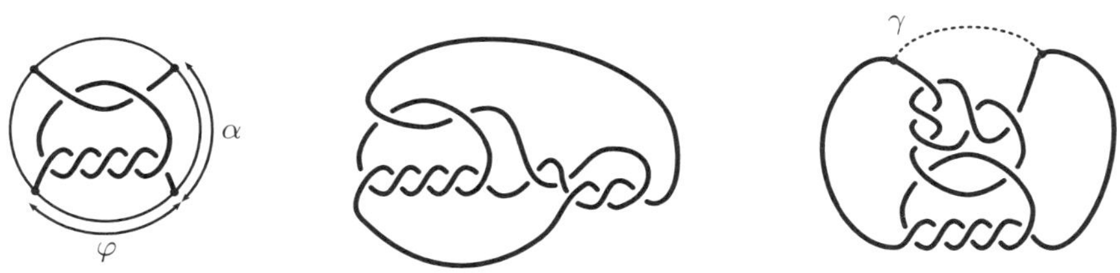

FIGURE 11. A tangle (left) with Seifert fibred two-fold branched cover (by abuse, the arcs $\gamma_{\frac{1}{0}}$ and γ_0 have been labelled by their respective lifts), the branch set associated to $7\alpha + 3\varphi$ Dehn filling giving rise to the knot 12^n_{500} (centre), and another view of 12^n_{500} with dashed arc γ that lifts to a knot in the cover $\widetilde{\gamma}$ isotopic to a regular fibre.

First consider the tangle $T = (B^3, \tau)$ shown in Figure 11. The Seifert fibration in the cover has base orbifold $D^2(2, 5)$ (notice that the tangle is the sum of two rational tangles). Moreover, this is a quasi-alternating framing for T: $\tau(0)$ is a connect sum of two-bridge links (the Hopf link and the cinqfoil) with $\det(\tau(0)) = (2)(5)$, while $\tau(\frac{1}{0})$ is a two-bridge knot (the trefoil) with $\det(\tau(\frac{1}{0})) = 3$. Indeed, one may check that $\tau(1)$ is the knot 7_3 with $\det(\tau(1)) = 13 = 3 + (5)(2)$ as required.

As an application of Theorem 3.2, it follows that $\tau(r)$ is a quasi-alternating link for every $r \geq 0$. For example, the quasi-alternating knot $\tau(\frac{7}{3}) \simeq 12^n_{500}$ is shown in Figure 11, noting that $\frac{7}{3} = [2, 3]$. This gives the branch set associated to Dehn filling $\partial\left(\mathbf{\Sigma}(B^3, \tau)\right)$ along the slope $7\alpha + 3\varphi$. Here, $\det(\tau(\frac{7}{3})) = 7(3) + 3(5)(2) = 51$ which decomposes as $\det(\tau(\frac{5}{2})) + \det(\tau(2)) = (5(3) + 2(10)) + (2(3) + 1(10)) = 35 + 16$ by resolving the final crossing added by the continued fraction. Note that one must verify that $c_M = 1$, though this may be easily determined from $\det(\tau(0))$ and $\det(\tau(\frac{1}{0}))$.

Given that 12^n_{500} is a quasi-alternating knot, with small Seifert fibred two-fold branched cover, we may repeat the above process forming a new quasi-alternating

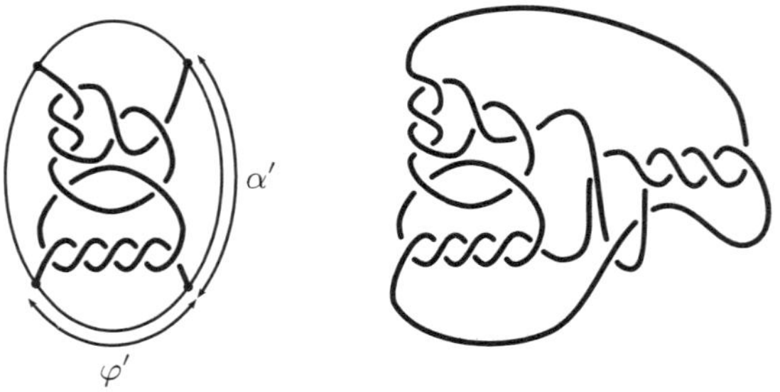

FIGURE 12. A quasi-alternating tangle (left) and a quasi-alternating Montesinos knot (right).

tangle $T' = (B^3, \tau')$. By removing a neighbourhood of the arc γ shown in Figure 11 we obtain the framed quasi-alternating tangle shown in Figure 12. By construction, $\tau(0)$ gives a connect sum of two-bridge knots (with $\det(\tau(0)) = (2)(5)(7)$) and branch set for Dehn filling along φ', while $\tau(\frac{1}{0}) \simeq 12^n_{500}$ is the branch set for Dehn filling along α'. In fact, γ lifts to a knot isotopic to φ in $\mathbf{\Sigma}(S^3, 12^n_{500})$.

Again, every link $\tau'(r)$ is quasi-alternating for $r \geq 0$ as a result of Theorem 3.2. A particular example, corresponding to filling along the slope $13\alpha' + 9\varphi'$, is shown in Figure 12. It may be easily verified that $\frac{13}{9} = [1, 2, 4]$ and $\det(\tau(\frac{13}{9})) = 13(51) + 9(2)(5)(7) = 1293$.

This process may now be iterated *ad infinitum* to obtain further infinite families of quasi-alternating Montesinos links. We remark that every quasi-alternating Montesinos link L is contained in such an infinite family: it suffices to identify an embedded arc γ with endpoints on L whose lift in $\mathbf{\Sigma}(S^3, L)$ is isotopic to a regular fibre, and repeat the construction above. As a result, while this observation does not immediately serve to enumerate quasi-alternating Montesinos links, it does ensure that every quasi-alternating Montesinos link arises through this iterative construction.

Acknowledgements. The author thanks Michel Boileau and Steve Boyer for helpful discussions.

References

[Ber] John Berge, *Some knots with surgeries yielding lens spaces.*, Unpublished manuscript.

[Boy02] Steven Boyer, *Dehn surgery on knots*, Handbook of geometric topology, North-Holland, Amsterdam, 2002, pp. 165–218.

[CK09] Abhijit Champanerkar and Ilya Kofman, *Twisting quasi-alternating links*, Proc. Amer. Math. Soc. **137** (2009), no. 7, 2451–2458.

[FS80] Ronald Fintushel and Ronald J. Stern, *Constructing lens spaces by surgery on knots*, Math. Z. **175** (1980), no. 1, 33–51.

[Gor83] C. McA. Gordon, *Dehn surgery and satellite knots*, Trans. Amer. Math. Soc. **275** (1983), no. 2, 687–708.

[Gre10] Joshua Greene, *Homologically thin, non-quasi-alternating links.*, Math. Res. Lett. **17** (2010), no. 1, 39–49.

[Hed09] Matthew Hedden, *On knot Floer homology and cabling: 2.*, Int. Math. Res. Not. **2009** (2009), no. 12, 2248–2274.

[Hei74] Wolfgang Heil, *Elementary surgery on Seifert fiber spaces*, Yokohama Math. J. **22** (1974), 135–139.

[HR85] Craig Hodgson and J. H. Rubinstein, *Involutions and isotopies of lens spaces*, Knot theory and manifolds (Vancouver, B.C., 1983), Lecture Notes in Math., vol. 1144, Springer, Berlin, 1985, pp. 60–96.

[Kho00] Mikhail Khovanov, *A categorification of the Jones polynomial*, Duke Math. J. **101** (2000), no. 3, 359–426.

[Man07] Ciprian Manolescu, *An unoriented skein exact triangle for knot Floer homology*, Math. Res. Lett. **14** (2007), no. 5, 839–852.

[Mat00] Thomas Mattman, *The Culler-Shalen seminorms of pretzel knots*, Ph.D. thesis, McGill University, 2000.

[MO07] Ciprian Manolescu and Peter Ozsváth, *On the Khovanov and knot Floer homologies of quasi-alternating links*, Proceedings of the 14th Gökova Geometry-Topology Conference, 2007, pp. 61–80.

[Mon75] José M. Montesinos, *Surgery on links and double branched covers of* S^3, Knots, groups, and 3-manifolds (Papers dedicated to the memory of R. H. Fox), Princeton Univ. Press, Princeton, N.J., 1975, pp. 227–259. Ann. of Math. Studies, No. 84.

[Mon76] ———, *Revêtements ramifiés de nœds, espaces fibré de Seifert et scindements de Heegaard*, 1976, Lecture notes, Orsay.

[Mos71] Louise Moser, *Elementary surgery along a torus knot*, Pacific J. Math. **38** (1971), 737–745.

[ORS] Peter Ozsváth, Jacob Rasmussen, and Zoltán Szabó, *Odd Khovanov homology*, Preprint, 2007, available at arXiv:0710.4300.

[OS04] Peter Ozsváth and Zoltán Szabó, *Holomorphic disks and knot invariants*, Adv. Math. **186** (2004), no. 1, 58–116.

[OS05a] ———, *On knot Floer homology and lens space surgeries*, Topology **44** (2005), no. 6, 1281–1300.

[OS05b] ———, *On the Heegaard Floer homology of branched double-covers*, Adv. Math. **194** (2005), no. 1, 1–33.

[OS11] ———, *Knot Floer homology and rational surgeries.*, Algebr. Geom. Topol. **11** (2011), 1–68.

[Osb81] Richard P. Osborne, *Knots with Heegaard genus 2 complements are invertible*, Proc. Amer. Math. Soc. **81** (1981), no. 3, 501–502.

[Ras03] Jacob Rasmussen, *Floer homology and knot complements*, Ph.D. thesis, Harvard University, 2003.

[Sch24] Otto Schreier, *Über die Gruppen* $A^aB^b = 1$, Abh. Math. Sem. Univ. Hamburg **3** (1924), 167–169.

[Wat] Liam Watson, *Surgery obstructions from Khovanov homology.*, Preprint, 2008, available at arXiv:0807.1341v3.

[Wid09] Tamara Widmer, *Quasi-alternating Montesinos links*, J. Knot Theory Ramifications **18** (2009), no. 10, 1459–1469.

Département de Mathématiques, Université du Québec à Montréal, Montréal Canada.

Current address: Department of Mathematics, UCLA, 520 Portola Plaza, Los Angeles, CA 90095.

E-mail address: `lwatson@math.ucla.edu`

Proceedings of Symposia in Pure Mathematics
Volume **82**, 2011

Thurston norm and cosmetic surgeries

Yi Ni

ABSTRACT. Two Dehn surgeries on a knot are called cosmetic if they yield homeomorphic manifolds. For a null-homologous knot with certain conditions on the Thurston norm of the ambient manifold, if the knot admits cosmetic surgeries, then the surgery coefficients are equal up to sign.

1. Introduction

Heegaard Floer homology is a powerful theory introduced by Ozsváth and Szabó [**OSz1**]. One important aspect of Heegaard Floer homology is that it behaves well under Dehn surgeries. In fact, if one knows about the knot Floer complex of a knot, then one can compute the Heegaard Floer homology of any surgery on the knot as in Ozsváth–Szabó [**OSz3, OSz5**] and Rasmussen [**R**]. This makes Heegaard Floer homology very useful in the study of Dehn surgery.

In this paper, we will use Heegaard Floer homology to study cosmetic surgeries. We first recall the definition of cosmetic surgeries.

DEFINITION 1.1. If two Dehn surgeries on a knot yield homeomorphic manifolds, then these two surgeries are *cosmetic*.

Cosmetic surgeries are very rare. More precisely, one has the following Cosmetic Surgery Conjecture.

CONJECTURE 1.2. [**K**, Problem 1.81] *Suppose K is a knot in a closed manifold Y. If two surgeries on K yield manifolds which are homeomorphic via an orientation preserving homeomorphism, then there is a homeomorphism of $Y - K$ which takes the slope of one surgery to the slope of the other.*

The main theorem of this paper is an analogue of Ozsváth–Szabó [**OSz5**, Theorem 9.7] and Ni [**Ni**, Theorem 1.5]. See also Wu [**W**].

All manifolds in this paper are oriented, unless otherwise stated.

THEOREM 1.3. *Suppose Y is a closed 3-manifold with $b_1(Y) > 0$. Let K be a null-homologous knot in Y, then the inclusion map $Y - K \to Y$ induces an*

2010 *Mathematics Subject Classification.* Primary 57M27.
Key words and phrases. Thurston norm, cosmetic surgeries, Heegaard Floer homology.
The author is partially supported by an AIM Five-Year Fellowship and NSF grant number DMS-0805807.

isomorphism $H_2(Y-K)\cong H_2(Y)$, *so we can identify* $H_2(Y)$ *with* $H_2(Y-K)$. *Suppose* $r\in\mathbb{Q}\cup\{\infty\}$, *let* $Y_r(K)$ *be the manifold obtained by* r*–surgery on* K. *Suppose* (Y,K) *satisfies that*

$$x_Y(h)<x_{Y-K}(h),\quad \textit{for any nonzero element}\quad h\in H_2(Y). \tag{1.1}$$

Here x_M *is the Thurston norm* [**T**] *in* M. *The conclusion is, if two rational numbers* r,s *satisfy that* $Y_r(K)\cong\pm Y_s(K)$, *then* $r=\pm s$.

Sometimes the condition (1.1) can be weakened if there is a certain additional condition. For example, we can prove the following theorem.

THEOREM 1.4. *Suppose* Y *is a closed* 3*–manifold with* $b_1(Y)>0$. *Suppose* K *is a null-homologous knot in* Y. *Suppose* $x_Y\equiv 0$, *while the restriction of* x_{Y-K} *on* $H_2(Y)$ *is nonzero. Then we have the same conclusion as Theorem 1.3. Namely, if two rational numbers* r,s *satisfy that* $Y_r(K)\cong\pm Y_s(K)$, *then* $r=\pm s$.

2. Non-triviality theorems

In this section, we will state some non-triviality theorems in Heegaard Floer homology. We first set up some notations we will use in this paper.

Let Y be a closed 3–manifold. Suppose $\mathfrak{S}$ is a subset of $\mathrm{Spin}^c(Y)$, let

$$HF^\circ(Y,\mathfrak{S})=\bigoplus_{\mathfrak{s}\in\mathfrak{S}}HF^\circ(Y,\mathfrak{s}),$$

where HF° is one of $\widehat{HF},HF^\infty,HF^+,HF^-$. Furthermore, if $h\in H_2(Y)$, then

$$HF^\circ(Y,h,i)=\bigoplus_{\mathfrak{s}\in\mathrm{Spin}^c(Y),\langle c_1(\mathfrak{s}),h\rangle=2i}HF^\circ(Y,\mathfrak{s}).$$

Similarly, if F is a Seifert surface for a knot $K\subset Y$, then

$$\widehat{HFK}(Y,K,[F],i)=\bigoplus_{\xi\in\underline{\mathrm{Spin}^c}(Y,K),\langle c_1(\xi),\widehat{F}\rangle=2i}\widehat{HFK}(Y,K,\xi),$$

see Ozsváth–Szabó [**OSz3**] for more details. Following Kronheimer and Mrowka [**KM10**], let

$$HF^\circ(Y|h)=HF^\circ(Y,h,\frac{1}{2}x(h)).$$

A very important feature of Heegaard Floer homology is that it detects the Thurston norm of a 3–manifold. In Ozsváth–Szabó [**OSz2**], this result is stated for universally twisted Heegaard Floer homology. Nevertheless, this result should also hold if one uses untwisted coefficients. In fact, the analogous result for Monopole Floer homology is stated with untwisted coefficients in Kronheimer–Mrowka [**KM07**, Corollary 41.4.2]. In order to state our results, we first recall two definitions.

DEFINITION 2.1. Suppose M is a compact 3–manifold, a properly embedded surface $S\subset M$ is *taut* if $x(S)=x([S])$ in $H_2(M,\partial S)$, no proper subsurface of S is null-homologous, and if any component of S lies in a homology class that is represented by an embedded sphere then this component is a sphere. Here $x(\cdot)$ is the Thurston norm.

DEFINITION 2.2. Suppose K is a null-homologous knot in a closed 3–manifold Y. An oriented surface $F \subset Y$ is a *Seifert-like surface* for K, if $\partial F = K$. When F is connected, we say that F is a *Seifert surface* for K. We also view a Seifert-like surface as a proper surface in $Y - \overset{\circ}{\nu}(K)$.

As in the proof of Hedden–Ni [**HN**, Theorem 2.2], using the known non-triviality results for twisted coefficients stated in Ni [**Ni**] and the Universal Coefficients Theorem, we can prove the following theorems. (The same results can also be proved via the approach taken by Juhász [**J**] and Kronheimer–Mrowka [**KM10**].)

THEOREM 2.3. *Suppose Y is a closed 3–manifold, $h \in H_2(Y)$, then*

$$HF^+(Y|h) \otimes \mathbb{Q} \neq 0, \quad \widehat{HF}(Y|h) \otimes \mathbb{Q} \neq 0.$$

THEOREM 2.4. *Suppose K is a null-homologous knot in a closed 3–manifold Y. Let F be a taut Seifert-like surface for K. Then*

$$\widehat{HFK}(Y, K, [F], \frac{x(F)+1}{2}) \otimes \mathbb{Q} \neq 0.$$

3. A surgery formula

Suppose $K \subset Y$ is a null-homologous knot. Let $Y_{p/q}(K)$ denote the manifold obtained by $\frac{p}{q}$–surgery on K. Note that there is a natural identification

$$\mathrm{Spin}^c(Y_{p/q}(K)) \cong \mathrm{Spin}^c(Y) \times \mathbb{Z}/p\mathbb{Z}.$$

Let $\pi\colon \mathrm{Spin}^c(Y_{p/q}(K)) \to \mathrm{Spin}^c(Y)$ be the projection to the first factor.

The goal of this section is to prove the following theorem, which is a (much easier) analogue of Ozsváth–Szabó [**OSz5**, Theorem 1.1].

THEOREM 3.1. *Suppose $K \subset Y$ is a null-homologous knot. If $\widehat{HF}(Y, \mathfrak{s}) = 0$, then there exists a constant $C = C(Y, K, \mathfrak{s})$, such that*

$$\mathrm{rank}\, \widehat{HF}(Y_{p/q}(K), \pi^{-1}(\mathfrak{s})) = qC.$$

3.1. Large surgeries on rationally null-homologous knots. Suppose $K \subset Y$ is a rationally null-homologous knot. Construct a Heegaard diagram $(\Sigma, \boldsymbol{\alpha}, \boldsymbol{\beta}, w, z)$ for (Y, K), such that $\beta_1 = \mu$ is a meridian of K. Moreover, w, z are two base points associated with a marked point on β_1 as in Ozsváth–Szabó [**OSz3**]. There is a curve $\lambda \subset \Sigma$ which gives rise to the knot K. Doing oriented cut-and-pastes to λ and m parallel copies of μ, we get a connected simple closed curve supported in a small neighborhood of $\mu \cup \lambda$. We often denote this curve by $m\mu + \lambda$. The m parallel copies of μ are supported in a small neighborhood of μ. We call this neighborhood the *winding region* for $m\mu + \lambda$. $(\Sigma, \boldsymbol{\alpha}, \boldsymbol{\gamma}, z)$ is a diagram for $Y_{m\mu+\lambda}(K)$, where $\gamma_1 = m\mu + \lambda$ and all other γ_i's are small Hamiltonian translations of β_i's.

DEFINITION 3.2. As in Ozsváth–Szabó [**OSz5**, Section 4], one defines a map

$$\Xi\colon \mathrm{Spin}^c(Y_{m\mu+\lambda}(K)) \to \underline{\mathrm{Spin}^c}(Y, K)$$

as follows. If $\mathfrak{t} \in \mathrm{Spin}^c(Y_{m\mu+\lambda}(K))$ is represented by a point $\mathbf{y}$ supported in the winding region, let $\mathbf{x} \in \mathbb{T}_\alpha \cap \mathbb{T}_\beta$ be the "nearest point", and let $\psi \in \pi_2(\mathbf{y}, \Theta, \mathbf{x})$ be a small triangle. Then

$$\Xi(\mathfrak{t}) = \underline{\mathfrak{s}}_{w,z}(\mathbf{x}) + \big(n_w(\psi) - n_z(\psi)\big) \cdot \mu. \tag{3.1}$$

When we construct the Heegaard triple diagram

$$(\Sigma, \boldsymbol{\alpha}, \boldsymbol{\beta}, \boldsymbol{\gamma}, w, z),$$

the position of the meridian β_1 relative to the points in $\lambda \cap \gamma_1$ may vary. Our next lemma says that the choice of the position of β_1 does not affect the definition of Ξ.

LEMMA 3.3. *Suppose we have two Heegaard triple diagrams as above*

$$\Gamma_1 = (\Sigma, \boldsymbol{\alpha}, \boldsymbol{\beta}^1, \boldsymbol{\gamma}, w^1, z^1), \quad \Gamma_2 = (\Sigma, \boldsymbol{\alpha}, \boldsymbol{\beta}^2, \boldsymbol{\gamma}, w^2, z^2).$$

The two sets $\boldsymbol{\beta}^1$ and $\boldsymbol{\beta}^2$ differ at the meridian, where the meridian $\beta_1^2 \in \boldsymbol{\beta}^2$ is a parallel translation of the meridian $\beta_1^1 \in \boldsymbol{\beta}^1$, still supported in the winding region. The two base points are moved together with the meridian.

Using these two diagrams, we can define two maps

$$\Xi^1, \Xi^2 \colon \operatorname{Spin}^c(Y_{m\mu+\lambda}(K)) \to \underline{\operatorname{Spin}^c}(Y, K).$$

Then $\Xi^1 = \Xi^2$.

PROOF. Without loss of generality, we may assume there is only one intersection point of $\lambda \cap \gamma_1$ between β_1^1 and β_1^2. See Figure 1 for an illustration.

Suppose $\mathbf{y}^1, \mathbf{y}^2 \in \mathbb{T}_\alpha \cap \mathbb{T}_\gamma$ are two intersection points supported in the winding region, and suppose their γ_1–coordinates are y^1, y^2, respectively. Assume $\mathfrak{s}_{w^1}(\mathbf{y}^1) = \mathfrak{s}_{w^2}(\mathbf{y}^2) = \mathfrak{t}$, we want to prove that $\Xi^1(\mathfrak{t}) = \Xi^2(\mathfrak{t})$.

By Ozsváth–Szabó [**OSz1**, Lemma 2.19],

$$\begin{aligned} \mathfrak{s}_{w^1}(\mathbf{y}^1) - \mathfrak{s}_{w^1}(\mathbf{y}^2) &= \operatorname{PD}(\varepsilon(\mathbf{y}^2, \mathbf{y}^1)), \\ \mathfrak{s}_{w^2}(\mathbf{y}^2) - \mathfrak{s}_{w^1}(\mathbf{y}^2) &= \operatorname{PD}(\mu). \end{aligned}$$

Hence $\varepsilon(\mathbf{y}^2, \mathbf{y}^1) = \mu$. Let $\tilde{\mathbf{y}}^1 \in \mathbb{T}_\alpha \cap \mathbb{T}_\gamma$ be the point whose coordinates coincide with the coordinates of $\mathbf{y}^1$, except that its γ_1–coordinate is the next intersection point to y^1 on the same α–curve, denoted $\tilde{y}^1$. Then $\varepsilon(\tilde{\mathbf{y}}^1, \mathbf{y}^1) = \mu$, so $\tilde{\mathbf{y}}^1$ is in the same equivalence class as $\mathbf{y}^2$.

Now we only need to prove that

$$\Xi^1(\mathfrak{s}_{w^1}(\mathbf{y}^1)) = \Xi^2(\mathfrak{s}_{w^2}(\tilde{\mathbf{y}}^1)). \tag{3.2}$$

Let $\mathbf{x}^1 \in \mathbb{T}_\alpha \cap \mathbb{T}_{\beta^1}, \tilde{\mathbf{x}}^1 \in \mathbb{T}_\alpha \cap \mathbb{T}_{\beta^2}$ be the nearest points to $\mathbf{y}^1, \tilde{\mathbf{y}}^1$, respectively. It is clear that $\underline{\mathfrak{s}}_{w^1,z^1}(\mathbf{x}^1) = \underline{\mathfrak{s}}_{w^2,z^2}(\tilde{\mathbf{x}}^1)$. Moreover, the small triangle for $\tilde{\mathbf{y}}^1$ in Γ_2 is just a translation of the small triangle for $\mathbf{y}^1$ in Γ_1, so they contribute the same $n_w(\psi) - n_z(\psi)$ term in (3.1). So (3.2) follows. $\square$

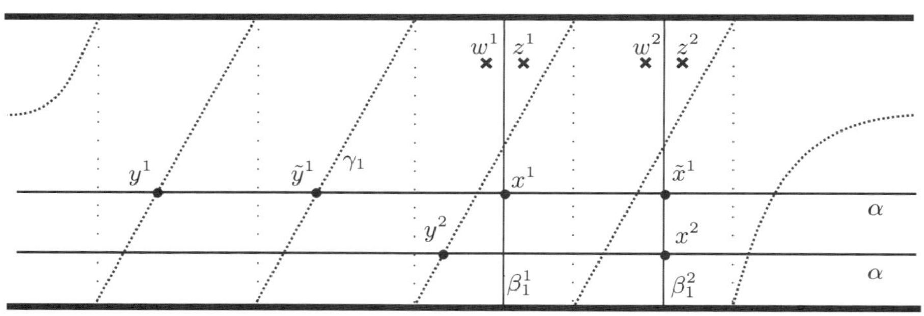

FIGURE 1. Local picture of the two triple Heegaard diagrams

REMARK 3.4. In Ozsváth–Szabó [**OSz5**], in order to define $\Xi(\mathfrak{t})$, one places the meridian in a position such that the equivalence class of intersection points representing $\mathfrak{t}$ is supported in the winding region. The above lemma removes this restriction.

LEMMA 3.5. *Suppose $\xi \in \underline{\mathrm{Spin}^c}(Y,K)$. For all sufficiently large m, there exists $\mathfrak{t} \in \mathrm{Spin}^c(Y_{m\mu+\lambda}(K))$, such that $\Xi(\mathfrak{t}) = \xi$.*

PROOF. Let $\mathfrak{s} \in \mathrm{Spin}^c(Y)$ be the underlying Spin^c structure of ξ. We can choose a Heegaard diagram for (Y,K) such that some $\mathbf{x} \in \mathbb{T}_\alpha \cap \mathbb{T}_\beta$ represents $\mathfrak{s}$, then $\xi = \underline{\mathfrak{s}}_{w,z}(\mathbf{x}) + n \cdot \mu$ for some $n \in \mathbb{Z}$. Now our desired result follows from the definition of Ξ. □

The following proposition is a part of Ozsváth–Szabó [**OSz5**, Theorem 4.1].

PROPOSITION 3.6. *Let $K \subset Y$ be a rationally null-homologous knot in a closed, oriented three-manifold, equipped with a framing λ. Let*

$$\widehat{A}_\xi(Y,K) = C_\xi\{\max\{i,j\} = 0\},$$

*where $C_\xi = CFK^\infty(Y,K,\xi)$ as in Ozsváth–Szabó [**OSz5**]. Then, for all sufficiently large m and all $\mathfrak{t} \in \mathrm{Spin}^c(Y_{m\mu+\lambda}(K))$, there is an isomorphism*

$$\Psi_{\mathfrak{t},m} \colon \widehat{CF}(Y_{m\mu+\lambda}(K),\mathfrak{t}) \to \widehat{A}_{\Xi(\mathfrak{t})}(Y,K).$$

3.2. Rational surgeries on null-homologous knots. Let K be a null-homologous knot in Y. As in Ozsváth–Szabó [**OSz5**, Section 7], $Y_{\frac{p}{q}}(K)$ can be realized by a Morse surgery with coefficient a on the knot $K' = K \# O_{q/r} \subset Y' = Y \# L(q,r)$, where $O_{q/r}$ is a U–knot in $L(q,r)$, $p = aq + r$. Let

$$\Xi' \colon \mathrm{Spin}^c(Y'_{a\mu'+\lambda'}) \to \underline{\mathrm{Spin}^c}(Y',K')$$

be the map defined in Definition 3.2.

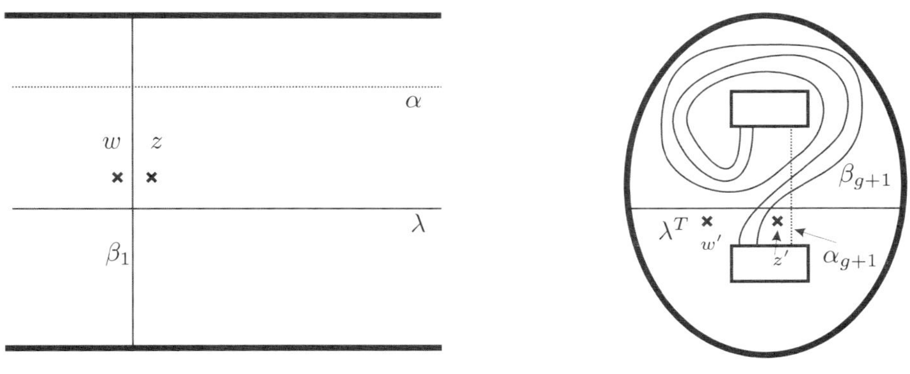

FIGURE 2. The left hand side is a piece of a Heegaard diagram for (Y,K). The right hand side is a genus 1 Heegaard diagram for $(L(q,r),O_{q/r})$. The boundary of the oval is capped off with a disk, and the boundaries of the two rectangles are glued together via a reflection. Here we choose $q=3, r=2$.

CONSTRUCTION 3.7. Let

$$(\Sigma, \boldsymbol{\alpha} = \{\alpha_1, \dots, \alpha_g\}, \boldsymbol{\beta} = \{\beta_1, \dots, \beta_g\}, w, z)$$

be a doubly-pointed Heegaard diagram for (Y, K), such that β_1 is a meridian for K and the two base points are induced from a marked point on β_1. Suppose $\lambda \subset \Sigma$ represents a longitude of K.

Let

$$(T, \{\alpha_{g+1}\}, \{\beta_{g+1}\}, w', z')$$

be a genus 1 Heegaard diagram for $(L(q, r), O_{q/r})$. As in Figure 2, β_{g+1} intersects α_{g+1} exactly q times and intersects the boundary of each rectangle exactly r times. Suppose $\lambda^T \subset T$ represents a longitude of $O_{q/r}$.

We perform the connected sum of Σ and T by identifying the neighborhoods of z and w', hence we get a new genus $(g+1)$ surface Σ'. Then

$$(\Sigma', \boldsymbol{\alpha}' = \boldsymbol{\alpha} \cup \{\alpha_{g+1}\}, \boldsymbol{\beta}' = \boldsymbol{\beta} \cup \{\beta_{g+1}\}, w, z')$$

is a Heegaard diagram for (Y', K'). The longitude λ' of K' is a connected sum of λ and λ^T. □

We define

$$\Pi_1 \colon \underline{\mathrm{Spin}^c}(Y', K') \to \underline{\mathrm{Spin}^c}(Y, K)$$

as follows. Given $\xi' \in \underline{\mathrm{Spin}^c}(Y', K')$, suppose $\mathbf{x}' \in \mathbb{T}_{\alpha'} \cap \mathbb{T}_{\beta'}$ represents the underlying Spin^c structure of ξ', then

$$\xi' = \underline{\mathfrak{s}}_{w,z'}(\mathbf{x}') + n \cdot \mu'$$

for some $n \in \mathbb{Z}$. Now let $\mathbf{x}$ be the projection of $\mathbf{x}'$ to $\mathbb{T}_\alpha \cap \mathbb{T}_\beta$, then

$$\Pi_1(\xi') = \underline{\mathfrak{s}}_{w,z}(\mathbf{x}) + n \cdot \mu.$$

The following proposition is obvious. (See also Ozsváth–Szabó [**OSz5**, Corollary 5.3].)

PROPOSITION 3.8. *For any* $\xi' \in \underline{\mathrm{Spin}^c}(Y', K')$, *we have*

$$CFK^\infty(Y', K', \xi') \cong CFK^\infty(Y, K, \Pi_1(\xi'))$$

as $\mathbb{Z} \oplus \mathbb{Z}$*–filtered chain complexes.*

LEMMA 3.9. *When* m *is sufficiently large, we have*

$$\pi = G_{Y,K} \circ \Pi_1 \circ \Xi'.$$

Here $G_{Y,K} \colon \underline{\mathrm{Spin}^c}(Y, K) \to \mathrm{Spin}^c(Y)$ *is the map defined in Ozsváth–Szabó* [**OSz5**, Section 2.2].

PROOF. We follow the notation in Construction 3.7. Since λ' intersects β_1 exactly once, we can slide β_{g+1} over β_1 r times to eliminate the intersection points in $\beta_{g+1} \cap \lambda'$. The new curve is denoted β'_{g+1} as in Figure 3. Then

$$(\Sigma', \boldsymbol{\alpha}', \boldsymbol{\beta}'' = \boldsymbol{\beta} \cup \{\beta'_{g+1}\}, w, z')$$

is also a Heegaard diagram for (Y', K'). Let $\gamma_1 = a\beta_1 + \lambda'$, then

$$(\Sigma', \boldsymbol{\alpha}', \boldsymbol{\gamma}_1 = \{\gamma_1, \beta_2, \dots, \beta_g, \beta'_{g+1}\}, w)$$

is a Heegaard diagram for $Y'_{a\mu'+\lambda'}(K')$.

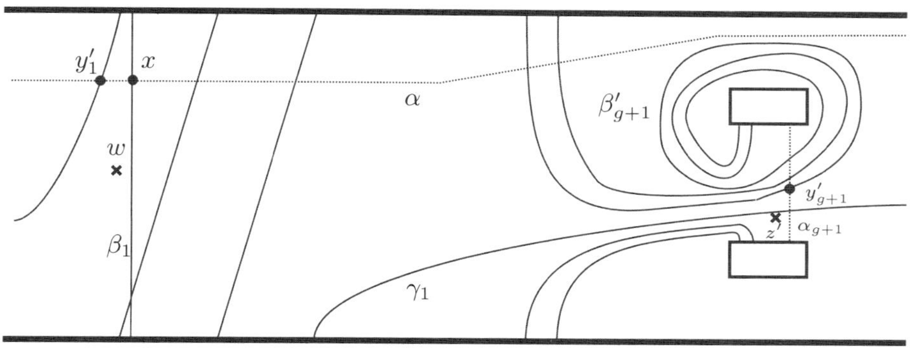

FIGURE 3. A Heegaard diagram for $Y'_{a\mu'+\lambda'}(K')$. Here we choose $a = 3$.

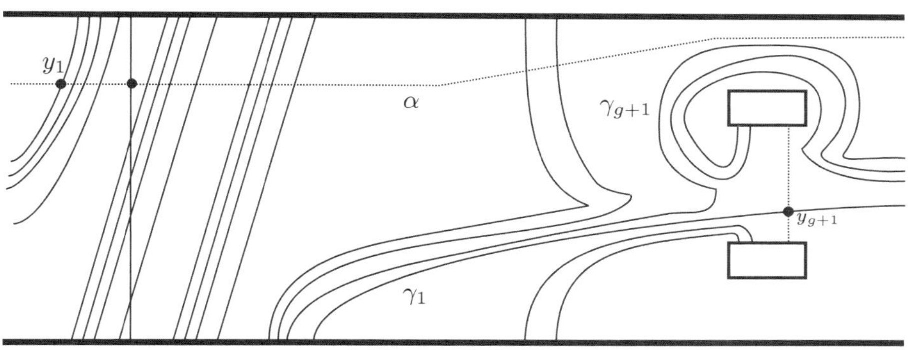

FIGURE 4. After q handleslides, we get a Heegaard diagram for $Y_{p/q}(K)$.

The curve α_{g+1} intersects γ_1 exactly once. We can slide β'_{g+1} over γ_1 q times to eliminate its q intersection points with α_{g+1}. The new curve is denoted γ_{g+1} as in Figure 4. Now

$$(\Sigma', \boldsymbol{\alpha}', \boldsymbol{\gamma}_2 = \{\gamma_1, \beta_2, \dots, \beta_g, \gamma_{g+1}\}, w)$$

is a Heegaard diagram for $Y'_{a\mu'+\lambda'}(K') = Y_{p/q}(K)$. Moreover, we may slide other α–curves over α_{g+1} to eliminate their intersection points with γ_1. A destabilization will remove α_{g+1} and γ_1. Now we get a diagram

$$(\Sigma^*, \boldsymbol{\alpha}^*, \boldsymbol{\gamma}^*, w)$$

which is isomorphic to

$$(\Sigma, \boldsymbol{\alpha}, \{\beta_2, \dots, \beta_g, \gamma^*_{g+1}\}, w),$$

where γ^*_{g+1} is the image of γ_{g+1} under the destabilization.

We want to show that γ^*_{g+1} is isotopic to $p\mu + q\lambda$, the curve obtained by doing cut-and-pastes to p parallel copies of μ and q parallel copies of λ. In fact, γ^*_{g+1} is supported in a small neighborhood of $\mu \cup \lambda$, so it must be isotopic to $p'\mu + q'\lambda$ for some p', q'. It is easy to compute the intersection numbers of γ_{g+1} with λ and

$\mu = \beta_1$, which are $p = aq + r$ and q. The intersection numbers of γ^*_{g+1} with μ and λ remains the same, so $\gamma^*_{g+1} = p\mu + q\lambda$.

Suppose $\mathfrak{t} \in \mathrm{Spin}^c(Y_{p/q}(K))$. We want to prove

$$\pi(\mathfrak{t}) = G_{Y,K} \circ \Pi_1 \circ \Xi'(\mathfrak{t}). \tag{3.3}$$

We first consider the right hand side of (3.3). Let $\mathbf{y}'$ be a point in $\mathbb{T}_{\alpha'} \cap \mathbb{T}_{\gamma_1}$ which is supported in the winding region and represents $\mathfrak{t}$ (Figure 3). Suppose the γ_1–coordinate of $\mathbf{y}'$ is y'_1 and the β'_{g+1}–coordinate is y'_{g+1}.

Let $\mathbf{x}' \in \mathbb{T}_{\alpha'} \cap \mathbb{T}_{\beta''}$ be the nearest point to $\mathbf{y}'$, then (3.1) implies that

$$\Xi'(\mathfrak{t}) = \underline{\mathfrak{s}}_{w,z'}(\mathbf{x}') + n \cdot \mu'$$

for some $n \in \mathbb{Z}$. Let $\mathbf{x}$ be the projection of $\mathbf{x}'$ to $\mathbb{T}_\alpha \cap \mathbb{T}_\beta$, then

$$\Pi_1 \circ \Xi'(\mathfrak{t}) = \underline{\mathfrak{s}}_{w,z}(\mathbf{x}) + n \cdot \mu.$$

Hence

$$G_{Y,K} \circ \Pi_1 \circ \Xi'(\mathfrak{t}) = \mathfrak{s}_w(\mathbf{x}).$$

Now we consider the left hand side of (3.3). As in Figure 4, we get another Heegaard diagram for $Y_{p/q}(K)$ by q handle slides. In this diagram, we can find a point $\mathbf{y} \in \mathbb{T}_{\alpha'} \cap \mathbb{T}_{\gamma_2}$ which represents $\mathfrak{t}$ as $\mathbf{y}'$ does. In fact, since α_{g+1} intersects γ_1 exactly once and is disjoint from other γ–curves, $\mathbf{y}$ must contain the intersection point of α_{g+1} and γ_1, denoted y_{g+1}. The γ_1–coordinate of $\mathbf{y}$, called y_1, is determined by y'_1 and y'_{g+1}: it is one of the q intersection points on γ_{g+1} near y'_1, and the choice among these q points is specified by the position of y'_{g+1}. Other coordinates of $\mathbf{y}$ are the same as $\mathbf{y}'$.

After handleslides and one destabilization, we get a point $\mathbf{y}^* \in \mathbb{T}_{\alpha^*} \cap \mathbb{T}_{\gamma^*}$ whose coordinates are the same as $\mathbf{x}$ except that its γ_1–coordinate is y_1. So its nearest point in $\mathbb{T}_\alpha \cap \mathbb{T}_\beta$ is $\mathbf{x}$, hence $\mathbf{x}$ represents $\pi(\mathfrak{t})$. This proves (3.3). □

LEMMA 3.10. *Let $H(\widehat{A}_\xi(Y,K))$ be the homology of $\widehat{A}_\xi(Y,K)$. For a fixed ξ, when $|n| \gg 0$,*

$$H(\widehat{A}_{\xi+n\cdot\mu}(Y,K)) \cong \widehat{HF}(Y, G_{Y,K}(\xi)).$$

PROOF. By the definitions

$$\begin{aligned} \widehat{A}_{\xi+n\cdot\mu}(Y,K) &= C_{\xi+n\cdot\mu}\{\max\{i,j\} = 0\} \\ &= C_\xi\{\max\{i, j-n\} = 0\}. \end{aligned}$$

By the adjunction inequality, $H(C_\xi\{i,j\}) = 0$ when $|i-j| \gg 0$. So

$$H(C_\xi\{\max\{i,j-n\} = 0\}) \cong H(C_\xi\{i = 0\})$$

when $n \gg 0$. The latter group is isomorphic to $\widehat{HF}(Y, G_{Y,K}(\xi))$ by Ozsváth–Szabó [**OSz5**, Proposition 3.2].

When $n \ll 0$, we have

$$H(C_\xi\{\max\{i,j-n\} = 0\}) \cong H(C_\xi\{j = n\}) \cong H(C_\xi\{j = 0\}),$$

which is isomorphic to $\widehat{HF}(Y, G_{Y,-K}(\xi))$ by Ozsváth and Szabó [**OSz5**, Proposition 3.2]. Now by [**OSz5**, Equation (4)] and the fact that K is null-homologous, we have $G_{Y,K}(\xi) = G_{Y,-K}(\xi)$. □

LEMMA 3.11. *Suppose $\widehat{HF}(Y,\mathfrak{s}) = 0$, then $H(\widehat{A}_{\xi'}(Y',K')) \neq 0$ for only finitely many $\xi' \in (G_{Y,K} \circ \Pi_1)^{-1}(\mathfrak{s})$.*

PROOF. For each $\xi \in \underline{\mathrm{Spin}^c}(Y,K)$, there are exactly q relative Spin^c structures in $\Pi_1^{-1}(\xi)$. Moreover, by Proposition 3.8, if $\xi' \in \Pi_1^{-1}(\xi)$, then

$$\widehat{A}_{\xi'}(Y',K') \cong \widehat{A}_{\xi}(Y,K).$$

Hence we only need to show that $H(\widehat{A}_{\xi}(Y,K)) \neq 0$ for only finitely many $\xi \in G_{Y,K}^{-1}(\mathfrak{s})$.

Pick any $\xi \in G_{Y,K}^{-1}(\mathfrak{s})$, then

$$G_{Y,K}^{-1}(\mathfrak{s}) = \{\xi + i\cdot\mu \mid i \in \mathbb{Z}\}.$$

By Lemma 3.10, $H(\widehat{A}_{\xi+i\cdot\mu}(Y,K))$ is isomorphic to $\widehat{HF}(Y,\mathfrak{s})$ when $|i|$ is large, hence is 0. This finishes the proof. □

PROPOSITION 3.12. *When m is sufficiently large,*

$$\begin{aligned}\widehat{HF}(Y'_{m\mu'+\lambda'}(K'),\pi^{-1}(\mathfrak{s})) &\cong \bigoplus_{\{\xi' \mid G_{Y,K}\circ\Pi_1(\xi')=\mathfrak{s}\}} H(\widehat{A}_{\xi'}(Y',K')) \\ &\cong \bigoplus^{q} \bigoplus_{\{\xi \mid G_{Y,K}(\xi)=\mathfrak{s}\}} H(\widehat{A}_{\xi}(Y,K)).\end{aligned}$$

PROOF. By Proposition 3.6, when m is sufficiently large

$$\widehat{HF}(Y'_{m\mu'+\lambda'}(K'),\pi^{-1}(\mathfrak{s})) \cong \bigoplus_{\mathfrak{t}\in\pi^{-1}(\mathfrak{s})} H(\widehat{A}_{\Xi'(\mathfrak{t})}(Y',K')).$$

By Lemma 3.9,

$$\Xi'(\pi^{-1}(\mathfrak{s})) = \Xi'\left(\Xi'^{-1}\circ(G_{Y,K}\circ\Pi_1)^{-1}(\mathfrak{s})\right) \subset (G_{Y,K}\circ\Pi_1)^{-1}(\mathfrak{s}).$$

Consider the map

$$\Xi'_{\mathfrak{s}}\colon\ \pi^{-1}(\mathfrak{s}) \to (G_{Y,K}\circ\Pi_1)^{-1}(\mathfrak{s}).$$

By Ni [**Ni**, Lemma 2.4], $\Xi'_{\mathfrak{s}}$ is injective. Moreover, by Lemmas 3.5 and 3.11, when m is sufficiently large, the range of $\Xi'_{\mathfrak{s}}$ contains all $\xi' \in (G_{Y,K}\circ\Pi_1)^{-1}(\mathfrak{s})$ satisfying $H(\widehat{A}_{\xi'}(Y',K')) \neq 0$. This proves the first equality.

In order to prove the second equality, we note that for each $\xi \in \underline{\mathrm{Spin}^c}(Y,K)$, there are exactly q relative Spin^c structures in $\Pi_1^{-1}(\xi)$. Moreover, by Proposition 3.8, if $\xi' \in \Pi_1^{-1}(\xi)$, then

$$\widehat{A}_{\xi'}(Y',K') \cong \widehat{A}_{\xi}(Y,K).$$

So the second equality easily follows. □

PROOF OF THEOREM 3.1. Let

$$C = \operatorname{rank} \bigoplus_{\{\xi \mid G_{Y,K}(\xi)=\mathfrak{s}\}} H(\widehat{A}_{\xi}(Y,K)).$$

By Proposition 3.12,

$$\operatorname{rank} \widehat{HF}(Y_{p/q},\pi^{-1}(\mathfrak{s})) = qC$$

when p is sufficiently large.

Since $\widehat{HF}(Y,\mathfrak{s}) = 0$, we have $\widehat{HF}(Y',\mathfrak{s}') = 0$ for any $\mathfrak{s}'$ that extends $\mathfrak{s}$. By Ozsváth–Szabó [**OSz2**, Theorem 9.12], we have the long exact sequence

$$\begin{array}{ccc} \widehat{HF}(Y', P_1^{-1}(\mathfrak{s})) & \longrightarrow & \widehat{HF}(Y'_{m\mu'+\lambda'}(K'), \pi_m^{-1}(\mathfrak{s})) \ , \\ \uparrow & \swarrow & \\ \widehat{HF}(Y'_{(m+1)\mu'+\lambda'}(K'), \pi_{m+1}^{-1}(\mathfrak{s})) & & \end{array}$$

where

$$P_1 \colon \operatorname{Spin}^c(Y') \to \operatorname{Spin}^c(Y),$$
$$\pi_m \colon \operatorname{Spin}^c(Y'_{m\mu'+\lambda'}(K')) \to \operatorname{Spin}^c(Y)$$

are the natural projection maps. Since $\widehat{HF}(Y', P_1^{-1}(\mathfrak{s})) = 0$, we have

$$\widehat{HF}(Y'_{a\mu'+\lambda'}(K'), \pi_a^{-1}(\mathfrak{s})) \cong \widehat{HF}(Y'_{m\mu'+\lambda'}(K'), \pi_m^{-1}(\mathfrak{s}))$$

for m sufficiently large. Hence its rank is always qC. □

4. Cosmetic surgeries

Proof of Theorem 1.3. Assume there are two rational numbers $\frac{p_1}{q_1}, \frac{p_2}{q_2}$ satisfying that there is a homeomorphism

$$f \colon Y_{\frac{p_1}{q_1}} \to \pm Y_{\frac{p_2}{q_2}},$$

then $|p_1| = |p_2|$ for homological reasons. If $\frac{p_1}{q_1} \neq \pm\frac{p_2}{q_2}$, then we can assume

$$0 < q_1 < q_2.$$

Without loss of generality, we may assume $Y - K$ is irreducible. By (1.1) and the adjunction inequality, we conclude that $\widehat{HF}(Y, h, \frac{1}{2}x_{Y-K}(h)) = 0$. It then follows from Theorem 3.1 that there is a constant C_h, such that

$$\operatorname{rank} \widehat{HF}(Y_{p/q}(K), h, \frac{1}{2}x_{Y-K}(h)) = qC_h.$$

Since (1.1) holds, a result of Gabai [**G**, Corollary 2.4] implies that

$$x_{Y-K}(h) = x_{Y_{p/q}(K)}(h)$$

for any nonzero $h \in H_2(Y)$ and $\frac{p}{q} \in \mathbb{Q}$. Theorem 2.3 then implies that

$$\operatorname{rank} \widehat{HF}(Y_{p/q}(K)|h) = qC_h \neq 0.$$

Since K is null-homologous, the inclusion maps $Y - K \to Y_r$ induce isomorphisms on H_2 for each $r \in \mathbb{Q} \cup \{\infty\} \backslash \{0\}$. Hence we can identify $H_2(Y_r(K))$ with $H_2(Y)$. Now $f_* \colon H_2(Y_{\frac{p_1}{q_1}}) \to H_2(Y_{\frac{p_2}{q_2}})$ can be regarded as a map

$$f_* \colon H_2(Y) \to H_2(Y).$$

Fix a nonzero $h \in H_2(Y)$, we have

$$\operatorname{rank} \widehat{HF}(Y_{\frac{p_1}{q_1}}|f_*^n(h)) = \frac{q_1}{q_2} \operatorname{rank} \widehat{HF}(Y_{\frac{p_2}{q_2}}|f_*^n(h)) \neq 0$$

for any $n \in \mathbb{Z}$. Moreover, since $f \colon Y_{\frac{p_1}{q_1}} \to \pm Y_{\frac{p_2}{q_2}}$ is a homeomorphism, we have

$$\operatorname{rank} \widehat{HF}(Y_{\frac{p_1}{q_1}}|f_*^{n-1}(h)) = \operatorname{rank} \widehat{HF}(Y_{\frac{p_2}{q_2}}|f_*^n(h)).$$

Thus we get

$$\operatorname{rank}\widehat{HF}(Y_{\frac{p_1}{q_1}}|f_*^n(h)) = \left(\frac{q_1}{q_2}\right)^n \operatorname{rank}\widehat{HF}(Y_{\frac{p_1}{q_1}}|h) \neq 0.$$

So $0 < \operatorname{rank}\widehat{HF}(Y_{\frac{p_1}{q_1}}|h) < 1$ when n is sufficiently large, which is impossible. □

PROOF OF THEOREM 1.4. Since $x_Y \equiv 0$, the adjunction inequality implies that $\widehat{HF}(Y, h, \frac{1}{2}x_{Y-K}(h)) = 0$ for any $h \in H_2(Y)$ satisfying $x_{Y-K}(h) \neq 0$. Using Theorems 3.1, 2.3 and Gabai [**G**, Corollary 2.4], we have

$$\operatorname{rank}\widehat{HF}(Y_{p/q}(K)|h) = qC_h$$

for some nonzero constant C_h. Now the argument is the same as in the proof of Theorem 1.3. □

References

[G] D. Gabai, *Foliations and the topology of 3–manifolds II*, J. Differential Geom. **26** (1987), no. 3, 461–478.

[HN] M. Hedden and Y. Ni, *Manifolds with small Heegaard Floer ranks*, Geom. Topol. **14** (2010), 1479–1501.

[JM] S. Jabuka and T. Mark, *Product formulae for Ozsváth–Szabó 4–manifold invariants*, Geom. Topol. **12** (2008), 1557–1651.

[J] A Juhász, *Floer homology and surface decompositions*, Geom. Topol. **12** (2008), 299–350.

[K] *Problems in low-dimensional topology*, Edited by Rob Kirby. AMS/IP Stud. Adv. Math., 2.2, Geometric topology (Athens, GA, 1993), 35–473, AMS, Providence, 1997.

[KM07] P. Kronheimer and T. Mrowka, *Monopoles and three-manifolds*, New Mathematical Monographs **10**, Cambridge University Press, Cambridge, 2007.

[KM10] P. Kronheimer and T. Mrowka, *Knots, sutures and excision*, J. Differential Geom. **84** (2010), no. 2, 301–364.

[Ni] Y. Ni, *Non-separating spheres and twisted Heegaard Floer homology*, preprint (2009), available at arXiv:0902.4034.

[OSz1] P. Ozsváth and Z. Szabó, *Holomorphic disks and topological invariants for closed three-manifolds*, Ann. of Math. (2), **159** (2004), no. 3, 1027–1158.

[OSz2] P. Ozsváth and Z. Szabó, *Holomorphic disks and three-manifold invariants: properties and applications*, Ann. of Math. (2), **159** (2004), no. 3, 1159–1245.

[OSz3] P. Ozsváth and Z Szabó, *Holomorphic disks and knot invariants.* Adv. Math. **186** (2004), no. 1, 58–116.

[OSz4] P. Ozsváth and Z Szabó, *Holomorphic disks and genus bounds*, Geom. Topol. **8** (2004), 311–334.

[OSz5] P. Ozsváth and Z. Szabó, *Knot Floer homology and rational surgeries*, Algebr. Geom. Topol. **11** (2011), 1–68.

[R] J. Rasmussen, *Floer homology and knot complements*, PhD Thesis, Harvard University (2003), available at arXiv:math.GT/0306378.

[T] W. Thurston, *A norm for the homology of 3–manifolds*, Mem. Amer. Math. Soc. **59** (1986), no. 339, i–vi and 99–130.

[W] Z. Wu, *Cosmetic Surgery in Integral Homology L-Spaces*, preprint (2009), available at arXiv:0911.5333.

DEPARTMENT OF MATHEMATICS, CALTECH, MC 253-37, 1200 E CALIFORNIA BLVD, PASADENA, CA 91125

E-mail address: `yini@caltech.edu`

Proceedings of Symposia in Pure Mathematics
Volume **82**, 2011

On the relative Giroux correspondence

Tolga Etgü and Burak Ozbagci

Abstract. Recently, Honda, Kazez and Matić described an adapted partial open book decomposition of a compact contact 3-manifold with convex boundary by generalizing the work of Giroux in the closed case. They also implicitly established a one-to-one correspondence between isomorphism classes of partial open book decompositions modulo positive stabilization and isomorphism classes of compact contact 3-manifolds with convex boundary. In this expository article we explicate the relative version of Giroux correspondence.

1. Introduction

Let (M, Γ) be a balanced sutured 3-manifold and let ξ be a contact structure on M with convex boundary whose dividing set on ∂M is isotopic to Γ. Recently, Honda, Kazez and Matić [**HKM09**] introduced an invariant of the contact structure ξ which lives in the sutured Floer homology group defined by Juhász [**Ju**]. This invariant is a relative version of the contact class in Heegaard Floer homology in the closed case as defined by Ozsváth and Szabó [**OzSz**] and reformulated in [**HKM07**]. Both the original definition in [**OzSz**] and the reformulation of the contact class by Honda, Kazez and Matić are based on the so called Giroux correspondence [**Gi02**] which is a one-to-one correspondence between open book decompositions modulo positive stabilization and isotopy classes of contact structures on closed 3-manifolds.

In order to adapt their reformulation [**HKM07**] of the contact class to the case of a contact manifold (M, ξ) with convex boundary, Honda, Kazez and Matić described in [**HKM09**], a partial open book decomposition of M (adapted to ξ) by generalizing the work of Giroux in the closed case. This description coupled with Theorem 1.2 (and the subsequent discussion) in [**HKM09**] induces a map from isomorphism classes of compact contact 3-manifolds with convex boundary to isomorphism classes of partial open book decompositions modulo positive stabilization. Here we spell out the inverse of this map, by describing a compact contact

2010 *Mathematics Subject Classification.* Primary 53D35; Secondary 57M50, 57R17.

Key words and phrases. partial open book decomposition, contact three manifold with convex boundary, sutured manifold, compatible contact structure.

The first author was partially supported by a GEBIP grant of the Turkish Academy of Sciences and a CAREER grant of the Scientific and Technological Research Council of Turkey.

The second author was partially supported by the research grant 107T053 of the Scientific and Technological Research Council of Turkey and the Marie Curie International Outgoing Fellowship 236639.

3-manifold with convex boundary compatible with an *abstract* partial open book decomposition. To define a contact structure compatible with an abstract partial open book decomposition we chose to mimic the analogous result of Torisu [**To**] (rather than adapting the construction of Thurston and Winkelnkemper [**ThWi**]) which conveniently allowed us to keep track of the dividing set on the boundary. Consequently, one obtains a relative version of Giroux correspondence which is due to Honda, Kazez and Matić.

THEOREM 1.1. *There is a one-to-one correspondence between isomorphism classes of partial open book decompositions modulo positive stabilization and isomorphism classes of compact contact* 3*-manifolds with convex boundary.*

The relative Giroux correspondence helps understand the geometric properties of contact 3-manifolds using partial open books, e.g. if the monodromy of a corresponding partial open book is not right-veering, then the contact structure is overtwisted. It also plays a critical role in the definition of the (relative) contact invariant in sutured Floer homology which helps to analyze the contact invariant of a closed manifolds in terms of the relative contact invariants of certain compact pieces. In [**GHV**], it is proved that the contact invariant vanishes in the presence of Giroux torsion using some properties of the relative invariant.

The paper is organized as follows: In Section 2 we give the definition of an abstract partial open book decomposition (S, P, h), construct a balanced sutured manifold (M, Γ) associated to (S, P, h), and construct a (unique) *compatible* contact structure ξ on M which makes ∂M convex with a dividing set isotopic to Γ. In Section 3 we prove Theorem 1.1 after reviewing the related results due to Honda, Kazez and Matić [**HKM09**]. In the last section we provide examples of abstract partial open books compatible with some basic contact 3-manifolds with boundary.

The reader is advised to turn to Etnyre's notes [**Etn**] for the related material on contact topology of 3-manifolds.

ACKNOWLEDGEMENTS. We would like to thank András Stipsicz, Sergey Finashin and John Etnyre for valuable comments on a draft of this paper. We also thank the anonymous referee for helpful remarks, especially for the remark in the footnote for Proposition 2.6.

2. Partial open books, sutured manifolds and contact structures

DEFINITION 2.1. An abstract partial open book decomposition is a triple (S, P, h) satisfying the following conditions:

(1) S is a compact oriented connected surface with $\partial S \neq \emptyset$,

(2) $P = P_1 \cup P_2 \cup \ldots \cup P_r$ is a proper (not necessarily connected) subsurface of S such that S is obtained from $\overline{S \setminus P}$ by successively attaching 1-handles $P_1, P_2, \ldots, P_r$,

(3) $h : P \to S$ is an embedding such that $h|_A =$ identity, where $A = \partial P \cap \partial S$.

REMARK. Figures 1 and 2 present simple examples of partial open book decompositions. It follows from the definition that A is a 1-manifold with nonempty boundary (but it may have closed components as in Figure 4) and $\overline{\partial P \setminus A}$ is a nonempty set consisting of some arcs (but no closed components). The connectedness condition on S is not essential, but simplifies the discussion.

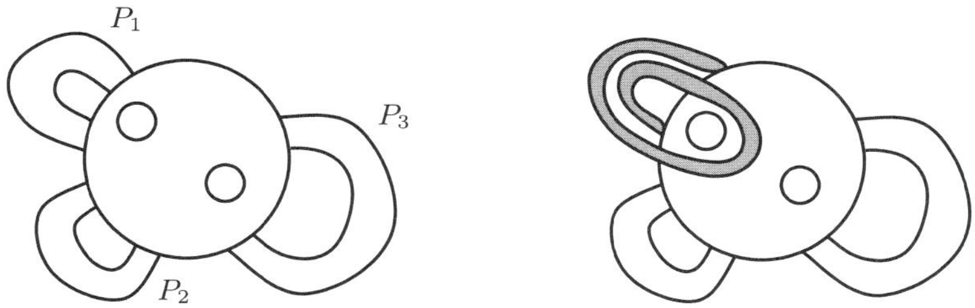

FIGURE 1. An example of S and P satisfying the conditions in Definition 2.1: $\overline{S \setminus P}$ is a twice punctured disk, $r = 3$, and h is the embedding which is identity on P_2 and P_3, and the image of P_1 is the shaded region indicated in the figure on the right.

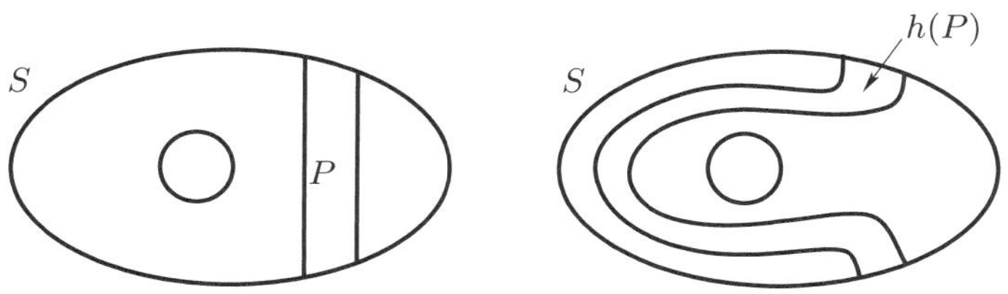

FIGURE 2. Another example of an abstract partial open book.

We now briefly turn our attention to sutured manifolds which was introduced by Gabai [**Ga**] to study foliations. A sutured manifold (M, Γ) is a compact oriented 3-manifold with nonempty boundary, together with a compact subsurface $\Gamma = A(\Gamma) \cup T(\Gamma) \subset \partial M$, where $A(\Gamma)$ is a union of pairwise disjoint annuli and $T(\Gamma)$ is a union of tori. Moreover each component of $\partial M \setminus \Gamma$ is oriented, subject to the condition that whether or not the orientation agrees with the orientation induced as the boundary of M changes every time we nontrivially cross $A(\Gamma)$. Let $R_+(\Gamma)$ (resp. $R_-(\Gamma)$) be the open subsurface of $\partial M \setminus \Gamma$ on which the orientation agrees with (resp. is the opposite of) the boundary orientation on ∂M. A sutured manifold (M, Γ) is balanced if M has no closed components, $\pi_0(A(\Gamma)) \to \pi_0(\partial M)$ is surjective, and $\chi(R_+(\Gamma)) = \chi(R_-(\Gamma))$ on every component of M. It turns out that if (M, Γ) is balanced, then $\Gamma = A(\Gamma)$ and every component of ∂M nontrivially intersects Γ. Since all the sutured manifolds that we will deal with in this paper are balanced, we will think of Γ as a set of *oriented curves* on ∂M by identifying each annulus in Γ with its core circle. Here we orient Γ as the boundary of $R_+(\Gamma)$.

We now emphasize the relation between dividing sets and sutures. Let ξ be a contact structure on a compact oriented 3-manifold M whose dividing set on the convex boundary ∂M is denoted by Γ. Then it is fairly easy to see that (M, Γ) is a *balanced* sutured manifold (with annular sutures) via the identification we mentioned above. Conversely, given a balanced sutured manifold (M, Γ), there exists a contact structure ξ on M which makes ∂M convex and realizes Γ as its diving set on ∂M. However one should keep in mind that the contact structure is not uniquely determined and cannot always be chosen to be tight.

Given a partial open book decomposition (S, P, h), we construct a sutured manifold (M, Γ) as follows: Let

$$H = (S \times [-1, 0])/\sim$$

where $(x, t) \sim (x, t')$ for $x \in \partial S$ and $t, t' \in [-1, 0]$. It is easy to see that H is a solid handlebody whose oriented boundary is the surface $S \times \{0\} \cup -S \times \{-1\}$ (modulo the relation $(x, 0) \sim (x, -1)$ for every $x \in \partial S$). Similarly let

$$N = (P \times [0, 1])/\sim$$

where $(x, t) \sim (x, t')$ for $x \in A$ and $t, t' \in [0, 1]$. Since P is not necessarily connected N is not necessarily connected. Observe that each component of N is also a solid handlebody. The oriented boundary of N can be described as follows: Let the arcs $c_1, c_2, \ldots, c_n$ denote the connected components of $\overline{\partial P \setminus A}$. Then, for $1 \leq i \leq n$, the disk $D_i = (c_i \times [0, 1])/\sim$ belongs to ∂N. Thus part of ∂N is given by the disjoint union of D_i's. The rest of ∂N is the surface $P \times \{1\} \cup -P \times \{0\}$ (modulo the relation $(x, 0) \sim (x, 1)$ for every $x \in A$).

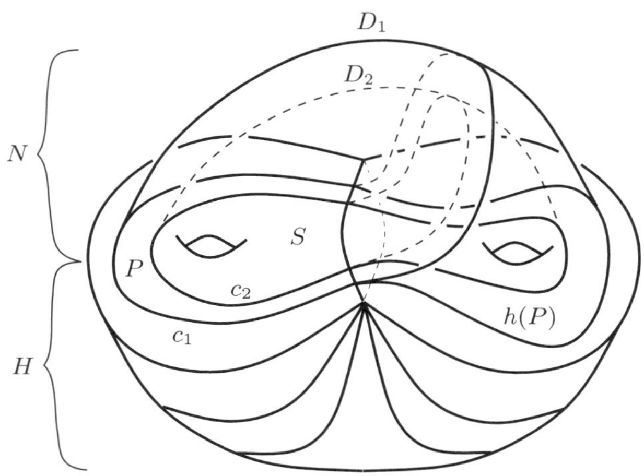

FIGURE 3. A partial open book decomposition: M as the union of N and H

Let $M = N \cup H$ where we glue these manifolds by identifying $P \times \{0\} \subset \partial N$ with $P \times \{0\} \subset \partial H$ and $P \times \{1\} \subset \partial N$ with $h(P) \times \{-1\} \subset \partial H$. Since the gluing identification is orientation reversing M is a compact oriented 3-manifold with oriented boundary

$$\partial M = (S \setminus P) \times \{0\} \cup -(S \setminus h(P)) \times \{-1\} \cup (\overline{\partial P \setminus A}) \times [0, 1]$$

(modulo the identifications given above).

DEFINITION 2.2. If a compact 3-manifold M with boundary is obtained from (S, P, h) as discussed above, then we call the triple (S, P, h) a partial open book decomposition of M.

We define the suture Γ on ∂M as the set of closed curves (see Remark 2) obtained by gluing the arcs $c_i \times \{1/2\} \subset \partial N$, for $1 \leq i \leq n$, with the arcs in

$(\overline{\partial S \setminus \partial P}) \times \{0\} \subset \partial H$, hence as an oriented simple closed curve and modulo identifications

$$\Gamma = (\overline{\partial S \setminus \partial P}) \times \{0\} \cup -(\overline{\partial P \setminus A}) \times \{1/2\} .$$

REMARK. If a sutured manifold (M, Γ) has only annular sutures, then it is convenient to refer to the set of core circles of these annuli as Γ.

DEFINITION 2.3. The sutured manifold (M, Γ) obtained from a partial open book decomposition (S, P, h) as described above is called the sutured manifold associated to (S, P, h).

DEFINITION 2.4 ([**Ju**]). A sutured manifold (M, Γ) is balanced if M has no closed components, $\pi_0(A(\Gamma)) \to \pi_0(\partial M)$ is surjective, and $\chi(R_+(\Gamma)) = \chi(R_-(\Gamma))$ on every component of M.

REMARK. It follows that if (M, Γ) is balanced, then $\Gamma = A(\Gamma)$ and every component of ∂M nontrivially intersects the suture Γ.

LEMMA 2.5. *The sutured manifold (M, Γ) associated to a partial open book decomposition (S, P, h) is balanced.*

PROOF. It is clear that M is connected since we assumed that S is connected. We observe that $\partial M \neq \emptyset$ since P is a proper subset of S by our definition. In fact, ∂M can be described starting from the connected surface $\partial H = S \times \{0\} \cup -S \times \{-1\}$: Let κ_j be $a_j \cup h(a_j)$, where a_j is the cocore of the 1-handle P_j in P (see Figure 4 for suitable a_j's). Then ∂M is obtained by cutting ∂H along κ_j's and capping off each resulting boundary by a disk $D_i = (c_i \times [0,1])/\sim$ for some i. From this description it is clear that every component of ∂M contains a $c_i \times \{1/2\} \subset \Gamma$ and therefore $\pi_0(A(\Gamma)) \to \pi_0(\partial M)$ is surjective. Now let $R_+(\Gamma)$ be the open subsurface in ∂M obtained by gluing

$$((S \setminus \partial S) \setminus P) \times \{0\} \subset \partial H \quad \text{and} \quad \cup_{i=1}^n \ (c_i \times [0, 1/2))/\sim \ \subset \partial N$$

and $R_-(\Gamma)$ be the open subsurface in ∂M obtained by gluing

$$((S \setminus \partial S) \setminus h(P)) \times \{-1\} \subset \partial H \quad \text{and} \quad \cup_{i=1}^n \ (c_i \times (1/2, 1])/\sim \ \subset \partial N$$

under the gluing map that is used to construct M. Since $h : P \to S$ is an embedding we have $\chi(P) = \chi(h(P))$ and it follows that $\chi(R_+(\Gamma)) = \chi(R_-(\Gamma))$. □

The following result is inspired by Torisu's work [**To**] in the closed case.

PROPOSITION 2.6. *Let (M, Γ) be the balanced sutured manifold associated to a partial open book decomposition (S, P, h). Then there exists a contact structure ξ on M satisfying the following conditions:*

(1) *ξ is tight when restricted to H and N,*
(2) *∂H is a convex surface in (M, ξ) whose dividing set is $\partial S \times \{0\}$,*
(3) *∂N is a convex surface in (M, ξ) whose dividing set is $\partial P \times \{1/2\}$.*

Moreover such ξ is unique up to isotopy.

PROOF. We will prove that there is a unique tight contact structure (up to isotopy) on H and N with the given boundary conditions, using arguments along similar lines.[1] Once we have these contact structures on H and N, since the dividing

[1]In fact, one can prove a general existence and uniqueness theorem using an explicit contact form $\lambda + dt$ on $\Sigma \times [0,1]/\sim$, for any surface Σ with boundary, where λ is a primitive of a volume form on Σ that is standard near the boundary. It can be argued that this contact form gives a tight contact structure making the boundary convex with dividing set $\partial\Sigma \times \{1/2\}$.

sets on ∂H and ∂N agree on the subsurface along which we glue H and N, we obtain a unique contact structure (up to isotopy) on M satisfying the above conditions, by gluing together the contact structures on these pieces.

To prove the existence of tight contact structures on H and N with prescribed dividing sets we simply consider H and N embedded in the closed contact 3-manifold (Y, ξ') supported by the open book (S, id) and appeal to the closed case (see [**To**] and [**Etn**, Lemma 4.4]). For H, observe that

$$H = (S \times [-1, 0])/\sim \ \subset (S \times [-1, 1])/\sim \ = Y \ ,$$

where the equivalence relation $\sim$ is given by, $(x, t) \sim (x, t')$ for $x \in \partial S$ and $t, t' \in [-1, 1]$, and $(s, -1) \sim (s, 1)$ for $s \in S$. The contact structure ξ' is Stein fillable by [**Gi02**], hence tight by [**ElGr**], and therefore its restriction to H is also tight. In fact, ∂H is convex with respect to ξ' with dividing set $\partial S \times \{0\}$ (see Lemma 4.4 in [**Etn**]). Similarly, N trivially embeds in H since $\partial P \times \{1/2\}$ is the union of $A \times \{0\}$ and the arcs $c_i \times \{1/2\}$, for $1 \leq i \leq n$. So ξ' restricts to a tight contact structure on N. To identify its dividing set we first observe that the dividing set on $P \times \{1\} \cup -P \times \{0\} = \partial N \cap \partial H$ is the set $A \times \{0\} = \partial N \cap (\partial S \times \{0\})$. The rest of ∂N consists of the disks $D_i = (c_i \times [0, 1])/\sim$. Each one of these disks can be made convex so that the dividing set is a single arc since its boundary intersects the dividing set twice. It follows that the dividing set on ∂N is as required after rounding the edges.

In order to prove the uniqueness for H, as in Lemma 4.4 in [**Etn**], we take a set $\{d_1, \ldots, d_p\}$ of properly embedded pairwise disjoint arcs in S whose complement is a single disk. (It follows that the set $\{d_1, d_2, \ldots, d_p\}$ represents a basis of $H_1(S, \partial S)$.) For $1 \leq k \leq p$, let δ_k denote the closed curve on ∂H which is obtained by gluing the arc d_k on $S \times \{0\}$ with the arc d_k on $S \times \{-1\}$. Then we observe that $\{\delta_1, \delta_2, \ldots, \delta_p\}$ is a set of homologically linearly independent closed curves on ∂H so that δ_k bounds a compressing disk $D_k^\delta = (d_k \times [0, -1])/\sim$ in H. It is clear that when we cut H along D_k^δ's (and smooth the corners) we get a 3-ball B^3. Moreover δ_k intersects the dividing set twice by our construction. Now we put each δ_k into Legendrian position (by the Legendrian realization principle [**H00**]) and make the compressing disk D_k^δ convex [**Gi91**]. The dividing set on D_k^δ will be an arc connecting two points on $\partial D_k^\delta = \delta_k$. Then we cut along these disks and round the edges (see [**H00**]) to get a connected dividing set on the remaining B^3. Consequently, Theorem 2.7 due to Eliashberg (although stated in different terms in [**El**]) implies the uniqueness of a tight contact structure on H with the assumed boundary conditions. Recall that a standard contact 3-ball is a tight contact 3-ball with convex boundary whose dividing set is connected.

THEOREM 2.7 (Eliashberg). *There is a unique standard contact* 3*-ball.*

The proof of the uniqueness of such a tight contact structure on N follows a similar line. Instead of a basis of $H_1(S, \partial S)$ we take suitable cocores $\{a_1, \ldots, a_r\}$ of the 1-handles P_j's in P to get a basis of $H_1(P, A)$ (see Figure 4 for an example). Then one can proceed as in the proof given above for the handlebody H. □

Proposition 2.6 leads to the following definition of compatibility of a contact structure and a partial open book decomposition.

DEFINITION 2.8. Let (M, Γ) be the balanced sutured manifold associated to a partial open book decomposition (S, P, h). A contact structure ξ on (M, Γ) is

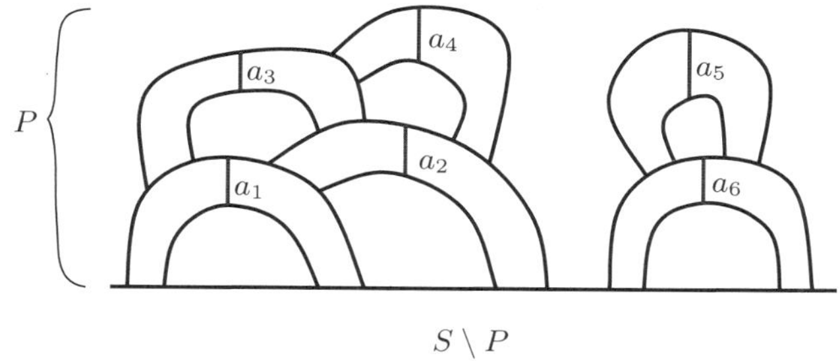

FIGURE 4. A basis of $H_1(P, A)$: cocores $a_1, a_2, \ldots, a_6$ of the 1-handles in P

said to be compatible with (S, P, h) if it is isotopic to a contact structure satisfying conditions (1), (2) and (3) stated in Proposition 2.6.

DEFINITION 2.9. Two partial open book decompositions (S, P, h) and $(\widetilde{S}, \widetilde{P}, \widetilde{h})$ are isomorphic if there is a diffeomorphism $f : S \to \widetilde{S}$ such that $f(P) = \widetilde{P}$ and $\widetilde{h} = f \circ h \circ (f^{-1})|_{\widetilde{P}}$.

REMARK. It follows from Proposition 2.6 that every partial open book decomposition has a unique compatible contact structure, up to isotopy, on the balanced suture manifold associated to it, such that the dividing set of the convex boundary is isotopic to the suture. Moreover if (S, P, h) and $(\widetilde{S}, \widetilde{P}, \widetilde{h})$ are isomorphic partial open book decompositions, then the associated compatible contact 3-manifolds (M, Γ, ξ) and $(\widetilde{M}, \widetilde{\Gamma}, \widetilde{\xi})$ are also isomorphic.

DEFINITION 2.10. Let (S, P, h) be a partial open book decomposition. A partial open book decomposition (S', P', h') is called a positive stabilization of (S, P, h) if there is a properly embedded arc s in S such that

- S' is obtained by attaching a 1-handle to S along ∂s,
- P' is defined as the union of P and the attached 1-handle,
- $h' = R_\sigma \circ h$, where the extension of h to P' by the identity is also denoted by h, and R_σ denotes the right-handed Dehn twist along the closed curve σ which is the union of s and the core of the attached 1-handle.

The effect of positively stabilizing a partial open book decomposition on the associated sutured manifold and the compatible contact structure is taking a connected sum with (S^3, ξ_{std}) away from the boundary. We will prove this statement in Lemma 2.11 and the notion of sutured Heegaard diagram will be helpful in our argument. So we digress to review basic definitions and properties of Heegaard diagrams of sutured manifolds (cf. [**Ju**]).

A sutured Heegaard diagram is given by $(\Sigma, \boldsymbol{\alpha}, \boldsymbol{\beta})$, where the Heegaard surface Σ is a compact oriented surface with nonempty boundary and $\boldsymbol{\alpha} = \{\alpha_1, \alpha_2, \ldots, \alpha_m\}$ and $\boldsymbol{\beta} = \{\beta_1, \beta_2, \ldots, \beta_n\}$ are two sets of pairwise disjoint simple closed curves in $\Sigma \setminus \partial\Sigma$. Every sutured Heegaard diagram $(\Sigma, \boldsymbol{\alpha}, \boldsymbol{\beta})$, uniquely defines a sutured manifold (M, Γ) as follows: Let M be the 3-manifold obtained from $\Sigma \times [0, 1]$ by attaching 3-dimensional 2-handles along the curves $\alpha_i \times \{0\}$ and $\beta_j \times \{1\}$ for $i = 1, \ldots, m$ and $j = 1, \ldots, n$. The suture Γ on ∂M is defined by the set of curves $\partial\Sigma \times \{1/2\}$ (see Remark 2).

In [**Ju**], Juhász proved that if (M,Γ) is defined by $(\Sigma,\boldsymbol{\alpha},\boldsymbol{\beta})$, then (M,Γ) is balanced if and only if $|\boldsymbol{\alpha}| = |\boldsymbol{\beta}|$, the surface Σ has no closed components and both $\boldsymbol{\alpha}$ and $\boldsymbol{\beta}$ consist of curves linearly independent in $H_1(\Sigma,\mathbb{Q})$. Hence a sutured Heegaard diagram $(\Sigma,\boldsymbol{\alpha},\boldsymbol{\beta})$ is called balanced if it satisfies the conditions listed above. We will abbreviate balanced sutured Heegaard diagram as balanced diagram.

A partial open book decomposition of (M,Γ) gives a sutured Heegaard diagram $(\Sigma,\boldsymbol{\alpha},\boldsymbol{\beta})$ of $(M,-\Gamma)$ as follows: Let

$$\Sigma = P\times\{0\}\cup -S\times\{-1\}/\sim\ \subset \partial H$$

be the Heegaard surface. Observe that, modulo identifications,

$$\partial\Sigma = (\overline{\partial P\setminus A})\times\{0\}\cup -(\overline{\partial S\setminus \partial P})\times\{-1\}\simeq -\Gamma\ .$$

As in the proof of Proposition 2.6, let $a_1,a_2,\ldots,a_r$ be properly embedded pairwise disjoint arcs in P with endpoints on A such that $S\setminus\cup_j a_j$ deformation retracts onto $\overline{S\setminus P}$. Then define two families $\boldsymbol{\alpha}=\{\alpha_1,\alpha_2,\ldots,\alpha_r\}$ and $\boldsymbol{\beta}=\{\beta_1,\beta_2,\ldots,\beta_r\}$ of simple closed curves in the Heegaard surface Σ by

$$\alpha_j = a_j\times\{0\}\cup a_j\times\{-1\}/\sim \quad\text{and}\quad \beta_j = a_j\times\{0\}\cup h(a_j)\times\{-1\}/\sim\ .$$

$(\Sigma,\boldsymbol{\alpha},\boldsymbol{\beta})$ is a sutured Heegaard diagram of $(M,-\Gamma)$. Here the suture is $-\Gamma$ since $\partial\Sigma$ is isotopic to $-\Gamma$.

LEMMA 2.11. *The balanced sutured manifold associated to a partial open book decomposition and the compatible contact structure are invariant under positive stabilization.*

PROOF. Let (S,P,h) be a partial open book decomposition of (M,Γ), s be a properly embedded arc in S, and (S',P',h') be the corresponding positive stabilization of (S,P,h). Consider the sutured Heegaard diagram $(\Sigma,\boldsymbol{\alpha},\boldsymbol{\beta})$ of $(M,-\Gamma)$ given by (S,P,h) using properly embedded disjoint arcs $a_1,a_2,\ldots,a_r$ in P.

Let a_0 be the cocore of the 1-handle attached to S during stabilization. The endpoints of a_0 are on $A'=\partial P'\cap\partial S'$ and $S'\setminus\cup_{j=0}^r a_j$ deformation retracts onto $S'\setminus P' = S\setminus P$. Using the properly embedded disjoint arcs $a_0,a_1,a_2,\ldots,a_r$ in P' we get a sutured Heegaard diagram $(\Sigma',\boldsymbol{\alpha}',\boldsymbol{\beta}')$ of $(M',-\Gamma')$, where (M',Γ') is the sutured manifold associated to (S',P',h'). Observe that $\boldsymbol{\alpha}' = \{\alpha_0\}\cup\boldsymbol{\alpha}$, $\boldsymbol{\beta}'=\{\beta_0\}\cup\boldsymbol{\beta}$, and

$$\Sigma' = P'\times\{0\}\cup -S'\times\{-1\}/\!\sim\ \cong T^2\#\Sigma\ .$$

Since h' is a right-handed Dehn twist along σ composed with the extension of h which is identity on $P'\setminus P$, α_0 is disjoint from every β_j with $j>0$. Therefore $(\Sigma',\boldsymbol{\alpha}',\boldsymbol{\beta}')$ is a stabilization of the Heegaard diagram $(\Sigma,\boldsymbol{\alpha},\boldsymbol{\beta})$, and consequently $(M',\Gamma')\cong(M,\Gamma)$. The contact structure ξ' compatible with (S',P',h') is contactomorphic to ξ since ξ' is obtained from ξ by taking a connected sum with (S^3,ξ_{std}) away from the boundary. This can be seen as in the closed case, and holds essentially because of the fact that the abstract open book with an annulus page and monodromy given by a right-handed Dehn twist (which is the one that gives the genus-1 Heegaard decomposition with a single $\boldsymbol{\alpha}$ -curve that intersects the single $\boldsymbol{\beta}$ -curve geometrically once) is compatible with the standard contact structure on S^3. □

3. Relative Giroux correspondence

The following theorem is the key to obtaining a description of a partial open book decomposition of (M, Γ, ξ) in the sense of Honda, Kazez and Matić.

THEOREM 3.1 ([**HKM09**], Theorem 1.1). *Let (M, Γ) be a balanced sutured manifold and let ξ be a contact structure on M with convex boundary whose dividing set $\Gamma_{\partial M}$ on ∂M is isotopic to Γ. Then there exist a Legendrian graph $K \subset M$ whose endpoints lie on $\Gamma \subset \partial M$ and a regular neighborhood $N(K) \subset M$ of K which satisfy the following:*

(A) (i) $T = \overline{\partial N(K) \setminus \partial M}$ is a convex surface with Legendrian boundary.
(ii) For each component γ_i of ∂T, $\gamma_i \cap \Gamma_{\partial M}$ has two connected components.
(iii) There is a system of pairwise disjoint compressing disks D_j^α for $N(K)$ so that ∂D_j^α is a curve on T intersecting the dividing set Γ_T of T at two points and each component of $N(K) \setminus \cup_j D_j^\alpha$ is a standard contact 3-ball, after rounding the edges.

(B) (i) Each component H of $\overline{M \setminus N(K)}$ is a handlebody (with convex boundary).
(ii) There is a system of pairwise disjoint compressing disks D_k^δ for H so that each ∂D_k^δ intersects the dividing set $\Gamma_{\partial H}$ of ∂H at two points and $H \setminus \cup_k D_k^\delta$ is a standard contact 3-ball, after rounding the edges.

Based on Theorem 3.1, Honda, Kazez and Matić describe a partial open book decomposition on (M, Γ) in [**HKM09**, Section 2]. In this paper, for the sake of simplicity and without loss of generality, we will assume that M is connected. As a consequence $M \setminus N(K)$ in Theorem 3.1 is also connected.

We claim that the description in [**HKM09**] gives a partial open book decomposition (S, P, h); that the balanced sutured manifold associated to (S, P, h) is isotopic to (M, Γ); and that ξ is compatible with (S, P, h) — all in the sense that we defined in this paper. In the rest of this section we prove these claims and Lemma 3.3 to obtain a proof of Theorem 1.1.

The tubular portion T of $-\partial N(K)$ in Theorem 3.1(A)(i) is split by its dividing set into positive and negative regions, with respect to the orientation of $\partial(M \setminus N(K))$. Let P be the positive region. Note that the negative region $T \setminus P$ is diffeomorphic to P. Since (M, Γ) is assumed to be a (balanced) sutured manifold, ∂M is divided into $R_+(\Gamma)$ and $R_-(\Gamma)$ by the suture Γ. Let $R_+ = R_+(\Gamma) \setminus \cup_i D_i$, where D_i's are the components of $\partial N(K) \cap \partial M$ and let S be the surface which is obtained from $\overline{R_+}$ by attaching the positive region P. If we denote the dividing set of T by $A = \partial P \cap \partial S$, then it is easy to see that

$$N(K) \cong (P \times [0,1])/\sim$$

where $(x, t) \sim (x, t')$ for $x \in A$ and $t, t' \in [0, 1]$, such that the dividing set of $\partial N(K)$ is given by $\partial P \times \{1/2\}$.

In [**HKM09**], Honda, Kazez and Matić observed that

$$\overline{M \setminus N(K)} \cong (S \times [-1, 0])/\sim$$

where $(x, t) \sim (x, t')$ for $x \in \partial S$ and $t, t' \in [-1, 0]$, such that the dividing set of $\overline{M \setminus N(K)}$ is given by $\partial S \times \{0\}$.

Moreover the embedding $h : P \to S$ which is obtained by first pushing P across $N(K)$ to $T \setminus P \subset \partial(M \setminus N(K))$, and then following it with the identification of $\overline{M \setminus N(K)}$ with $(S \times [-1, 0])/\sim$ is called the monodromy map in the Honda-Kazez-Matić description of a partial open book decomposition.

In conclusion, we see that the triple (S, P, h) satisfies the conditions in Definition 2.1:

(1) The compact oriented surface S is connected since we assumed that M is connected and it is clear that $\partial S \neq \emptyset$.

(2) The surface P is a proper subsurface of S such that S is obtained from $\overline{S \setminus P}$ by successively attaching 1-handles by construction.

(3) The monodromy map $h : P \to S$ is an embedding such that h fixes $A = \partial P \cap \partial S$ pointwise.

Next we observe that $N(K)$ (resp. $\overline{M \setminus N(K)}$) corresponds to N (resp. H) in our construction of the balanced sutured manifold associated to a partial open book decomposition proceeding Definition 2.1. The monodromy map h amounts to describing how $N = N(K)$ and $H = \overline{M \setminus N(K)}$ are glued together along the appropriate subsurface of their boundaries. This proves that the balanced sutured manifold associated to (S, P, h) is diffeomorphic to (M, Γ).

LEMMA 3.2. *The contact structure ξ in Theorem 3.1 is compatible with the partial open book decomposition (S, P, h) described above.*

PROOF. We have to show that the contact structure ξ in Theorem 3.1 satisfies the conditions (1), (2) and (3) stated in Proposition 2.6 with respect to the partial open book decomposition (S, P, h) described above. We already observed that $N = N(K)$ and $H = \overline{M \setminus N(K)}$. Then

(1) The restrictions of the contact structure ξ onto $N(K)$ and $\overline{M \setminus N(K)}$ are tight by conditions (A)(iii) and (B)(ii) of Theorem 3.1, respectively. This is because in either case one obtains a standard contact 3-ball or a disjoint union of standard contact 3-balls by cutting the manifold along a collection of compressing disks each of whose boundary geometrically intersects the dividing set exactly twice, and hence the dividing set of each of these compressing disks is a single boundary-parallel arc (see [**H02**, Corollary 2.6 (2)]).

(2) $\partial H = \partial(M \setminus N(K)) = (\partial M \setminus \cup_i D_i) \cup T$ is convex by the convexity of ∂M and the convexity of T (condition (A)(i) in Theorem 3.1). Its dividing set is the union of those of $\partial M \setminus \cup_i D_i$ and T, hence it is isotopic to $(\overline{\partial S \setminus \partial P}) \times \{0\} \cup A \times \{0\} = \partial S \times \{0\}$.

(3) $\partial N = \partial N(K) = \cup_i D_i \cup T$ is convex by the convexity of $D_i \subset \partial M$ and the convexity of T. Its dividing set is the union of those of D_i's and T, hence it is isotopic to $(\overline{\partial P \setminus \partial S}) \times \{1/2\} \cup A \times \{0\} = \partial P \times \{1/2\}$. □

The following lemma is the only remaining ingredient in the proof of Theorem 1.1.

LEMMA 3.3. *Let (S, P, h) be a partial open book decomposition, (M, Γ) be the balanced sutured manifold associated to it, and ξ be a compatible contact structure. Then (S, P, h) is given by the Honda-Kazez-Matić description.*

PROOF. Consider the graph K in P that is obtained by gluing the core of each 1-handle in P (see Figure 5 for example).

It is clear that P retracts onto K. We will denote $K \times \{1/2\} \subset P \times \{1/2\}$ also by K. We can first make $P \times \{1/2\}$ convex and then Legendrian realize K

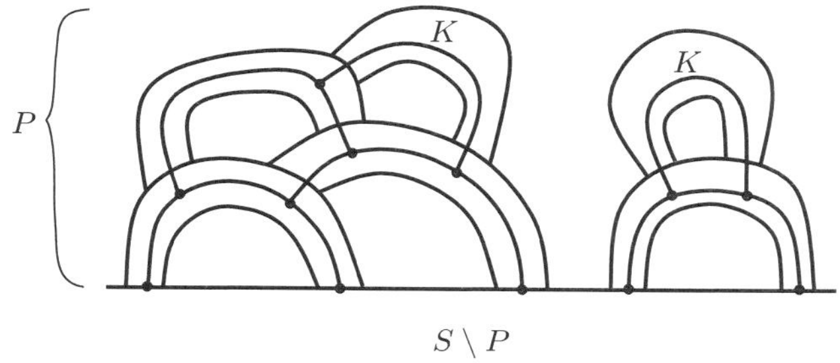

FIGURE 5. Legendrian graph K in P

with respect to the compatible contact structure ξ on $N \subset M$. This is because each component of the complement of K in P contains a boundary component (see [**Etn**, Remark 4.30]). Hence K is a Legendrian graph in (M, ξ) with endpoints in $\partial P \times \{1/2\} \setminus \partial S \times \{0\} \subset \Gamma \subset \partial M$ such that $N = P \times [0,1]/\sim$ is a neighborhood $N(K)$ of K in M. Then all the conditions except (A)(i) in Theorem 3.1 on $N(K) = N$ and $\overline{M \setminus N(K)} = H$ are satisfied because of the way we constructed ξ in Proposition 2.6. Since ∂N is convex T is also convex. It remains to check that the boundary of the tubular portion T of N is Legendrian. Note that each component of this boundary $\partial D_i = \partial(c_i \times [0,1]) \subset \partial N$ is identified with $\gamma_i = c_i \times \{0\} \cup h(c_i) \times \{-1\}$ in the convex surface $\partial H = S \times \{0\} \cup -S \times \{-1\}$. Since each γ_i intersects the dividing set $\Gamma_{\partial H} = S \times \{0\}$ of ∂H transversely at two points $\partial c_i \times \{0\}$, the set $\{\gamma_1, \gamma_2, \ldots, \gamma_n\}$ is non-isolating in ∂H and hence we can use the Legendrian Realization Principle to make each γ_i Legendrian. □

Proof of Theorem 1.1. By Proposition 2.6 each partial open book decomposition is compatible with a unique compact contact 3-manifold with convex boundary up to contact isotopy. This gives a map from the set of all partial open book decompositions to the set of all compact contact 3-manifolds with convex boundary and by Remark 2 this map descends to a map from the set of isomorphism classes of all partial open book decompositions to the set of isomorphism classes of all compact contact 3-manifolds with convex boundary. Moreover by Lemma 2.11 this gives a well-defined map Ψ from the isomorphism classes of all partial open book decompositions modulo positive stabilization to that of isomorphism classes of compact contact 3-manifolds with convex boundary. On the other hand, Honda-Kazez-Matić description gives a well-defined map Φ in the reverse direction by [**HKM09**, Theorems 1.1 and 1.2]. Furthermore, $\Psi \circ \Phi$ is identity by Lemma 3.2 and $\Phi \circ \Psi$ is identity by Lemma 3.3. □

4. Examples

Below we provide examples of abstract partial open books which correspond to some basic contact 3-manifolds with boundary. These examples were previously appeared in [**EtOz**] where their contact invariants were calculated.

EXAMPLE 4.1. Let S be an annulus, P be a regular neighborhood of r disjoint arcs connecting the two distinct boundary components of S as in Figure 6, and the monodromy h be the inclusion of P into S. The partial open book (S, P, h) is compatible with the contact structure obtained by removing r disjoint standard contact open 3-balls from the unique (up to isotopy) tight contact structure ξ_{std} on $S^1 \times S^2$.

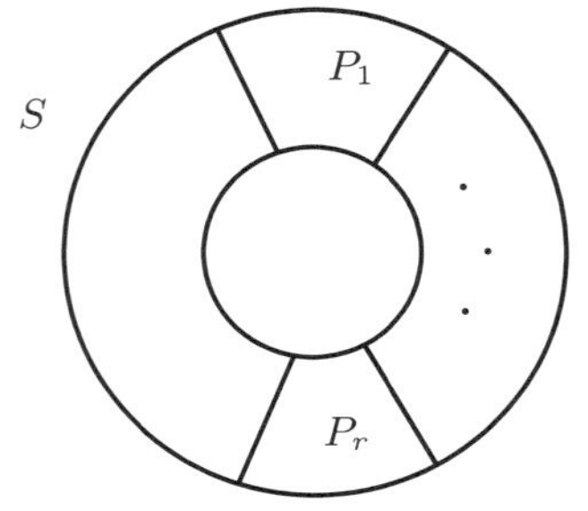

FIGURE 6. The annulus S, r components $P_1, \ldots, P_r$ of P in Example 4.1.

EXAMPLE 4.2 (**Standard contact 3-ball**). Let S and P be as in Example 4.1 for $r = 1$, and the monodromy h be the restriction (to P) of a *right*-handed Dehn twist along the core of S. The contact 3-manifold (M, Γ, ξ) compatible with this partial open book is the standard contact 3-ball. Here the Legendrian graph K which satisfies the conditions in Theorem 3.1 is a single arc in B^3 connecting two distinct points on Γ as depicted in Figure 7. The complement H of a regular neighborhood $N = N(K)$ in the standard contact 3-ball B^3 is a solid torus with two parallel dividing curves (see Figure 8) on ∂H which are homotopically nontrivial inside H. Here a meridional disk in H will serve as the required compressing disk D_1^δ for H in Theorem 3.1 (B). On the other hand, N is already a standard contact 3-ball. This shows in particular that the standard contact 3-ball can be obtained from a tight solid torus H by attaching a tight 2-handle N.

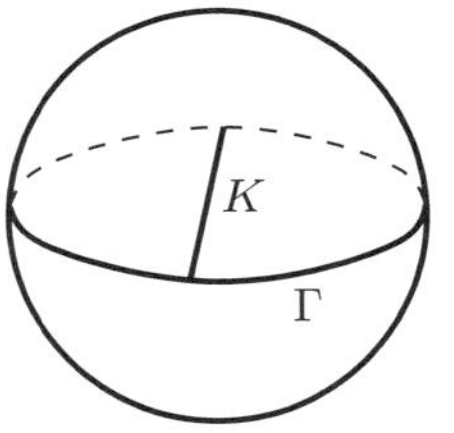

FIGURE 7. The Legendrian arc K in the standard contact 3-ball.

EXAMPLE 4.3 (**Standard neighborhood of an overtwisted disk**). Let (S, P, h) be the partial open book decomposition shown in Figure 2. This is the partial open book considered in [**HKM09**, Example 1] which is compatible with the standard neighborhood of an overtwisted disk.

Here we observe that by Proposition 2.6, (M, Γ, ξ) is obtained by gluing a pair of compact connected contact 3-manifolds with convex boundaries, namely $(H, \Gamma_{\partial H}, \xi|_H)$ and $(N, \Gamma_{\partial N}, \xi|_N)$, along parts of their boundaries. We know that

$$H = (S \times [-1, 0])/\sim$$

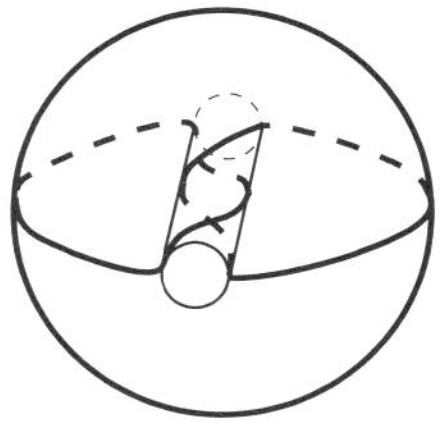

FIGURE 8. The dividing curves on ∂H.

where S is an annulus and $(x,t) \sim (x,t')$ for $x \in \partial S$ and $t,t' \in [-1,0]$. There is a unique (up to isotopy) compatible tight contact structure on H whose dividing set $\Gamma_{\partial H}$ on ∂H is $\partial S \times \{0\}$ (cf. Proposition 2.6). Hence $(H, \Gamma_{\partial H}, \xi|_H)$ is a solid torus carrying a tight contact structure where $\Gamma_{\partial H}$ consists of two parallel curves on ∂H which are homotopically nontrivial in H. We observe that when we cut H along a compressing disk we get a standard contact 3-ball B^3 with its connected dividing set $\Gamma_{\partial B^3}$ on its convex boundary. Note that $\Gamma_{\partial B^3}$ is obtained by "gluing" $\Gamma_{\partial H}$ and the dividing set on the compressing disk. Similarly we know that $N = (P \times [0,1])/\sim$, where $(x,t) \sim (x,t')$ for $x \in A$ and $t,t' \in [0,1]$. There is a unique (up to isotopy) compatible tight contact structure on N whose dividing set $\Gamma_{\partial N}$ on ∂N is $\partial P \times \{1/2\}$ (cf. Proposition 2.6). We observe that $(N, \Gamma_{\partial N}, \xi|_N)$ is the standard contact 3-ball.

References

[El] Y. Eliashberg, *Contact 3-manifolds twenty years since J. Martinet's work,* Ann. Inst. Fourier (Grenoble) **42** (1992), no. 1-2, 165–192.

[ElGr] Y. Eliashberg and M. Gromov, *Convex symplectic manifolds,* Several complex variables and complex geometry, Part 2 (Santa Cruz, CA, 1989), 135–162, Proc. Sympos. Pure Math., **52**, Part 2, Amer. Math. Soc., Providence, RI, 1991.

[EtOz] T. Etgü and B. Ozbagci, *Partial open book decompositions and the contact class in sutured Floer homology*, Turkish J. Math. **33** (2009), 295-312.

[Etn] J. B. Etnyre, *Lectures on open book decompositions and contact structures,* Floer homology, gauge theory, and low-dimensional topology, 103–141, Clay Math. Proc., **5**, Amer. Math. Soc., Providence, RI, 2006.

[Ga] D. Gabai, *Foliations and the topology of 3-manifolds*, J. Differential Geom. **18** (1983), no. 3, 445–503.

[Gi91] E. Giroux, *Convexité en topologie de contact,* Comment. Math. Helv. **66** (1991), no. 4, 637–677.

[Gi02] E. Giroux, *Géometrie de contact: de la dimension trois vers les dimensions supérieures,* Proceedings of the International Congress of Mathematicians (Beijing 2002), Vol. II, 405–414.

[GHV] P. Ghiggini, K. Honda, and J. Van Horn-Morris, *The vanishing of the contact invariant in the presence of torsion*, preprint, arXiv:0706.1602v2.

[H00] K. Honda, *On the classification of tight contact structures. I*, Geom. Topol. **4** (2000), 309–368.

[H02] K. Honda, *Gluing tight contact structures*, Duke Math. J. **115** (2002), 435–478.

[HKM07] K. Honda, W. Kazez, and G. Matić, *Right-veering diffeomorphisms of compact surfaces with boundary,* Invent. Math., **169** (2007), no. 2, 427–449.

[HKM09] K. Honda, W. Kazez, and G. Matić, *The contact invariant in the sutured Floer homology,* Invent. Math., **176** (2009), no. 3, 637–676.

[Ju] A. Juhász, *Holomorphic discs and sutured manifolds*, Algebr. Geom. Topol. **6** (2006), 1429–1457.

[OzSt] B. Ozbagci and A. I. Stipsicz, *Surgery on contact 3-manifolds and Stein surfaces*, Bolyai Soc. Math. Stud., Vol.13, Springer, 2004.

[OzSz] P. Ozsváth and Z. Szabó, *Heegaard Floer homology and contact structures,* Duke Math. J. **129** (2005), no. 1, 39–61.

[To] I. Torisu, *Convex contact structures and fibered links in* 3*-manifolds*, Internat. Math. Res. Notices **2000**, no. 9, 441–454.

[ThWi] W. Thurston and H. Winkelnkemper, *On the existence of contact forms,* Proc. Amer. Math. Soc. **52** (1975), 345–347.

DEPARTMENT OF MATHEMATICS, KOÇ UNIVERSITY, ISTANBUL, TURKEY
E-mail address: tetgu@ku.edu.tr

DEPARTMENT OF MATHEMATICS, KOÇ UNIVERSITY, ISTANBUL, TURKEY
E-mail address: bozbagci@ku.edu.tr

Proceedings of Symposia in Pure Mathematics
Volume **82**, 2011

A note on the support norm of a contact structure

John A. Baldwin and John B. Etnyre

ABSTRACT. In this note we observe that the no two of the three invariants defined for contact structures in [**EO**] – that is, the support genus, binding number and support norm – determine the third.

In [**EO**], the second author and B. Ozbagci define three invariants of contact structures on closed, oriented 3-manifolds in terms of supporting open book decompositions. These invariants are the support genus, binding number and support norm. There are obvious relationships between these invariants, but [**EO**] leaves open the question of whether any two of them determine the third. We show in this note that this is not the case.

Recall that an open book decomposition (L, π) of a 3–manifold M consists of an oriented link L in M and a fibration $\pi : (M - L) \to S^1$ of the complement of L whose fibers are Seifert surfaces for L. The fibers $\pi^{-1}(\theta)$ of π are called *pages* of the open book and L is called the *binding*. It is often convenient to record an open book decomposition (L, π) by a pair (Σ, ϕ), where Σ is a compact surface which is homeomorphic to the closure of a page of (L, π), and $\phi : \Sigma \to \Sigma$ is the monodromy of the fibration π. A contact structure ξ on M is said to be supported by the open book decomposition (L, π) if ξ is the kernel of a 1-form α which evaluates positively on tangent vectors to L that agree with the orientation of L, and for which $d\alpha$ restricts to a positive volume form on each page of (L, π). According to a result of Giroux [**Gi**] every contact structure is supported by some open book decomposition.

With this in mind, we may describe the three invariants defined in [**EO**]. The *support genus* of a contact structure ξ on M is defined to be

$$\mathrm{sg}(\xi) = \min\{g(\pi^{-1}(\theta)) \mid (L, \pi) \text{ supports } \xi\},$$

where θ is any point in S^1 and $g(\pi^{-1}(\theta))$ is the genus of the page $\pi^{-1}(\theta)$. The *binding number* of ξ is defined to be

$$\mathrm{bn}(\xi) = \min\{|L| \mid (L, \pi) \text{ supports } \xi \text{ and } \mathrm{sg}(\xi) = g(\pi^{-1}(\theta))\},$$

2010 *Mathematics Subject Classification.* 57R17, 53D10.

Key words and phrases. contact structure, open book decompsition, support genus, support norm, binding number.

The first author was supported by an NSF Postdoctoral Fellowship and NSF Grant DMS-0635607. The second author was partially supported by NSF Grant DMS-0804820.

where $|L|$ denotes the number of components of L (or, equivalently, the number of boundary components of any page of (L, π)). And the *support norm* of ξ is defined to be

$$\mathrm{sn}(\xi) = \min\{-\chi(\pi^{-1}(\theta)) \mid (L, \pi) \text{ supports } \xi\},$$

where $\chi(\pi^{-1}(\theta))$ denotes the Euler characteristic of any page $\pi^{-1}(\theta)$. It is a simple observation that $\mathrm{sn}(\xi) \geq -1$, with equality if and only if ξ is the standard tight contact structure on S^3.

Since, for any surface Σ, we have the equality

$$-\chi(\Sigma) = 2g(\Sigma) + |\partial\Sigma| - 2,$$

it is immediately clear that

$$\mathrm{sn}(\xi) \leq 2\,\mathrm{sg}(\xi) + \mathrm{bn}(\xi) - 2.$$

Moreover, if the support norm of ξ is achieved by an open book whose pages have genus $g > \mathrm{sg}(\xi)$ and whose binding has m components, then $\mathrm{sn}(\xi) = 2g + m - 2$, which is at least $2\,\mathrm{sg}(\xi) + 1$. The following lemma from [**EO**] summarizes these bounds.

LEMMA 1.1. *For any contact structure ξ on a closed, oriented 3–manifold,*

$$\min\{2\,\mathrm{sg}(\xi) + \mathrm{bn}(\xi) - 2,\ 2\,\mathrm{sg}(\xi) + 1\} \leq \mathrm{sn}(\xi) \leq 2\,\mathrm{sg}(\xi) + \mathrm{bn}(\xi) - 2.$$

Thus, for contact structures with $\mathrm{bn}(\xi) \leq 3$, it follows that that $\mathrm{sn}(\xi) = 2\,\mathrm{sg}(\xi) + \mathrm{bn}(\xi) - 2$. Yet, the results in [**EO**] do not resolve whether the upper bound on the support norm in Lemma 1.1 can ever be a strict inequality. Our main result is that this bound can indeed be a strict inequality; that is, the support genus and binding number do not, in general, wholly determine the support norm.

For the rest of this note, Σ will denote the genus one surface with one boundary component. Let $\phi_{\mathbf{n},m}$ be the diffeomorphism of Σ given by

$$\phi_{\mathbf{n},m} = D_\delta^m \cdot D_x D_y^{-n_1} \cdots D_x D_y^{-n_k},$$

where x, y and δ are the curves pictured in Figure 1, D_c denotes a right handed Dehn twist about the curve c, and $\mathbf{n} = (n_1, \ldots, n_k)$ is a k-tuple of non-negative integers for which some $n_i \neq 0$. Let $\xi_{\mathbf{n},m}$ denote the contact structure supported by the open book $(\Sigma, \phi_{\mathbf{n},m})$, and let $M_{\mathbf{n},m}$ denote the 3-manifold with this open book decomposition.

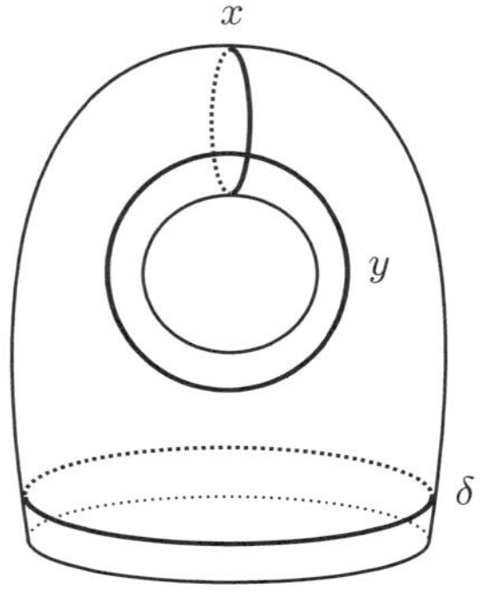

FIGURE 1. The surface Σ and the curves x, y and δ.

THEOREM 1.2. *For $m \leq 0$, the contact structure $\xi_{\mathbf{n},m}$ satisfies*

$$\mathrm{sg}(\xi_{\mathbf{n},m}) = 0.$$

For any fixed tuple $\mathbf{n}$, *there is a finite subset* $E_{\mathbf{n}}$ *of the integers such that*

$$\mathrm{bn}(\xi_{\mathbf{n},m}) > 3 \text{ and } \mathrm{sn}(\xi_{\mathbf{n},m}) = 1$$

for all $m \leq 0$ which are not in $E_{\mathbf{n}}$. In particular,

$$\mathrm{sn}(\xi_{\mathbf{n},m}) < 2\,\mathrm{sg}(\xi_{\mathbf{n},m}) + \mathrm{bn}(\xi_{\mathbf{n},m}) - 2$$

for all $m \leq 0$ which are not in $E_{\mathbf{n}}$.

In contrast, the support genus $\mathrm{sg}(\xi_{\mathbf{n},m}) = 1$ when $m > 0$, [**B2**]. Therefore, for $m > 0$, $\mathrm{bn}(\xi_{\mathbf{n},m}) = 1$ and $\mathrm{sn}(\xi_{\mathbf{n},m}) = 1$, and, hence,

$$\mathrm{sn}(\xi_{\mathbf{n},m}) = 2\,\mathrm{sg}(\xi_{\mathbf{n},m}) + \mathrm{bn}(\xi_{\mathbf{n},m}) - 2.$$

That is, the upper bound in Lemma 1.1 is achieved for $\xi = \xi_{\mathbf{n},m}$ when $m > 0$.

PROOF OF THEOREM 1.2. One can easily see that, for $m \leq 0$, the open book $(\Sigma, \phi_{\mathbf{n},m})$ is not *right-veering* (cf. [**B1**, Section 4]); therefore, for such m, the contact structure $\xi_{\mathbf{n},m}$ is overtwisted [**HKM**] and $\xi_{\mathbf{n},m}$ is supported by an open book with planar pages [**Et**].

Observe that if a contact structure ξ on M is supported by an open book with planar pages and the binding number of ξ is three or less, then M must be a Seifert fibered space. More precisely, if the binding number of ξ is two then M is a lens space, and if the binding number is three then M is a small Seifert fibered space. One can see this by drawing a surgery picture corresponding to the open book supporting ξ which realizes the binding number.

The diffeomorphism $\phi_{\mathbf{n},0} = D_x D_y^{-n_1} \cdots D_x D_y^{-n_k}$ is pseudo-Anosov since the trace of the induced map on $H_1(\Sigma; \mathbb{Z})$ is greater than two in absolute value (for instance, $\phi_{(1),0}$ is the monodromy of the figure eight knot in S^3). Therefore, the binding of the open book given by $(\Sigma, \phi_{\mathbf{n},0})$ is a hyperbolic knot, and the manifold $M_{\mathbf{n},m}$ is obtained from $M_{\mathbf{n},0}$ via $-\frac{1}{m}$ surgery on this knot. Thurston's Dehn Surgery Theorem then implies that there is some finite subset $E_{\mathbf{n}}$ of the integers for which $M_{\mathbf{n},m}$ is hyperbolic for all m not in $E_{\mathbf{n}}$ [**Th**]. In particular, $M_{\mathbf{n},m}$ is not a Seifert fibered space, except, perhaps, for some of the m in the exceptional set $E_{\mathbf{n}}$. Hence, the binding number of $\xi_{\mathbf{n},m}$ must be greater than three for all $m \leq 0$ which are not in $E_{\mathbf{n}}$.

We are left to check that the support norm of $\xi_{\mathbf{n},m}$ is one when $m \leq 0$ and $M_{\mathbf{n},m}$ is hyperbolic. If the support norm were not one, then it would be zero (the support norm must be non-negative since $\xi_{\mathbf{n},m}$ is not the tight contact structure on S^3). But the only surface with boundary which has Euler characteristic zero is the annulus, and the only 3-manifolds with open book decompositions whose pages are annuli are lens spaces. □

It is natural to ask if the difference between $\mathrm{sn}(\xi)$ and $2\,\mathrm{sg}(\xi) + \mathrm{bn}(\xi) - 2$ can be arbitrarily large. While we cannot answer this question we do note the following.

THEOREM 1.3. *For a fixed* $\mathbf{n}$ *the difference between* $\mathrm{sn}(\xi_{\mathbf{n},m})$ *and* $2\,\mathrm{sg}(\xi_{\mathbf{n},m}) + \mathrm{bn}(\xi_{\mathbf{n},m}) - 2$ *is bounded independent of $m < 0$.*

Before we prove this theorem we estimate the binding numbers $\mathrm{bn}(\xi_{\mathbf{n},m})$ in some special cases.

PROPOSITION 1.4. The binding number of $\xi_{(1),-1}$ satisfies $3 \leq \text{bn}(\xi_{(1),-1}) \leq 9$. For each $m < -1$, the binding number of $\xi_{(1),m}$ satisfies $4 \leq \text{bn}(\xi_{(1),m}) \leq 9$.

The manifold $M_{(1),-1}$ is the Brieskorn sphere $\Sigma(2,3,7)$. Since $\Sigma(2,3,7)$ is not a lens space, it does not admit an open book decomposition with planar pages and two or fewer binding components. Therefore, $\text{bn}(\xi_{(1),-1}) \geq 3$. It is well-known that the only exceptional surgeries on the figure eight are integral surgeries [**Th**]. Therefore, $E_{(1)} = \{-1, 0, 1\}$. So, from Theorem 1.2, we know that $\text{bn}(\xi_{(1),m}) > 3$ for all $m < -1$. To prove Proposition 1.4, we construct an open book decomposition of $M_{(1),m}$ with planar pages and nine binding components and we show that it supports $\xi_{(1),m}$ for $m < 0$.

Recall that overtwisted contact structures on a 3-manifold M are isotopic if and only if they are homotopic as 2-plane fields. Moreover, the homotopy type of a 2-plane field ξ is uniquely determined by its induced *Spin*^{c} structure $\mathbf{t}_\xi$ and its *3-dimensional invariant* $d_3(\xi)$. Therefore, in order to show that the open book decomposition we construct actually supports $\xi_{(1),m}$ (for $m \leq 0$), we need only prove that the contact structure it supports is overtwisted and has the same 3-dimensional invariant as $\xi_{(1),m}$ (their *Spin*^{c} structures automatically agree since $H_1(M_{(1),m};\mathbb{Z}) = 0$). Below, we describe how to compute these invariants from supporting open book decompositions. For more details, see the exposition in [**EO**].

Suppose that ϕ is a product of Dehn twists around homologically non-trivial curves $\gamma_1, \ldots, \gamma_k$ in some genus g surface S with n boundary components. The open book (S, id) supports the unique tight contact structure on $\#^{2g+n-1}(S^1 \times S^2)$, and the γ_i may be thought of as Legendrian curves in this contact manifold. The contact manifold (M, ξ) supported by the open book (S, ϕ) bounds an achiral Lefschetz fibration X, which is constructed from $\natural^{2g+n-1}(S^1 \times D^3)$ by attaching 2-handles along these Legendrian curves. Each 2-handle is attached with contact framing $+1$ or -1 depending on whether the corresponding Dehn twist in ϕ is left- or right-handed, respectively. As long as $c_1(\mathbf{t}_\xi)$ is torsion in $H^2(M;\mathbb{Z})$, $d_3(\xi)$ is an element of $\mathbb{Q}$ and may be computed according to the formula,

$$d_3(\xi) = \frac{1}{4}(c^2(X) - 2\chi(X) - 3\sigma(X)) + q. \tag{1.1}$$

Here, q is the number of left-handed Dehn twists in the factorization ϕ. The number $c^2(X)$ is the square of the class $c(X) \in H^2(X;\mathbb{Z})$ which is Poincaré dual to

$$\sum_{i=1}^{k} \text{rot}(\gamma_i) C_i \in H_2(X, M; \mathbb{Z}),$$

where C_i is the cocore of the 2-handle attached along γ_i, and $\text{rot}(\gamma_i)$ is the rotation number of γ_i. The class $c(X)$ restricts to $c_1(\mathbf{t}_\xi)$ in $H^2(M;\mathbb{Z})$. Since we have assumed that $c_1(\mathbf{t}_\xi)$ is torsion, some multiple $k \cdot c(X)$ is sent to zero by the map $i^* : H^2(X;\mathbb{Z}) \to H^2(M;\mathbb{Z})$, and, hence, comes from a class $c_r(X)$ in $H^2(X, M;\mathbb{Z})$, which can be squared. So, by $c^2(X)$, we mean $\frac{1}{k^2} c_r^2(X)$.

LEMMA 1.5. *For $m < 0$, the 3-dimensional invariant $d_3(\xi_{(1),m}) = 1/2$.*

PROOF OF LEMMA 1.5. The manifold $M_{(1),m}$ can also be described as the result of $-\frac{1}{m}$ surgery on the figure eight knot. It is therefore a rational homology 3-sphere, and the 3-dimensional invariant $d_3(\xi_{(1),m})$ is a well-defined element of $\mathbb{Q}$.

Observe that the Dehn twist D_δ is isotopic to the composition $(D_x D_y)^6$. As described above, the contact manifold supported by the open book $(\Sigma, \phi_{(1),m})$ bounds an achiral Lefschetz fibration X, constructed from $\natural^2(S^1 \times D^3)$ by attaching $12|m|+2$ 2-handles corresponding to the Dehn twists in the factorization

$$\phi_{(1),m} = (D_x D_y)^{6m} \cdot D_x D_y^{-1}.$$

From the discussion in [**EO**, Section 6.1], it follows that $\mathrm{rot}(x) = \mathrm{rot}(y) = 0$; hence, $c(X) = 0$. Moreover, $\chi(X) = 12|m| + 1$ and $q = 12|m| + 1$. Therefore, the formula in (1.1) gives

$$d_3(\xi_{(1),m}) = \frac{12|m| + 1}{2} - \frac{3\sigma(X)}{4}.$$

The achiral Lefschetz fibration associated to the monodromy $(D_x D_y)^{2m}$ gives a well-known Milnor fiber with the reverse orientation which has signature is $8|m|$. One may easily check (via Kirby calculus or gluing formulas for the signature or computations of the degree of related Heegaard-Floer contact invariants) that the two extra 2–handles attached to the Milnor fiber (with reversed orientation) to create X do not change the signature, so $\sigma(X) = 8|m|$. Thus $d_3(\xi_{(1),m})) = 1/2$. □

Proposition 1.4 follows if we can find a planar open book with nine binding components which supports an overtwisted contact structure on $M_{(1),m}$ with $d_3 = 1/2$. The figure eight knot K is pictured in Figure 2. The knot K can be embedded as a homologically non-trivial curve on the surface S obtained by plumbing together two positive Hopf bands and two negative Hopf bands, as shown on the left in Figure 3.

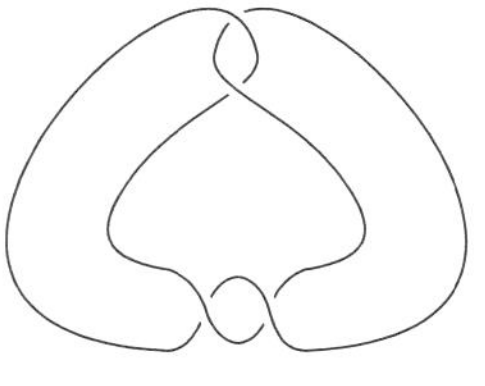

FIGURE 2. The figure eight, drawn here as a twist knot.

Topologically, S is an embedded copy of the planar surface P with five boundary components shown on the right in Figure 3. Moreover, S is a page of the open book decomposition of $M_{(1),0} \cong S^3$ given by (P, ϕ), where ϕ is the product of right-handed Dehn twists around the curves γ_1 and γ_3 and left-handed Dehn twists around the curves γ_2 and γ_4. The knot K is the image, under this embedding, of the curve $r \subset P$.

Since the Seifert framing of K agrees with the framing induced by S, $(P, D_r^m \cdot \phi)$ is an open book decomposition for $-\frac{1}{m}$ surgery on K. Let ξ_m denote the contact structure on $M_{(1),m}$ which is supported by this open book. It is easy to check that the open book $(P, D_r^m \cdot \phi)$ is not right-veering for $m \leq 0$. (This can be seen by taking, for example, the horizontal arc connecting the right most boundary components of the surface on the right of Figure 3.) Therefore, the corresponding ξ_m are overtwisted [**HKM**].

LEMMA 1.6. *For $m \leq 0$, the 3-dimensional invariant is $d_3(\xi_m) = 3/2$.*

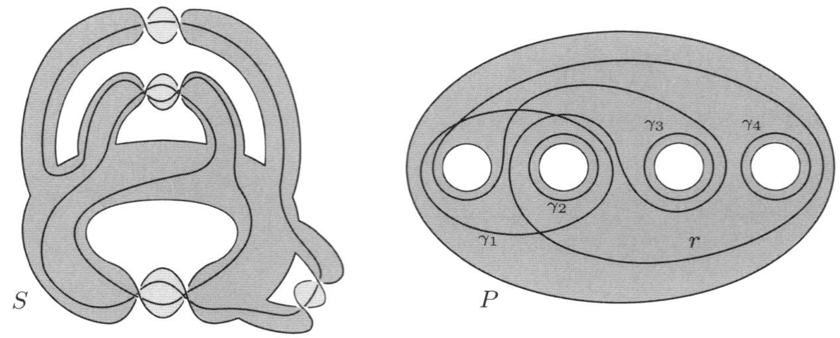

FIGURE 3. The figure eight knot, embedded on a planar surface with five boundary components.

PROOF OF LEMMA 1.6. Figure 4 shows another illustration of P, on the left; the four topmost horizontal segments are identified with the four bottommost horizontal segments to form 1-handles. As discussed above, we can think of these curves as knots in $\#^4(S^1 \times S^2) = \partial(\natural^4(S^1 \times D^3))$. The contact manifold supported by $(P, D_r^m \cdot \phi)$ bounds the achiral Lefschetz fibration X obtained from $\natural^4(S^1 \times D^3)$ by attaching 2-handles along the curves γ_2 and γ_4 with framing $+1$, along the curves γ_1 and γ_3 with framing -1, and along $|m|$ parallel copies of r (with respect to the blackboard framing) with framing $+1$, as indicated on the right in Figure 4.

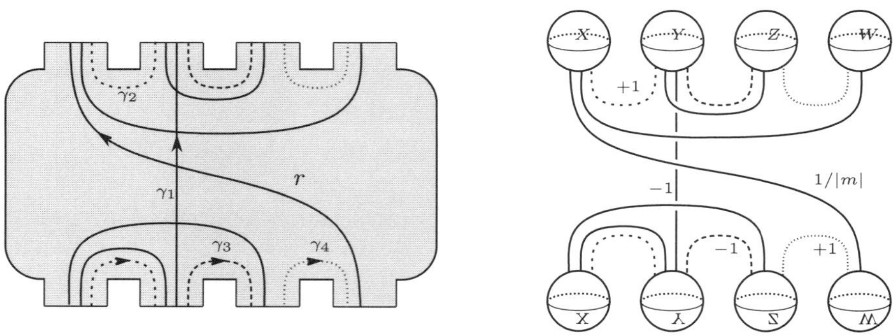

FIGURE 4. On the left, the surface P. On the right, a Kirby diagram for the achiral Lefschetz fibration corresponding to the open book $(P, D_r^m \cdot \phi)$. The label $1/|m|$ indicates that we attach 2-handles along $|m|$ parallel copies of the curve r with framing $+1$.

Let X, Y, Z and W denote the 1-handles attached to D^4 to form $\natural^4(S^1 \times D^3)$, as shown in Figure 4. Furthermore, let $S_{\gamma_1}, \dots, S_{\gamma_4}$ and $S_{r_1}, \dots, S_{r_{|m|}}$ denote the cores of the 2-handles attached to the curves $\gamma_1, \dots, \gamma_4$ and the $|m|$ parallel copies $r_1, \dots, r_{|m|}$ of r, and let $C_{\gamma_1}, \dots, C_{\gamma_4}$ and $C_{r_1}, \dots, C_{r_{|m|}}$ denote the cocores of these 2-handles. These cores form a basis for the group of 2-chains $C_2(X;\mathbb{Z})$; X, Y, Z and W for a basis for the 1-chains $C_1(X;\mathbb{Z})$; and the boundary map $d_2 : C_2(X;\mathbb{Z}) \to$

$C_1(X;\mathbb{Z})$ sends

$$\begin{aligned} d_2(S_{\gamma_1}) &= Y, \\ d_2(S_{\gamma_2}) &= X - Y, \\ d_2(S_{\gamma_3}) &= Y - Z, \\ d_2(S_{\gamma_4}) &= Z - W, \\ d_2(S_{r_i}) &= -Z + Y + W. \end{aligned}$$

The homology $H_2(X;\mathbb{Z})$ is therefore generated by $h_1, \ldots, h_{|m|}$, where

$$h_i = S_{r_i} + S_{\gamma_4} - S_{\gamma_1}.$$

By construction, X may also be obtained from D^4 by attaching 2-handles along $|m|$ parallel copies of the figure eight with framing $+1$, so the intersection matrix Q_X is simply the $|m| \times |m|$ identity matrix with respect to the corresponding basis. Since the curves r_i are parallel, it is clear that $h_i \cdot h_j = 0$ for $i \neq j$. It follows that $h_i \cdot h_i = 1$ for $i = 1, \ldots, |m|$.

Recall that the class $c(X)$ is Poincaré dual to

$$\sum_{i=1}^{4} \operatorname{rot}(\gamma_i) \cdot C_{\gamma_i} + \sum_{i=1}^{|m|} \operatorname{rot}(r_i) \cdot C_{r_i}.$$

Via the discussion in [**EO**, Section 3.1], we calculate that $\operatorname{rot}(\gamma_4) = \operatorname{rot}(\gamma_2) = \operatorname{rot}(\gamma_3) = -1$ and $\operatorname{rot}(\gamma_1) = \operatorname{rot}(r_i) = 0$. Therefore, $\langle c(X), h_i\rangle = -1$ for $i = 1, \ldots, |m|$. So, thought of as a class in $H^2(X, \partial X;\mathbb{Z})$, $c(X)$ is Poincaré dual to

$$-h_1 - \cdots - h_{|m|}.$$

Hence, $c^2(X) = |m|$. In addition, $\chi(X) = 1 + |m|$, $\sigma(X) = |m|$ and $q = 2 + |m|$. From the formula in (1.1), we have

$$d_3(\xi_m) = \frac{1}{4}(|m| - 2(1 + |m|) - 3|m|) + 2 + |m| = 3/2.$$

This completes the proof of Lemma 1.6. □

Proof of Proposition 1.4. Recall that

$$d_3(\xi \# \xi') = d_3(\xi) + d_3(\xi') + 1/2$$

for any two contact structures ξ and ξ'. (This can easily be see by recalling that Equation (1.1) defines d_3 using any almost complex 4–manifold X giving the plane field on the boundary as the set of complex tangencies, in this case $q = 0$. Then for the connect sum formula one simply attaches a "Stein" 1–handle to the almost complex manifold used to compute d_3 of ξ and ξ'.) Let $m < 0$. Since $d_3(\xi_{(1),m}) = 1/2$, $d_3(\xi_m) = 3/2$, and both $\xi_{(1),m}$ and ξ_m are overtwisted, it follows that $\xi_{(1),m}$ is isotopic to $\xi_m \# \xi'$, where ξ' is the unique (overtwisted) contact structure on S^3 with $d_3(\xi') = -3/2$. In [**EO**], Ozbagci and the second author show that $\operatorname{bn}(\xi') \leq 5$. In particular, ξ is supported by the open book $(P', D_b^{-1}D_a^{-1} \cdot \psi)$, where a and b are the curves on the surface P' shown in Figure 5 and ψ is a composition of right-handed Dehn twists around the four unlabeled curves.

Then, the planar open book $(P \#_b P', D_r^m \cdot \phi \cdot D_b^{-1}D_a^{-1} \cdot \psi)$ with nine binding components supports $\xi_{(1),m} \simeq \xi_m \# \xi'$, and the proof of Proposition 1.4 is complete. (Here, $\#_b$ denotes boundary connected sum.) □

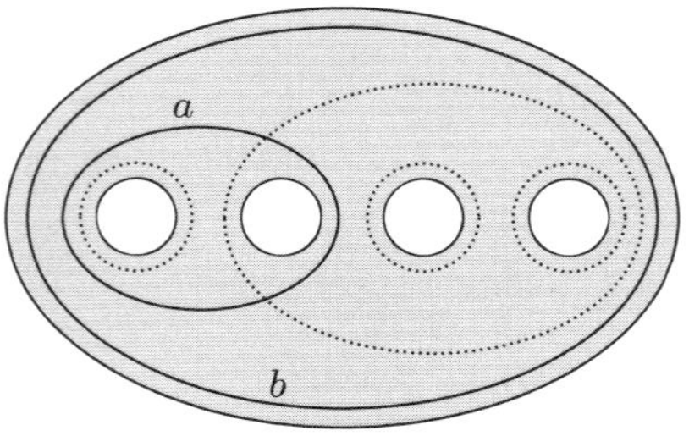

FIGURE 5. The surface P'.

The table below summarizes what we know of the support genus, binding number and support norm for the contact structures $\xi_{(1),m}$.

	sg	bn	sn
$m > 0$	1	1	1
$m = 0$	0	1	-1
$m = -1$	0	$[3, 9]$	1
$m < -1$	0	$[4, 9]$	1

TABLE 1. Values of sg, bn and sn for $\xi_{(1),m}$.

PROOF OF THEOREM 1.3. We first observe that $M_{\mathbf{n},m}$ is a rational homology sphere. This can be seen by noticing that $M_{\mathbf{n},0}$ can be obtained as the 2–fold cover of S^3 branched over an alternating (non-split) link (in fact, closure of a 3–braid). Thus the determinant of the link is non-zero and hence the cardinality of the first homology of the cover is finite. Since $M_{\mathbf{n},m}$ can be obtained from $M_{\mathbf{n},0}$ by $-1/m$ surgery on a null-homologous knot it has the same first homology.

Let K be the binding of the open book $(\Sigma, \phi_{\mathbf{n},0})$ in $M_{\mathbf{n},0}$. If we fix an overtwisted contact structure on $M_{\mathbf{n},0}$ we can find a Legendrian knot L in the knot type K with Thurston-Bennequin invariant 0 and overtwisted complement. In [**On**] it was shown that there is a planar open book (Σ', ϕ') for this overtwisted contact structure that contains L on a page so that the page framing is 0.

Notice that $M_{\mathbf{n},m}$ can be obtained from $M_{\mathbf{n},0}$ by composing ϕ' with a $+1$ Dehn twist along m copies of L on the page of the open book. Thus each $M_{\mathbf{n},m}$ has an overtwisted contact structure supported by a planar open book with the same number of binding components. The number of $Spin^c$ structures on $M_{\mathbf{n},m}$ is finite and independent of m. We can get from the constructed overtwisted contact structure on $M_{\mathbf{n},m}$ to an overtwisted contact structure realizing any $Spin^c$ structure by a bounded number of Lutz twists along generators of $H_1(M_{\mathbf{n},m}; \mathbb{Z})$ all of which lie on a page of the open book. As shown in [**Et**] we may positively stabilize the open book a bounded number of times and then compose its monodromy with extra Dehn twists to achieve these Lutz twists. The number of these stabilizations depends on the number of $Spin^c$ structures on $M_{\mathbf{n},m}$ and thus is independent of m. We now have planar open books realizing overtwisted contact structures representing all $Spin^c$

structures on $M_{\mathbf{n},m}$ with the number of binding components bounded independent of m.

To get an open book representing any overtwisted contact structure on $M_{\mathbf{n},m}$ we can take these and connect sum with overtwisted contact structures on S^3. In [**EO**] it was shown that all overtwisted contact structures on S^3 have $\mathrm{bn} \leq 6$. Thus we obtain a bound independent of m on the binding number for all overtwisted contact structures on $M_{\mathbf{n},m}$ and in particular on the $\xi_{\mathbf{n},m}, m < 0$. □

REMARK 1.7. One can also show, in a similar manner to the proof of Theorem 1.3, that the binding number of $\xi_{\mathbf{n},m}$ is bounded by a constant depending only on the length of $\mathbf{n}$.

As noted in Theorem 1.2, $\mathrm{sg}(\xi_{\mathbf{n},m}) = 0$ for $m < 0$ and $\mathrm{sn}(\xi_{\mathbf{n},m}) = 1$ for all but finitely many $m < 0$; that is, these two quantities do not depend (much) on $\mathbf{n}$ or m. While we do know that the binding number of $\xi_{\mathbf{n},m}$ is bounded independent of m it could depend on (the length of) $\mathbf{n}$. This suggests the following interesting question, which we leave unanswered.

QUESTION 1.8. Does there exist, for any positive integer n, a contact structure ξ such that $2\,\mathrm{sg}(\xi) + \mathrm{bn}(\xi) - 2 - \mathrm{sn}(\xi) = n$?

Note that $0 \leq 2\,\mathrm{sg}(\xi_{(1),m}) + \mathrm{bn}(\xi_{(1),m}) - 2 - \mathrm{sn}(\xi_{(1),m}) \leq 6$ for all $m \in \mathbb{Z}$, but we currently cannot prove that this difference is larger that 1. It would be very interesting to determine if there is an m such that this difference is greater than 1. In general, computing $2\,\mathrm{sg}(\xi_{\mathbf{n},m}) + \mathrm{bn}(\xi_{\mathbf{n},m}) - 2 - \mathrm{sn}(\xi_{\mathbf{n},m})$ could potentially provide a positive answer to this question.

Noticing that all our examples involve overtwisted contact structures, in the first version of this paper we asked the following question.

QUESTION 1.9. Is there a tight contact structure ξ such that

$$\mathrm{sn}(\xi) < 2\,\mathrm{sg}(\xi) + \mathrm{bn}(\xi) - 2 \;?$$

Within a few days Etgü and Lekili showed that the answer to this question was indeed YES by showing there were tight contact structures on the Seifert fibered space $M(-1; \frac{1}{2}, \frac{1}{2}, \frac{1}{m+2})$ supported by planar open books with more than 3 binding components but also supported by a genus one open books with one boundary component. See [**EL**] for the details.

References

[B1] J. A. Baldwin, *Tight contact structures and genus one fibered knots*, Algebr. Geom. Topol. **7** (2007) 701–735.

[B2] J. A. Baldwin, *Capping off open books and the Ozsváth-Szabó contact invariant*, (2009), arXiv:0901.3797.

[EL] T. Etgü and Y. Lekili, *Examples of planar tight contact structures with support norm one*, Int. Math. Res. Not. **2010** (2010), 3723–3728.

[Et] J. B. Etnyre, *Planar open book decompositions and contact structures*, Int. Math. Res. Not. **2004** (2004), 4255–4267.

[EO] J. B. Etnyre and B. Ozbagci, *Invariants of contact structures from open books*, Trans. Amer. Math. Soc. **360** (2008) 3133–3151.

[Gi] E. Giroux, *Géometrie de contact: de la dimension trois vers les dimensions supérieures*, Proceedings of the International Congress of Mathematicians (Beijing 2002), Vol. II, 405–414.

[HKM] K. Honda, W. Kazez, and G. Matić, *Right-veering diffeomorphisms of a compact surface with boundary*, Inv. Math. **169** (2007), 427–449.

[On] S. C. Onaran, *Invariants of Legendrian knots from open book decompositions*, Int. Math. Res. Not. **2010** (2010), 1831–1859.

[Th] W. Thurston, *The geometry and topology of three-manifolds*, Princeton University Press, 1997.

Department of Mathematics, Princeton University
E-mail address: baldwinj@math.princeton.edu

School of Mathematics, Georgia Institute of Technology
E-mail address: etnyre@math.gatech.edu

Proceedings of Symposia in Pure Mathematics
Volume **82**, 2011

Topological properties of Reeb orbits on boundaries of star-shaped domains in $\mathbb{R}^4$

Stefan Hainz and Ursula Hamenstädt

ABSTRACT. Let B^4 be the compact unit ball in $\mathbb{R}^4$ with boundary S^3. Let γ be a knot on S^3 which is transverse to the standard contact structure. We show that if there is an immersed symplectic disc $f\colon (D, \partial D) \to (B^4, \gamma)$ then $\text{lk}(\gamma) = 2\text{tan}(f) - 1$ where $\text{lk}(\gamma)$ is the self-linking number of γ and $\text{tan}(f)$ is the tangential self-intersection number of f. We also show that if $E \subset \mathbb{C}^2$ is compact and convex, with smooth boundary Σ, and if the principal curvatures of Σ are suitably pinched then the self-linking number of a periodic Reeb orbit on Σ of Maslov index 3 equals -1.

1. Introduction

Consider the four-dimensional euclidean space $\mathbb{R}^4$ with the standard *symplectic form* defined in standard coordinates by $\omega_0 = \sum_{i=1}^{2} dx_i \wedge dy_i$. This symplectic form is the differential of the one-form

$$\lambda_0 = \frac{1}{2}\sum_{i=1}^{2}(x_i dy_i - y_i dx_i).$$

For every bounded domain $\Omega \subset \mathbb{R}^4$ which is star-shaped with respect to the origin $0 \in \mathbb{R}^4$, with smooth boundary Σ, the restriction λ of λ_0 to Σ defines a smooth *contact form* on Σ. This means that $\lambda \wedge d\lambda$ is a volume form on Σ.

Let $\xi = \ker(\lambda)$ be the contact bundle. Each *transverse knot* γ on Σ, i.e. an embedded smooth closed curve on Σ which is everywhere transverse to ξ, admits a canonical orientation determined by the requirement that $\lambda(\gamma') > 0$. To such an oriented transverse knot γ we can associate its *self-linking number* $\text{lk}(\gamma)$ which is defined as follows. Let $S \subset \Sigma$ be a *Seifert surface* for γ, i.e. S is a smooth embedded oriented surface in Σ whose oriented boundary equals γ. Since γ is transverse to ξ, there is a natural identification of the restriction to γ of the oriented normal bundle of S in Σ with a real line subbundle N_S of the contact bundle $\xi|\gamma$. Then N_S defines a trivialization of the oriented two-plane bundle $\xi|\gamma$. The self-linking number $\text{lk}(\gamma)$

2010 *Mathematics Subject Classification.* Primary 57R17; Secondary 53D12, 57M25.
Key words and phrases. Transverse knots, Reeb orbits, self-linking number, Maslov index.
Partially supported by DFG-SPP 1154.

of γ is the winding number with respect to N_S of a trivialization of ξ over γ which extends to a trivialization of ξ on Σ.

Eliashberg [**Eli93**] showed that the self-linking number of a transverse knot in Σ is always an odd integer. If g denotes the *Seifert genus* of γ, i.e. the smallest genus of a Seifert surface for γ, then we have $\text{lk}(\gamma) \leq 2g - 1$ ([**Eli93**, Theorem 4.1.1]). Eliashberg also constructed for every $k \geq 1$ a transverse unknot ζ of self-linking number $\text{lk}(\zeta) = -2k - 1$ in the standard unit three-sphere $S^3 \subset \mathbb{R}^4$. This indicates that in general we can not expect additional relations between the self-linking number of a transverse knot γ on Σ and purely topological invariants of γ.

Our first goal is to relate the self-linking number of a (canonically oriented) transverse knot γ on the unit three-sphere $S^3 \subset \mathbb{R}^4$ to the symplectic topology of the closed unit ball $B^4 \subset \mathbb{R}^4$. For this let $D \subset \mathbb{C}$ be the closed unit disc with oriented boundary ∂D and let $f\colon (D, \partial D) \to (B^4, S^3)$ be a smooth immersion with $f^{-1}(S^3) = \partial D$. If all self-intersections of $f(D)$ are transverse then the *tangential index* $\tan(f)$ of f is the number of self-intersection points of f counted with signs and multiplicities. The disc $f\colon D \to B^4$ is called *symplectic* if for every $x \in D$ the restriction of the symplectic form ω_0 to $df(T_xD)$ does not vanish and defines the usual orientation of D. We show

THEOREM 1.1. *Let γ be a transverse knot on the boundary S^3 of the compact unit ball $B^4 \subset \mathbb{R}^4$. If γ bounds an immersed symplectic disc $f\colon (D, \partial D) \to (B^4, \gamma)$ then* $\text{lk}(\gamma) = 2\tan(f) - 1$.

Now let Σ be the boundary of an arbitrary bounded domain $\Omega \subset \mathbb{R}^4$ which is star-shaped with respect to the origin, with smooth boundary. The *Reeb vector field* of the contact form λ is the smooth vector field X on Σ defined by $\lambda(X) = 1$ and $d\lambda(X, \cdot) = 0$. Rabinowitz [**Rab79**] (see also [**W79**]) showed that the *Reeb flow* on Σ generated by the Reeb vector field X admits periodic orbits. Dynamical properties of the Reeb flow on Σ are related to properties of Ω viewed as a symplectic manifold. The proof of the following corollary is similar to the proof of Theorem 1.1.

COROLLARY 1.2. *Let γ be a periodic Reeb orbit on the boundary Σ of a star-shaped domain $\Omega \subset \mathbb{R}^4$ with compact closure C. If γ bounds an immersed symplectic disc $f\colon (D, \partial D) \to (C, \gamma)$ then* $\text{lk}(\gamma) = 2\tan(f) - 1$.

Note that by a result of Hofer, Wysocki and Zehnder [**HWZ96**], there is always a periodic Reeb orbit of self-linking number -1 on Σ which is unknotted.

Even though the radial diffeomorphism $\Psi\colon S^3 \to \Sigma$ maps the contact bundle of S^3 to the contact bundle of Σ and hence maps a transverse knot γ on S^3 to a transverse knot $\Psi\gamma$ on Σ, the corollary is not immediate from Theorem 1.1. Namely, in general the radial diffeomorphism Ψ does not extend to a symplectomorphism $B^4 \to C$ and hence there is no obvious relation between symplectic immersions of discs in the unit ball and in the domain Ω.

In general, the existence of an immersed symplectic disc in C whose boundary is a given Reeb orbit γ on Σ does not seem to be known. However, we observe in Section 4 that such an immersed symplectic disc always exists if Σ is the boundary of a strictly convex domain in $\mathbb{R}^4$.

There is a second numerical invariant for a periodic Reeb orbit γ on Σ, the so-called *Maslov index* $\mu(\gamma)$. If Σ is the boundary of a compact strictly convex body $C \subset \mathbb{R}^4$ then the Maslov index of any periodic Reeb orbit on Σ is at least three [**HWZ98**]. The *action* of γ is defined to be $\int_\gamma \lambda > 0$, and the orbit is *minimal* if its action is minimal among the actions of all periodic orbits of the Reeb flow. Ekeland [**Eke90**] showed that the Maslov index of a minimal Reeb orbit on Σ equals precisely three.

Our second result relates the Maslov index to the self-linking number for periodic Reeb orbits on the boundary of compact strictly convex bodies with geometric control.

THEOREM 1.3. *Let C be a compact strictly convex body with smooth boundary Σ. If the principal curvatures $a \geq b \geq c$ of Σ satisfy the pointwise pinching condition $a \leq b + c$ then a periodic Reeb orbit γ on Σ of Maslov index 3 bounds an embedded symplectic disc in C. In particular, the self-linking number of γ equals -1.*

As an immediate consequence, if Σ is as in Theorem 1.3 then a periodic Reeb orbit γ on Σ of Maslov index 3 is a slice knot in Σ. In fact, with some additional effort it is possible to show that such an orbit is unknotted [**H07**].

The proofs of these results use mainly tools from differential topology and differential geometry. In Section 2 we begin to investigate topological properties of transverse knots on the three-sphere S^3. We define a self-intersection number for a (not necessarily immersed) disc in the closed unit ball B^4 with boundary γ which does not have self-intersections near the boundary and relate this self-intersection number to the self-linking number of γ. In Section 3 we study topological invariants of immersed discs in B^4 with boundary γ and show Theorem 1.1 and the corollary. In Section 4 we look at boundaries of strictly convex bodies in $\mathbb{R}^4$ and derive Theorem 1.3.

2. Self-intersection of surfaces

In this section we investigate topological invariants of smooth maps from an oriented bordered surface S with connected boundary ∂S into an arbitrary smooth oriented simply connected 4-dimensional manifold W (without boundary) whose restrictions to a neighborhood of ∂S are embeddings. For the main application, $W = \mathbb{C}^2 = \mathbb{R}^4$. We use this discussion to investigate maps from the closed unit disc $D \subset \mathbb{C}$ into the compact unit ball $B^4 \subset \mathbb{C}^2$. For maps which map the oriented boundary ∂D of D to a canonically oriented transverse knot γ on Σ we define a self-intersection number and relate this to the self-linking number of γ.

Let for the moment S be any compact oriented surface with connected boundary $\partial S = S^1$.

DEFINITION 2.1. A smooth map $f\colon S \to W$, i.e. a map which is smooth up to and including the boundary, is called *boundary regular* if the singular points of f are contained in the interior of S, i.e. if there is a neighborhood A of ∂S in S such that the restriction of f to $f^{-1}(f(A))$ is an embedding.

McDuff investigated in [**McD91**] boundary regular *pseudo-holomorphic* discs in *almost complex 4-manifolds* (W, J). By definition, such a pseudo-holomorphic disc is a smooth boundary regular map f from the closed unit disc $D \subset \mathbb{C}$ into W whose differential is complex linear with respect to the complex structure on D and the almost complex structure J. She defined a topological invariant for such boundary regular pseudo-holomorphic discs which depends on a trivialization of the normal bundle over the boundary circle.

Our first goal is to find a purely topological analog of this construction. For this we say that two boundary regular maps $f, g\colon S \to W$ *are contained in the same boundary class* if g coincides with f near the boundary and is homotopic to f with fixed boundary. This means that there is a homotopy $h\colon [0,1] \times S \to W$ connecting $h_0 = f$ to $h_1 = g$ with $h(s,z) = f(z)$ for all $s \in [0,1]$, all $z \in \partial S$. We do not require that each of the maps $h_s\colon z \to h_s(z) = h(s,z)$, $s \in [0,1]$, is boundary regular. In particular, if $\pi_2(W) = 0$ then any two boundary regular maps $f, g\colon S \to W$ which coincide near the boundary ∂S of S are contained in the same boundary class (recall that we require that W is simply connected).

There is also the following stronger notion of homotopy for boundary regular maps.

DEFINITION 2.2. A homotopy $h\colon [0,1] \times S \to W$ is called *boundary regular* if for each s the map h_s is boundary regular and coincides with h_0 near ∂S.

The set of boundary regular maps in the boundary class of a map $f\colon S \to W$ can naturally be partitioned into boundary regular homotopy classes.

A boundary regular map $f\colon S \to W$ is an embedding near ∂S. Since S is oriented by assumption, the normal bundle L of $f(S)$ over the embedded circle $f(\partial S)$ is an oriented real two-dimensional subbundle of $TW|f(\partial S)$.

For each trivialization ρ of this normal bundle, the self-intersection number $\mathrm{Int}(f,\rho) \in \mathbb{Z}$ is defined as follows [**McD91**]. Let $\overline{N}$ be a closed tubular neighborhood of $f(\partial S) = \gamma$ in W with smooth boundary ∂N such that $f(S) \cap \overline{N}$ is an embedded closed annulus A which intersects ∂N transversely. Let $E \subset W$ be an embedded submanifold with boundary which contains A and is diffeomorphic to an open disc bundle over A. One of the two connected components $(\partial E)_0$ of the boundary ∂E of E has a natural identification with the total space of the normal bundle L of $f(S)$ over γ. Remove $\overline{N} - E$ from W and glue to the boundary $(\partial E)_0$ of the resulting manifold the oriented real two-dimensional vector bundle $D \times \mathbb{C} \to D$ in such a way that $\partial D \times \{0\}$ is identified with the curve $f(\partial S) = \gamma \subset (\partial E)_0$ and that the fibres $\{x\} \times \mathbb{C}$ ($x \in \partial D$) match up with the trivialized normal bundle $L|f(\partial S)$ of $f(S)$ over γ. Up to diffeomorphism, the resulting 4-dimensional smooth manifold W_ρ only depends on the homotopy class of the trivialization ρ and of the boundary class of f. Let S_0 be the closed oriented surface obtained by glueing a disc to the boundary of S in the usual way. The map f naturally extends to a map f_0 of S_0 into W_ρ. The *self-intersection number* $\mathrm{Int}(f,\rho)$ is then defined to be the topological self-intersection number of $f_0(S_0)$ in W_ρ. Thus $\mathrm{Int}(f,\rho)$ is the number of intersections of $f(S)$ with a surface f' which is a generic perturbation of $f(S)$ and such that $f(\partial S)$ is pushed into the direction given by ρ.

In the next lemma we determine the boundary regular homotopy classes in a fixed boundary class.

LEMMA 2.3. *Let $f\colon S \to W$ be a smooth boundary regular map. Choose a trivialization ρ of the oriented normal bundle of $f(S)$ over $f(\partial S)$. Then the assignment which associates to a boundary regular homotopy class of maps in the boundary class of f its self-intersection number with respect to ρ is a bijection onto* $\mathrm{Int}(f,\rho) + 2\mathbb{Z}$. *Moreover, if f is an embedding then each such class can be represented by an embedding.*

PROOF. Let $f\colon S \to W$ be a smooth boundary regular map. Write $\gamma = f(\partial S)$ and let $u\colon S \to W$ be a smooth boundary regular map in the boundary class of f. This means that there is a homotopy $h\colon [0,1] \times S \to W$ connecting $h_0 = f$ to $h_1 = u$ with fixed boundary. The maps u, f are contained in the same boundary regular homotopy class if and only if this homotopy can be chosen in such a way that there is a tubular neighborhood N of γ such that the intersection of $h_s(S)$ with N is independent of s.

Choose such an open tubular neighborhood N of γ with smooth boundary ∂N which is sufficiently small that both $f(S)$ and $u(S)$ intersect N in a smooth annulus containing γ as one of its two boundary components. We may assume that there is a compact subsurface $C \subset S$ with smooth boundary ∂C such that $S - C$ is an annulus neighborhood of ∂S and that $f(S-C) = u(S-C) = f(S) \cap N = u(S) \cap N$. Then $u|_C$ and $f|_C$ can be combined to a map into $W - N$ of the closed oriented surface $\tilde{S}$ which we obtain from C by gluing two copies of C along the boundary with an orientation reversing boundary identification. This map is homotopic in $W - N$ to a constant map if and only if u and f are contained in the same boundary regular homotopy class.

Now W is simply connected by assumption and N is homeomorphic to a 3-ball-bundle over a circle, with boundary $\partial N \sim \gamma \times S^2$. Thus by van Kampen's theorem, $W - N$ is simply connected and the second homotopy group $\pi_2(W - N)$ coincides with the second homology group $H_2(W - N, \mathbb{Z})$ via the Hurewicz isomorphism. Since two boundary regular maps in the same boundary class are homotopic with fixed boundary, we conclude that the family of boundary regular homotopy classes of maps in the boundary class of f can be identified with the kernel of the natural homomorphism $H_2(W - N, \mathbb{Z}) \to H_2(W, \mathbb{Z})$.

To compute this group, we use the long exact homology sequence of the pair $(W, W - N)$ given by

$$\cdots \to H_3(W, \mathbb{Z}) \quad \to H_3(W, W - N, \mathbb{Z}) \quad \to H_2(W - N, \mathbb{Z}) \quad \to H_2(W, \mathbb{Z}) \quad \to \dots$$

Excision shows that $H_3(W, W - N, \mathbb{Z}) = H_3(\overline{N}, \partial N, \mathbb{Z})$ where $\overline{N}$ is the closure of N. Since $\overline{N} = \gamma \times B^3 = S^1 \times B^3$ where B^3 denotes the closed unit ball in $\mathbb{R}^3$, the group $H_3(W, W - N, \mathbb{Z})$ is infinite cyclic and generated by a ball $\{z\} \times (B^3, S^2)$ where $z \in \gamma$ is any fixed point.

Every singular homology class $v \in H_3(W, \mathbb{Z})$ can be represented by a piecewise smooth singular cycle σ whose image is nowhere dense in W. On the other hand, the curve γ is contractible in W and therefore there is a smooth isotopy of W which moves σ away from N. Thus the image of $H_3(W, \mathbb{Z})$ under the natural homomorphism $H_3(W, \mathbb{Z}) \to H_3(W, W - N, \mathbb{Z}) = H_3(\overline{N}, \partial N, \mathbb{Z})$ vanishes and hence by exactness, the kernel of the natural homomorphism $H_2(W - N, \mathbb{Z}) \to H_2(W, \mathbb{Z})$ is isomorphic to $\mathbb{Z}$ and generated by a sphere $e = \{z\} \times S^2 \sim 1 \in \pi_2(\partial N) = \mathbb{Z}$ for some $z \in \gamma$. As a consequence, every boundary regular homotopy class in the boundary class of f can uniquely be represented in the form $[f] + ke$ where $[f]$ denotes the boundary regular homotopy class of f and where $k \in \mathbb{Z}$.

Next we show that if f is an embedding then each of these classes can be represented by an embedding as well. For this note that after possibly replacing N by a smaller tubular neighborhood of γ we may assume that $f^{-1}(\overline{N})$ is a closed annulus neighborhood of ∂S and that for some $z \in \gamma$, the sphere $M = \{z\} \times S^2 \subset \partial N$ intersects $f(S)$ transversely in a single point x. The orientation of the surface S

defines uniquely an orientation of M such that $T_xW = T_x(f(S)) \oplus T_xM$ as oriented vector spaces. Using standard surgery near the transverse intersection point x we can attach the sphere M to the surface $f(S)$ as follows (see [**GS99**, p. 38]). There is a closed neighborhood V of x in $W - \gamma$ which is diffeomorphic to a closed ball and such that the intersections $f(S) \cap V$, $M \cap V$ are smooth discs which intersect transversely in the single point x. The boundaries of these discs are two disjoint oriented circles in the boundary $\partial V \sim S^3$ of V. These circles define the Hopf link in S^3 and therefore they form the oriented boundary of a smooth embedded annulus in ∂V. The surgery replaces $(f(S) \cup M) \cap V$ by such an annulus (which can be done smoothly). We obtain in this way a compact oriented bordered surface which can be represented by a boundary regular map $g\colon S \to W$ which coincides with $f(S)$ near the boundary. The surgery does not change relative homology classes ([**GS99**, p. 38]) and hence $g(S)$ is homologous to $[f] + e$ via an identification of M with a generator e of the kernel of the natural map $H_2(W - N, \mathbb{Z}) \to H_2(W, \mathbb{Z})$. In other words, the embedded surface which we just constructed represents the boundary regular homotopy class $[f] + e$ in the boundary class of f. In the same way we can also construct a surface which represents the boundary regular homotopy class $[f] - e$ by attaching to f a sphere equipped with the reverse orientation. Namely, we also can connect the boundaries of the discs $f(S) \cap V, M \cap V$ with an embedded cylinder whose oriented boundary is the union of the oriented boundary of $f(S) \cap V$ with the boundary of $M \cap V$ equipped with the reversed orientation. Repeating this procedure finitely many times with different basepoints we obtain an embedding in every boundary regular homotopy class of maps in the boundary class of f.

Let ρ be a trivialization of an oriented normal bundle of $f(S)$ along $\gamma = f(\partial S)$. We are left with showing that a boundary regular homotopy class in the boundary class of f is determined by its self-intersection number with respect to ρ. For this let again $M = \{z\} \times S^2 \subset \partial N$ be an oriented embedded sphere as above which intersects $f(S)$ transversely in a single point x. Assume that the index of intersection between $f(S)$ and M with respect to the given orientations is positive. Let $g\colon S \to W$ be the map constructed above with $[g] = [f] + e$. Using the above notations, it is enough to show that $\operatorname{Int}(g, \rho) = \operatorname{Int}(f, \rho) + 2$. However, this can be seen as follows.

As above, denote by W_ρ the manifold used for the definition of the self-intersection number $\operatorname{Int}(f, \rho)$. Recall that up to diffeomorphism, the manifold W_ρ only depends on ρ and the boundary class of f. In particular, we may assume that W_ρ contains the images Γ_f, Γ_g of the closed surface S_0 under the natural extensions of the maps f, g. The self-intersection numbers of the surfaces Γ_f, Γ_g in the manifold W_ρ can now be compared via

$$\begin{aligned}\operatorname{Int}(g, \rho) &= \Gamma_g \cdot \Gamma_g = (\Gamma_f + e) \cdot (\Gamma_f + e) \\ &= \Gamma_f \cdot \Gamma_f + 2\Gamma_f \cdot e + e \cdot e = \operatorname{Int}(f, \rho) + 2\end{aligned}$$

since the topological self-intersection of the sphere e in W_ρ vanishes. But this just means that the assignment which associates to a boundary regular homotopy class in the boundary class of f its self-intersection number with respect to ρ is a bijection onto $\operatorname{Int}(f, \rho) + 2\mathbb{Z}$. From this the lemma follows. □

From now on we assume that the 4-dimensional manifold W is equipped with a smooth almost complex structure J.

DEFINITION 2.4. A smooth boundary regular map $f\colon S \to W$ is called *boundary holomorphic* if for each $z \in \partial S$ the tangent plane of $f(S)$ at z is a complex line in (TW, J) whose orientation coincides with the orientation induced from the orientation of S.

If $f\colon S \to W$ is boundary regular and boundary holomorphic then the pull-back f^*TW under f of the tangent bundle of W is a 2-dimensional complex vector bundle over S. Since f is boundary holomorphic, the restriction to ∂S of the tangent bundle TS of S is naturally a complex line-subbundle of $f^*TW|_{\partial S}$. Then the normal bundle of $f(S)$ over γ can be identified with a complex line subbundle of $(TW, J)|f(\partial S)$ as well. Every trivialization ρ of this normal bundle defines as before a smooth manifold W_ρ. This manifold admits a natural almost complex structure extending the almost complex structure on the complement of a small tubular neighborhood of $f(\partial S)$ in W. In particular, if we denote as before by f_0 the natural extension of f to the closed surface S_0 then the pull-back bundle $f_0^*TW_\rho$ is a complex two-dimensional vector bundle over S_0.

Up to homotopy, this bundle only depends on ρ and the boundary class of f. Namely, any homotopy h_s of $f = h_0$ which is the identitity near the boundary induces a homotopy of the pull-back bundles $h_s^*TW_\rho$. Now if f_0, f_1 are the extensions of h_0, h_1 to the closed surface S_0 then since the homotopy h_s is the identity near the boundary, it determines a homotopy of the complex pull-back bundle $f_0^*TW_\rho$ to the complex pull-back bundle $f_1^*TW_\rho$. Let $c(\rho)$ be the evaluation on S_0 of the first Chern class of this bundle.

Changing the trivialization ρ by a full positive (negative) twist in the group $U(1) \subset GL(1, \mathbb{C})$ changes both the self-intersection number $\mathrm{Int}(f, \rho)$ and the Chern number $c(\rho)$ by 1 (-1) (see [**McD91**]). In particular, there is up to homotopy a unique trivialization ρ of the complex normal bundle of $f(S)$ over $f(\partial S)$ such that $c(\rho) = 2$. We call such a trivialization a *preferred* trivialization. By the above observation, a preferred trivialization only depends on the boundary class of f but not on the boundary regular homotopy class of f. Moreover, the complex normal bundle of a boundary holomorphic boundary regular map $f\colon S \to W$ only depends on the oriented boundary circle $f(\partial S)$.

DEFINITION 2.5. The *self-intersection number* $\mathrm{Int}(f)$ of a boundary holomorphic boundary regular map $f\colon S \to W$ is the self-intersection number $\mathrm{Int}(f, \rho)$ of f with respect to a preferred trivialization ρ of the complex normal bundle of $f(S)$ over $f(\partial S)$.

Lemma 2.3 implies

COROLLARY 2.6. *Let $f\colon S \to W$ be boundary regular and boundary holomorphic. Then a boundary regular homotopy class in the boundary class of f is uniquely determined by its self-intersection number.*

Now consider the standard unit sphere S^3 in $\mathbb{C}^2$ which bounds the standard open unit ball $B_0^4 \subset \mathbb{C}^2$. The *contact distribution* on S^3 is the unique smooth two-dimensional subbundle ξ of TS^3 which is invariant under the (integrable) complex structure J on $\mathbb{C}^2$. A smooth embedding $\gamma\colon S^1 \to S^3$ is *transverse* if its tangent γ' is everywhere transverse to ξ. Let N be the outer normal field of S^3. Then JN is tangent to S^3 and orthogonal to ξ. The tangent of the transverse knot γ can be written in the form

$$\gamma'(t) = a(t)JN(\gamma(t)) + B(t)$$

where $B(t) \in \xi$ for all t and where $a(t) \neq 0$. Assume that γ is oriented in such a way that $a(t)$ is positive for all t. If as in the introduction we denote by λ the restriction to S^3 of the radial one-form λ_0 on $\mathbb{C}^2$ then this orientation of γ is determined by the requirement that the evaluation of λ on γ' is positive, and we call it *canonical.* If the transverse knot γ is canonically oriented then $J\gamma'$ points inside the ball B_0^4. Thus if $B^4 = B_0^4 \cup S^3$ denotes the closed unit ball then for every canonically oriented transverse knot γ on S^3 there is a smooth boundary regular boundary holomorphic map $f\colon S \to B^4$ with $f(\partial S) = \gamma$ and $f^{-1}(S^3) = \partial S$ whose restriction to a neighborhood of the boundary is an embedding: Just choose a smooth embedding of a closed annulus A into B^4 which maps one of the boundary circles ζ of A diffeomorphically onto γ and whose tangent plane at a point in ζ is J-invariant. In particular, A meets S^3 transversely along γ and hence we may assume that $A \cap S^3 = \gamma$. Extend this embedding in an arbitrary way to a smooth map of the surface S (with the annulus A as a neighborhood of ∂S) into B^4 which is always possible since B^4 is contractible.

Define a boundary regular map $f\colon (S, \partial S) \to (B^4, S^3)$ to be *boundary transverse* if f is transverse to S^3 along the boundary. Since B^4 is contractible, the above observation implies that every boundary regular boundary transverse map $f\colon (S, \partial S) \to (B^4, \gamma)$ can be homotoped within the family of such maps to a boundary regular boundary holomorphic map $f'\colon (S, \partial S) \to (B^4, \gamma)$. In particular, the oriented normal bundle of f over γ is naturally homotopic to the oriented normal bundle of f' over γ. The map f' is used to calculate the preferred trivialization of this normal bundle. Then the self-intersection number $\mathrm{Int}(f)$ can be defined as the self-intersection number of f with respect to the induced trivialization of the oriented normal bundle of f. This self-intersection number coincides with the self-intersection number $\mathrm{Int}(f')$ of f' and by Corollary 2.6, it only depends on f.

More generally, let Σ be the boundary of a bounded domain $\Omega \subset \mathbb{C}^2$ which contains the origin 0 in its interior and which is star-shaped with respect to 0. The contact form is the restriction λ to Σ of the radial one-form λ_0 on $\mathbb{C}^2$.

Let N be the outer normal field of $\Sigma \subset \mathbb{C}^2$. Since the one-form λ_0 can also be written in the form $(\lambda_0)_p(Y) = \frac{1}{2}\langle Jp, Y\rangle$ ($p \in \mathbb{C}^2, Y \in T_p\mathbb{C}^2$ and where $\langle , \rangle$ is the euclidean inner product), the *Reeb vector field* X on Σ is given by

$$X(p) = \phi(p)JN(p)$$

where

$$\phi(p) = \frac{2}{\langle p, N(p)\rangle} > 0.$$

Namely, for $p \in \Sigma$ we have

$$d\lambda_p(X, \cdot) = \phi(p)\omega_0(JN(p), \cdot) = -\phi(p)\langle N(p), \cdot\rangle = 0$$

on $T_p\Sigma$ and

$$\lambda_p(X) = \frac{1}{2}\langle Jp, X\rangle = \frac{1}{2}\phi(p)\langle Jp, JN(p)\rangle = 1.$$

In particular, if γ is a Reeb orbit on Σ then $J\gamma'$ is transverse to Σ and points inside the domain Ω. As a consequence, as for transverse knots on S^3, if we denote by $C = \Omega \cup \Sigma$ the closure of Ω then for every boundary regular boundary transverse map $f\colon S \to C$ whose boundary $f(\partial S)$ is a periodic Reeb orbit on Σ, the self-intersection number $\mathrm{Int}(f)$ of f is well defined.

The next lemma shows that in both cases, the self-intersection number of such a map $f\colon (S, \partial S) \to (C, \gamma)$ only depends on γ. For a convenient formulation, let C be the closure of a bounded star-shaped domain in $\mathbb{C}^2$ and call a smoothly embedded closed curve γ in the boundary Σ of C *admissible* if either C is the unit ball, $\Sigma = S^3$ and γ is a canonically oriented transverse knot or if γ is a Reeb orbit on Σ.

LEMMA 2.7. *Let $\gamma \subset \Sigma$ be an admissible curve. Then any two boundary regular boundary transverse maps $f\colon (S, \partial S) \to (C, \gamma), g\colon (S', \partial S') \to (C, \gamma)$ have the same self-intersection number.*

PROOF. Let $f\colon S \to C, g\colon S' \to C$ be any two boundary regular boundary transverse maps with boundary an admissible closed curve γ. The inner normals of the surfaces $f(S), g(S')$ along $\gamma = f(\partial S) = g(\partial S')$ point strictly inside the domain C. After a small deformation through boundary regular boundary transverse maps we may assume that there is a small annular neighborhood A of the boundary of S, an annular neighborhood A' of the boundary of S' and a homeomorphism $\phi\colon A \to A'$ which maps ∂S to $\partial S'$ and is such that $g(\phi(x)) = f(x)$ for all $x \in A$. We may moreover assume that the restrictions of f, g to $f^{-1}(A), g^{-1}(A')$ are embeddings. Since f, g are boundary regular, after possibly modifying f, g once more with a small boundary regular homotopy which pushes interior intersection points of $f(S), g(S')$ with Σ into the interior Ω of C we may assume that there is a compact star-shaped set $K \subset \Omega$ such that $f(S - A) \subset K, g(S' - A) \subset K$. But then the restrictions of f, g to $S - A, S' - A'$ are maps of surfaces $S - A, S' - A'$ into K with the same boundary curve γ'. Now K is contractible and hence the maps $f|S - A, g|S' - A'$ define the same relative homology class in $H_2(K, \gamma'; \mathbb{Z})$. By Lemma 2.3 and its proof, this implies that the self-intersection numbers of f, g indeed coincide. □

As a consequence, we can define:

DEFINITION 2.8. Let $\gamma \subset \Sigma$ be an admissible curve. The *self-intersection number* $\mathrm{Int}(\gamma)$ of γ is the self-intersection number of a boundary regular boundary transverse map $f\colon S \to C$ with boundary $f(\partial S) = \gamma$.

The final goal of this section is to calculate the self-intersection number of an admissible curve γ on Σ. For this we begin with calculating the preferred trivialization of the normal bundle of the complex line subbundle of $T\mathbb{C}^2|\gamma$ spanned by the tangent γ' of γ. Note that this normal bundle can naturally be identified with the restriction of the contact bundle ξ to γ.

To this end let D be the closed unit disc in $\mathbb{C}$ and let $f\colon D \to C$ be a boundary holomorphic boundary regular *immersion* which maps ∂D diffeomorphically onto the canonically oriented admissible curve γ. The image under df of the inner normal of D along ∂D points strictly inside C. Let

$$\hat{M}\colon (z_1, z_2) \to (-\overline{z}_2, \overline{z}_1)$$

be a J-orthogonal $\langle \cdot, \cdot \rangle$-compatible almost complex structure on $\mathbb{C}^2$ where as usual, $z \to \overline{z}$ is complex conjugation. Then we obtain a trivialization of the complex vector bundle $(f^*T\mathbb{C}^2, J)$ over D by the sections $X_1 = df(\frac{\partial}{\partial x}), X_2 = \hat{M}df(\frac{\partial}{\partial x})$ (with a slight abuse of notation).

The trivialization of the tangent bundle $df(TD)|\gamma$ of $f(D)$ over γ defined by the tangent γ' of the admissible curve γ has rotation number one with respect to the trivialization $df(\frac{\partial}{\partial x})$. Since the tangent bundle of the two-sphere S^2 has Chern number 2 and is obtained by glueing the tangent bundles of two standard discs D_1, D_2 along the boundary using the trivializations defined by the tangent field of the boundary, the restriction of the section X_2 to γ defines the preferred trivialization of the complex normal bundle L over γ. Now $\hat{M}$ is complex anti-linear and therefore the trivialization of L over γ defined by the section $\hat{M} \circ \gamma'$ has rotation number -1 with respect to the preferred trivialization. Recall that the preferred trivialization of the complex normal bundle of γ only depends on the boundary class of an infinitesimally holomorphic immersion of a surface S into $\mathbb{C}^2$.

To the admissible curve γ we can also associate its *self-linking number* $\mathrm{lk}(\gamma)$ (see [**Eli92**] and the introduction). The following proposition relates these two numbers.

PROPOSITION 2.9. *Let γ be an admissible curve on Σ. Then the self-intersection number* $\mathrm{Int}(\gamma)$ *of γ equals* $\mathrm{lk}(\gamma) + 1$.

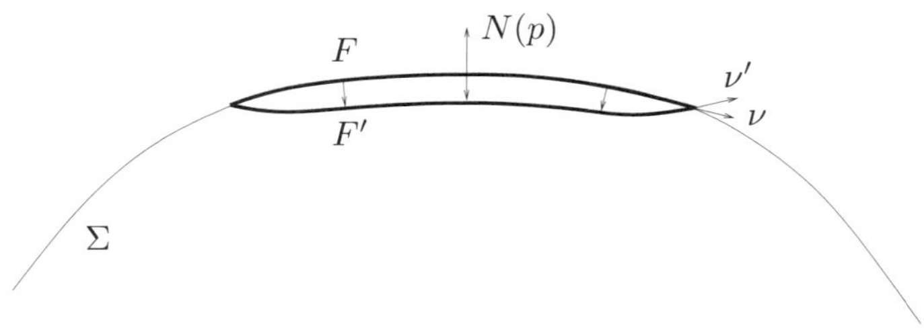

FIGURE 1. The surfaces F and F'

PROOF. Let N be the outer normal field of Σ and let L be the *complex subbundle* of $T\Sigma$, i.e. the 2-dimensional subbundle which is invariant under the complex structure J. If $\Sigma = S^3$ then this is just the contact bundle. The image of the outer normal N of Σ under the J-orthogonal $\langle , \rangle$-compatible almost complex structure $\hat{M}$ is a global section of the bundle L. Let $F \subset \Sigma$ be a *Seifert surface* for the admissible curve γ, i.e. F is an embedded oriented bordered surface in Σ with boundary γ. Let N_F be the oriented normal field of F in Σ with respect to the restriction of the euclidean metric $\langle , \rangle$. Since γ is transverse to L we may assume that for every $x \in \gamma$ the vector $N_F(x)$ is contained in the fibre L_x at x of the complex line bundle L. The self-linking number of γ is therefore the winding number of the section $x \to M(x) = \hat{M}N(x)$ of $L|\gamma$ with respect to the trivialization of $L|\gamma$ defined by the section $x \to N_F(x)$.

Let $F' \subset C$ be the embedded surface which we obtain by pushing F slightly in the direction $-N$ as in Figure 1. Then F' is an embedded surface in C which is boundary regular and boundary transverse. The restriction of N_F to γ extends to a global trivialization of the oriented normal bundle of the surface F and hence F' in $\mathbb{C}^2 - \gamma$. Thus the self-intersection number of F' with respect to the trivialization defined by N_F vanishes. Since by the above observation the winding number of the section M of $L|\gamma$ with respect to the preferred trivialization of $L|\gamma$ equals -1, the self-intersection number $\mathrm{Int}(\gamma)$ equals the winding number of M with respect to the trivialization of $L|\gamma$ defined by N_F plus one. This shows the proposition. □

3. Topological invariants of immersed discs

As in Section 2, we denote by S a compact oriented surface with connected boundary $\partial S = S^1$. Let (W, J) be a smooth simply connected 4-dimensional manifold equipped with a smooth almost complex structure J. In this section we investigate topological invariants of boundary holomorphic boundary regular *immersions* $f\colon S \to (W, J)$. For this we use the assumptions and notations from Section 2. Recall in particular the definition of the self-intersection number $\mathrm{Int}(f)$ of f.

Let $G(2, 4)$ be the Grassmannian of oriented (real) 2-planes in $\mathbb{R}^4 = \mathbb{C}^2$. This Grassmannian is just the homogeneous space $G(2, 4) = SO(4)/SO(2) \times SO(2) = S^2 \times S^2$, in particular the second homotopy group of $G(2, 4)$ coincides with its second homology group and is isomorphic to $\mathbb{Z} \oplus \mathbb{Z}$. Moreover, there are generators $\tilde{e}_1, \tilde{e}_2$ of $H_2(G(2, 4); \mathbb{Z})$ such that with respect to these generators, the homological intersection form ι is the symmetric form represented by the $(2, 2)$-matrix $(a_{i,j})$ with $a_{1,1} = a_{2,2} = 0$ and $a_{1,2} = a_{2,1} = \iota(\tilde{e}_1, \tilde{e}_2) = 1$.

The complex projective line $\mathbb{C}P^1 = S^2$ of *complex* oriented lines in $\mathbb{C}^2$ for the standard complex structure J is naturally embedded in $G(2, 4)$. Its homotopy class is the generator of an infinite cyclic subgroup Z_1 of $\pi_2(G(2, 4))$. We call this generator the *canonical generator* of Z_1 and denote it by e_1. The anti-holomorphic sphere of all complex oriented lines for the complex structure $-J$ is homotopic and hence homologous in $G(2, 4)$ to the complex projective line $\mathbb{C}P^1$. Namely, the complex structures $J, -J$ define the same orientation on $\mathbb{R}^4$ and hence J can be connected to $-J$ by a continuous curve of linear complex structures on $\mathbb{R}^4$. This curve then determines a homotopy of $\mathbb{C}P^1$ onto the anti-holomorphic sphere of complex oriented lines for $-J$. Since these two spheres are disjoint, the self-intersection number of the class in $H_2(G(2, 4); \mathbb{Z})$ defined by e_1 vanishes.

A second infinite cyclic subgroup Z_2 of $\pi_2(G(2, 4))$ is defined as follows. Let $S^2 \subset \mathbb{R}^3 \subset \mathbb{R}^4$ be the standard unit sphere. The map which associates to a point $y \in S^2$ the oriented tangent plane of S^2 at y, viewed as a 2-dimensional oriented linear subspace of $\mathbb{R}^4$, defines a smooth map of S^2 into $G(2, 4)$. Its homotopy class e_2 generates a subgroup Z_2 of $\pi_2(G(2, 4))$. We call e_2 the canonical generator of Z_2. Now the tangent bundle of $S^2 \subset \mathbb{R}^3 \subset \mathbb{C}^2$ intersects the complex projective line $\mathbb{C}P^1 \subset G(2, 4)$ in precisely one point (which is the tangent space of S^2 at $(0, 0, 1, 0)$). This intersection is transverse with positive intersection index. Therefore we have $\iota(e_1, e_2) = 1$ and hence the elements e_1, e_2 generate $\pi_2(G(2, 4))$.

Let W be a simply connected 4-dimensional manifold with smooth almost complex structure J. Equip the tangent bundle TW of W with a J-invariant Riemannian metric $\langle , \rangle$. Let $\mathcal{G} \to W$ be the smooth fibre bundle over W whose fibre at a point $x \in W$ consists of the Grassmannian of oriented 2-planes in T_xW. Let S be a compact oriented surface with connected boundary $\partial S \sim S^1$. Then every smooth boundary regular *immersion* $f\colon S \to W$ defines a smooth map Gf of S into the bundle $\mathcal{G}$ by assigning to a point $x \in S$ the oriented tangent space $df(T_xS)$ of $f(S)$ at $f(x)$. The complex pull-back bundle (f^*TW, J) over S admits a *complex* trivialization. This trivialization can be chosen to be of the form $df(X), V$ where X is a global nowhere vanishing section of the tangent bundle TS of S and V is a global section of the $\langle , \rangle$-orthogonal complement of the complex line subbundle of f^*TW which is spanned by the section $df(X)$. With respect to this complex trivialization of f^*TW, the pull-back $f^*\mathcal{G}$ of the bundle $\mathcal{G}$ can naturally be represented as a product $S \times G(2, 4)$. If f is boundary holomorphic, i.e. if for every $z \in \partial S$

the tangent space $df(T_zS) \subset T_{f(z)}W$ is J-invariant and if moreover its orientation coincides with the orientation induced by J, then in the above identification of $f^*\mathcal{G}$ with $S \times G(2,4)$ the circle of tangent planes of $f(S)$ over $f(\partial S)$ is given by the curve $\partial S \times L_0$ in $\partial S \times G(2,4)$ where $L_0 = \mathbb{C} \times \{0\} \subset \mathbb{C}^2$ is a fixed complex line. Thus in this case the map Gf can be viewed as a smooth map of the surface S into the Grassmannian $G(2,4)$ which maps the boundary ∂S of S to the single complex line L_0. In other words, if we let $\tilde{S}$ be the closed oriented surface obtained from S by collapsing ∂S to a point then Gf defines a map of $\tilde{S}$ into $G(2,4)$. This map then defines a homotopy class of maps $\tilde{S} \to G(2,4)$ and a homology class $[Gf] \in H_2(G(2,4),\mathbb{Z})$.

The following definition strengthens Definition 2.2.

Definition 3.1. A smooth homotopy $h\colon [0,1] \times S \to W$ is *regular* if h is boundary regular and if moreover for every $s \in [0,1]$ the map h_s is a boundary holomorphic immersion.

If $f, g\colon S \to W$ are two boundary regular boundary holomorphic immersions which are regularly homotopic, i.e. which can be connected by a regular homotopy, then the maps Gf and Gg are homotopic. Namely, if $h\colon [0,1] \times S \to W$ is a regular homotopy connecting $h_0 = f$ to $h_1 = g$, then there is a complex trivialization of the complex pull-back bundle (h^*TW, J) over $[0,1] \times S$ whose restriction to $[0,1] \times \partial S$ does not depend on $s \in [0,1]$ and is determined as before by the section $dh_s(X)$ where X is a global nowhere vanishing section of TS. This trivialization then defines an identification of the bundle $h^*\mathcal{G}$ with $[0,1] \times S \times G(2,4)$. For each $s \in [0,1]$ the tangent planes of the immersion h_s define a smooth section of the bundle $h^*\mathcal{G}$ over $\{s\} \times S$ and hence a smooth map of S into $G(2,4)$. This map depends smoothly on s and maps the boundary ∂S of S to a single point. Thus by continuity, the homotopy class of the tangent map of h_s is independent of $s \in [0,1]$ and hence it is an invariant of regular homotopy.

There is another way to obtain an invariant of regular homotopy.

Definition 3.2. The *tangential index* $\tan(f)$ of a boundary regular immersion f whose only self-intersection points are transverse double points is defined to be the number of self-intersection points of f counted with signs.

If a boundary regular immersion $f\colon S \to W$ has self-intersection points which are not transverse double points then it can be perturbed with a regular homotopy to an immersion whose only self-intersection points are transverse double points and whose tangential index is independent of the perturbation (see e.g. [**McD91**]). The tangential index is invariant under regular homotopy.

If we consider more specifically boundary regular boundary holomorphic immersions of *discs* then we can derive a more precise result. For this recall from Section 2 that for every boundary regular boundary holomorphic map $f\colon D \to W$ there is a preferred trivialization of the normal bundle of $f(D)$ over $f(\partial D)$. On the other hand, there is a trivialization N of the oriented normal bundle of $f(D)$ over $f(\partial D)$ which extends to a global trivialization of the oriented normal bundle of $f(D)$ in TW.

Definition 3.3. The *winding number* $\mathrm{wind}(f)$ of a boundary regular boundary holomorphic immersion $f\colon D \to W$ is the winding number of the preferred trivialization of the normal bundle of $f(D)$ over $f(\partial D)$ with respect to a trivialization which extends to a global trivialization of the normal bundle of $f(D)$.

For the formulation of the following version of the well known *adjunction formula* for immersed boundary regular boundary holomorphic discs, denote for a boundary regular boundary holomorphic immersion $f\colon D \to W$ by $\mathcal{C}_2(Gf)$ the component of $[Gf]$ in the subgroup Z_2 of $H_2(G(2,4),\mathbb{Z})$, viewed as an integer.

PROPOSITION 3.4. *For a boundary regular boundary holomorphic immersion* $f\colon D \to W$ *we have* $\mathrm{Int}(f) = \mathrm{wind}(f) + 2\tan(f)$, *and* $\mathrm{wind}(f) = 2\mathcal{C}_2(Gf)$.

PROOF. Let $f\colon D \to W$ be a boundary regular boundary holomorphic immersion. As in Section 2, let ρ be the preferred trivialization of the complex normal bundle of $f(D)$ over $f(\partial D)$ and use this trivialization to extend f to an immersion f_0 of the two-sphere S^2 into the almost complex manifold $(W_\rho, \tilde{J})$. Then $\mathrm{Int}(f)$ is the self-intersection number of $f_0(S^2)$ in W_ρ. Since f_0 is an immersion, this self-intersection number just equals $\chi(N) + 2\tan(f_0)$ where $\chi(N)$ is the Euler number of the normal bundle of $f_0(S^2)$ in W_ρ and where $\tan(f_0) = \tan(f)$ is the tangential index defined above (see e.g. [**McD91**, Lemma 4.2] or simply note that the formula is obvious if f_0 is an embedding and follows for immersions with only transverse double points by surgery at every double self-intersection point which increases the Euler class of the normal bundle by 2 if the double point has positive index and decreases it by 2 if the double point has negative index). By our definition of the winding number $\mathrm{wind}(f)$ of f, this is just the formula stated in the proposition.

To show that $\mathrm{wind}(f) = 2\mathcal{C}_2(Gf)$, note first that we have $\mathrm{wind}(f) = 0$ if $[Gf] \in Z_1$. Namely, using the above notations, recall that a preferred trivialization ρ of the normal bundle of the disc $f(D)$ over $f(\partial D) = \gamma$ is determined by the requirement that the evaluation of the first Chern class of the complex tangent bundle $(TW_\rho, \tilde{J})$ of W_ρ on the 2-sphere $f_0(S^2)$ equals two.

The tangent plane map of f can be viewed as a map $(D, \partial D) \to G(2,4)$ which maps the boundary ∂D of D to a single point and hence factors through a map $F\colon S^2 \to G(2,4)$. If $[Gf] \in Z_1$ then since $\pi_2(G(2,4)) = H_2(G(2,4),\mathbb{Z})$, the map F can be homotoped to a map $S^2 \to \mathbb{C}P^1$. By construction, this implies that up to homotopy, the complex vector bundle $(f_0^*TW, \tilde{J})$ decomposes as a direct sum $TS^2 \oplus N$ of two complex line bundles. The first Chern class of $(f_0^*TW, \tilde{J})$ is then the sum of the Chern classes of TS^2 and N. Therefore by our normalization, the first Chern class of the normal bundle $N = f_0^*TW/TS^2 \to S^2$ vanishes. As a consequence, the bundle $N \to S^2$ is trivial and hence the preferred trivialization of the normal bundle N over γ extends to a global trivialization of N over S^2. This shows that $\mathrm{wind}(f) = 0$ if $[Gf] \in Z_1$.

Arguing as in the proof of Lemma 2.3, if $g\colon D \to W$ is any boundary regular boundary holomorphic immersion with $\mathcal{C}_2(Gg) = ke_2$ for some $k \in \mathbb{Z}$ then there is a boundary regular boundary holomorphic immersion $f\colon D \to W$ with $f(\partial D) = g(\partial D)$, $[Gf] = [Gg] - ke_2 \in Z_1$ and such that $\mathrm{Int}(f) = \mathrm{Int}(g)$, $\tan(f) = \tan(g) + k$ and $\mathrm{wind}(f) = \mathrm{wind}(g) - 2k$. Namely, such an immersion f can be constructed as follows. Choose a point $z \in D$ such that there is a small ball $V \subset W$ about $g(z)$ which intersects $g(D)$ in an embedded disc B containing $g(z)$. Choose an embedded 2-sphere $\hat{S} \subset \partial V$ which intersects B transversely in precisely two points, one with positive and one with negative intersection index. The tangent bundle of the sphere is a generator e_2 of the subgroup Z_2 of $G(2,4)$. As in the proof of Lemma 2.3, attaching the sphere to $g(D)$ with surgery about the intersection point with positive intersection index results in a disc u which satisfies $\mathrm{Int}(u) = \mathrm{Int}(g)$, $[Gu] = [Gg] + e_2$ and $\tan(u) = \tan(g) - 1$. Similarly, attaching the sphere to $g(D)$

with surgery about the intersection point with negative intersection index results in a disc u' which satisfies $\mathrm{Int}(u') = \mathrm{Int}(g)$, $[Gu'] = [Gg] - e_2$ and $\tan(u') = \tan(g) + 1$. From this the proposition is immediate. □

A *complex point* of an immersed disc $f: D \to W$ is a point $z \in D$ such that the real two-dimensional subspace $df(T_zD)$ of TW is invariant under the almost complex structure J. The point is called *holomorphic* if the orientation of $df(T_zD)$ induced by the orientation of D coincides with the orientation induced by the almost complex structure J, and it is called *anti-holomorphic* otherwise.

COROLLARY 3.5. *Let $f: D \to W$ be a boundary regular boundary holomorphic immersion. If f does not have any anti-holomorphic points then* $\mathrm{Int}(f) = 2\tan(f)$.

PROOF. Let $f: D \to W$ be a boundary regular boundary holomorphic immersion without any anti-holomorphic point. Then the tangent map Gf of f does not intersect the anti-holomorphic sphere of complex lines in $\mathbb{C}^2$ equipped with the reverse of the orientation induced by the complex structure. Since the anti-holomorphic sphere is homologous in $G(2,4)$ to the complex projection line $\mathbb{C}P^1$ and has vanishing self-intersection (see the discussion at the beginning of this section), we have $[Gf] \in Z_1$ by consideration of intersection numbers. The corollary now is immediate from Proposition 3.4. □

As in the introduction, denote by ω_0 the standard symplectic form on $\mathbb{C}^2$. An immersion $f: D \to \mathbb{C}^2$ is called symplectic if for every $z \in D$ the restriction of $f^*\omega_0$ to the tangent plane T_zD does not vanish and defines the standard orientation of T_zD. As a consequence of Corollary 3.5 we obtain Theorem 1.1 and the corollary from the introduction.

COROLLARY 3.6. *Let γ either be a transverse knot on the standard three-sphere, the boundary of the standard unit ball $C \subset \mathbb{C}^2$, or a Reeb orbit on the boundary Σ of a domain in $\mathbb{C}^2$ which is star-shaped with respect to the origin, with compact closure C. If γ bounds a boundary regular immersed symplectic disc $f: (D, \partial D) \to (C, \gamma)$ then* $\mathrm{lk}(\gamma) = 2\tan(f) - 1$.

PROOF. By definition, symplectic immersions do not have any anti-holomorphic points. Thus if $f: (D, \partial D) \to (C, \gamma)$ is a boundary regular boundary holomorphic immersed symplectic disc then $\mathrm{lk}(\gamma) = 2\tan(f) - 1$ by Proposition 2.9 and Corollary 3.5.

Now if $f: (D, \partial D) \to (C, \gamma)$ is an arbitrary boundary regular immersed symplectic disc then f can be slightly modified with a smooth homotopy to a boundary transverse symplectic disc without changing the tangential index since being symplectic is an open condition. Locally near the boundary, this disc can be represented as a graph over an embedded symplectic annulus $A \subset C$ with γ as one of its boundary components whose tangent plane is J-invariant at every point in γ. Now for a fixed nonzero vector $X \in T\mathbb{C}^2$, the set of all nonzero vectors $Y \in T\mathbb{C}^2$ such that $\omega_0(X, Y) > 0$ is convex and hence contractible. Therefore locally near the boundary, this graph can be deformed to a graph which coincides with the annulus A near γ. The resulting map is a boundary regular and boundary holomorphic immersed symplectic disc with boundary γ whose tangential index coincides with the tangential index of f. □

4. Boundaries of compact convex bodies with controlled curvature

In this section we investigate periodic Reeb orbits on the boundary Σ of a compact strictly convex body $C \subset \mathbb{C}^2$. Our main goal is the proof of Theorem 1.3 from the introduction.

We begin with observing that Corollary 3.6 can be applied to periodic Reeb orbits on boundaries of compact convex bodies.

LEMMA 4.1. *Let γ be a periodic Reeb orbit on Σ. Then there is a boundary regular symplectic immersion $f\colon (D, \partial D) \to (C, \gamma)$.*

PROOF. Let γ be a periodic Reeb orbit on the boundary Σ of a compact strictly convex body $C \subset \mathbb{C}^2$. Choose two distinct points $a \neq b$ on γ and smooth parametrizations $\gamma_1, \gamma_2\colon [0, \pi] \to \gamma$ of the two subarcs of γ connecting a to b. We assume that the orientation of γ_2 coincides with the orientation of γ and that the parametrizations γ_1, γ_2 coincide near a, b with the parametrization of γ up to translation and reflection in the real line. Define a map $f\colon (D, \partial D) \to (C, \gamma)$ as follows. Let $\tilde{\gamma}_1, \tilde{\gamma}_2\colon [0, \pi] \to S^1$ be parametrizations by arc length of the two half-circles of the unit circle $S^1 \subset \mathbb{C}$ connecting 1 to -1, chosen in such a way that the orientation of $\tilde{\gamma}_2$ coincides with the orientation of ∂D. We require that f maps the line segment in D connecting $\tilde{\gamma}_1(t)$ to $\tilde{\gamma}_2(t)$ which is parametrized by arc length to the line segment in the convex body $C \subset \mathbb{C}^2$ connecting $\gamma_1(t)$ to $\gamma_2(t)$ and parametrized propotional to arc length on the same parameter interval. By construction, the map f is smooth, moreover it is symplectic near the points $1, -1$.

We claim that f is a symplectic immersion. For this let as before $\langle , \rangle$ be the usual euclidean inner product on $\mathbb{R}^4 = \mathbb{C}^2$. Let $t \in (0, \pi)$ and consider the straight line segment ℓ in C connecting $\gamma_1(t)$ to $\gamma_2(t)$. By strict convexity of C, the arc ℓ is contained in C and intersects Σ transversely at the endpoints. Let X, Y be the tangents of ℓ at the endpoints $\gamma_1(t), \gamma_2(t)$ and let as before N be the outer normal field of Σ. Then $\langle X, N(\gamma_1(t))\rangle < 0, \langle Y, N(\gamma_2(t))\rangle > 0$ and hence since $\gamma_1'(t) = -a_1 JN(\gamma(t)), \gamma_2'(t) = a_2 JN(\gamma_2(t))$ for some numbers $a_1 > 0, a_2 > 0$ we have $\omega_0(X, \gamma_1'(t)) > 0$ and $\omega_0(Y, \gamma_2'(t)) > 0$. Now with respect to the usual trivialization of $T\mathbb{C}^2$ we have $X = Y$. On the other hand, by the construction of the map f, for every point $s \in \ell$ the tangent space of $f(D)$ at s is spanned by $X = Y$ and a convex linear combination of $\gamma_1'(t), \gamma_2'(t)$. This shows that f is a symplectic immersion. Moreover f is clearly boundary regular whence the lemma. ☐

We call an immersion $f\colon D \to C$ as in Lemma 4.1 a *linear filling* of the Reeb orbit γ. By Corollary 3.6, if $\mathrm{lk}(\gamma) = -1$ then a linear filling f of γ satisfies $\tan(f) = 0$. However, an immersed symplectic disc may have transverse self-intersection points of negative intersection index, so there is no obvious relation between the tangential index of a boundary regular immersed symplectic disc and the number of its self-intersection points. On the other hand, if γ admits an *embedded* linear filling then Corollary 3.6 implies that $\mathrm{lk}(\gamma) = -1$.

Our final goal is to relate the Maslov index of a periodic Reeb orbit γ to the geometry of the hypersurface Σ. For this consider for the moment an arbitrary bounded domain $\Omega \subset \mathbb{C}^2$ with smooth boundary Σ which is star-shaped with respect to the origin. Write $C = \Omega \cup \Sigma$. As before, denote by J the usual complex structure on $\mathbb{C}^2$ and let $\langle , \rangle$ be the *euclidean* inner product. The restriction λ of the radial one-form λ_0 on $\mathbb{C}^2$ defined by $(\lambda_0)_p(Y) = \frac{1}{2}\langle Jp, Y\rangle$ $(p \in \mathbb{C}^2, Y \in T_p\mathbb{C}^2)$ defines a smooth contact structure on Σ.

Let N be the outer unit normal field of Σ. As in Section 2 write $\hat{M}(z_1, z_2) = (-\bar{z}_2, \bar{z}_1)$ and let M be the section of $T\Sigma$ defined by $M(p) = \hat{M} \circ N$. Its image is contained in the complex line subbundle L of the tangent bundle of Σ. The sections M, JM define a global trivialization of L which is symplectic with respect to the restriction of the symplectic form ω_0.

The kernel ξ of the contact form is a smooth real 2-dimensional subbundle of $T\Sigma$. Orthogonal projection P of $T\Sigma$ onto L defines a smooth bundle epimorphism whose kernel is the annihilator of the restriction of ω_0 to $T\Sigma$. Thus the morphism P preserves the restriction to $T\Sigma, L$ of the symplectic form and therefore its restriction to the subbundle ξ of $T\Sigma$ is a real symplectic bundle isomorphism. Its inverse $\pi\colon L \to \xi$ is a symplectic bundle morphism as well. Since by construction the sections M, JM of L form a symplectic basis of L we have

LEMMA 4.2. *The smooth sections $\pi \circ M, \pi \circ JM$ of the bundle ξ define a symplectic trivialization $T\colon \xi \to (\mathbb{R}^2, dx \wedge dy)$.*

In other words, for each $p \in \Sigma$ the restriction T_p of T to ξ_p is an area preserving linear map $T_p\colon (\xi_p, \omega_0) \to (\mathbb{R}^2, dx \wedge dy)$.

Recall from Section 2 that the Reeb vector field X on Σ is given by

$$X(p) = \phi(p) JN(p)$$

where

$$\phi(p) = \frac{2}{\langle p, N(p) \rangle} > 0.$$

Denote by $\Psi_t\colon \Sigma \to \Sigma$ the Reeb-flow of (Σ, λ) and let γ be a periodic orbit for Ψ_t of period $\chi > 0$. Using the above trivialization T of the bundle ξ we obtain a curve $\Phi\colon [0, \chi] \to SL(2, \mathbb{R})$ with $\Phi(0) = \mathrm{Id}$ by defining

$$\Phi(t)\colon \ = T_{\Psi_t(p)} \circ d\Psi_t(p) \circ T_p^{-1}.$$

where $p = \gamma(0)$. If the curve Φ is *non-degenerate*, which means that $\Phi(\chi)$ does not have one as an eigenvalue, then the Maslov index $\mu(\gamma)$ of γ is defined as the *μ-index* $\mu(\Phi)$ of the curve Φ as defined in [**HWZ95**] .

To estimate the μ-index of Φ define for a unit vector $X \in S^1 \subset \mathbb{R}^2$ the *rotation* of X with respect to the curve Φ as the total rotation angle $\mathrm{rot}(\Phi, X)$ (or the total winding) of the curve

$$t \to \frac{\Phi(t)X}{\|\Phi(t)X\|} \in S^1.$$

The following lemma is valid for *any* path in $SL(2, \mathbb{R})$ beginning at the identity. It uses an extension of the Maslov index to degenerate paths which is given in the proof of the lemma.

LEMMA 4.3. *Let $c\colon [0, \chi] \to SL(2, \mathbb{R})$ be a continuous arc with $c(0) = \mathrm{Id}$. Then $\mathrm{rot}(c, X) < (\mu(c) + 1)\pi$ for every $X \in S^1$.*

PROOF. We follow [**RS93**]. Assume that $\mathbb{R}^2$ is equipped with the standard symplectic form.

In standard euclidean coordinates let $V = \mathbb{R} \times \{0\} \subset \mathbb{R}^2$. The *Maslov cycle* determined by V is just V, viewed as a point in the real projective line $\mathbb{R}P^1$ of all one-dimensional subspaces of $\mathbb{R}^2$. A *crossing* of a smooth curve $\Lambda\colon [a, b] \to \mathbb{R}P^1$ is a number $t \in [a, b]$ such that $\Lambda(t) = V$. Then locally near t, we can write $\Lambda(s) = \{x + A(s)x\}$ where $A(s)\colon V \to V^\perp$ is linear and vanishes for $s = t$ (and where $V^\perp$ is the euclidean orthogonal complement of V in $\mathbb{R}^2$). With respect to the standard basis of $\mathbb{R}^2 = V \oplus V^\perp$ we can view $s \to A(s)$ as a real valued function. With this interpretation, the crossing is *non-degenerate* if $A'(t) \neq 0$. The *sign* $\operatorname{sign}\Gamma(\Lambda, V, t)$ of the crossing point t then equals the sign of $A'(t)$ ([**RS93**, p. 830]). The *Maslov index* of the curve $\Lambda\colon [a, b] \to \mathbb{R}P^1$ with only non-degenerate crossings is then defined to be

$$\mu(\Lambda, V) = \frac{1}{2}\operatorname{sign}\Gamma(\Lambda, V, a) + \sum_{a<t<b} \operatorname{sign}\Gamma(\Lambda, V, t) + \frac{1}{2}\operatorname{sign}\Gamma(\Lambda, V, b)$$

(see [**RS93**, p. 831]).

A smooth path $\Lambda\colon [a, b] \to \mathbb{R}P^1$ with $\Lambda(a) = V$ and only non-degenerate crossings lifts to a smooth path $\tilde{\Lambda}\colon [a, b] \to S^1 \subset \mathbb{C} = \mathbb{R}^2$ beginning at $(1, 0)$. Crossings of Λ are precisely those points $t \in [a, b]$ where $\tilde{\Lambda}(t) = (\pm 1, 0)$, and the sign of the crossing is the sign of the derivative of $\tilde{\Lambda}$ with respect to the usual counterclockwise orientation of S^1. As a consequence, we have $\mu(\Lambda, V) = p$ for $p \in \mathbb{Z}$ precisely if $\operatorname{rot}(\tilde{\Lambda}) = p\pi$, and $\mu(\Lambda, V) = p + \frac{1}{2}$ precisely if $\operatorname{rot}(\tilde{\Lambda}) \in (p\pi, (p+1)\pi)$. Here we write $\operatorname{rot}(\tilde{\Lambda})$ to denote the total rotation of the path $\tilde{\Lambda}$ in S^1.

For a curve $c\colon [0, \chi] \to SL(2, \mathbb{R})$, the Maslov index $\mu(c, V)$ of c with respect to V is defined to be the Maslov index of the curve $\Lambda\colon [0, \chi] \to \mathbb{R}P^1, t \to \Lambda(t) = c(t)V$ [**RS93**] and hence $\operatorname{rot}(c, (1, 0)) < (\mu(c, V) + \frac{1}{2})\pi$.

The above definition of a Maslov index for paths in $SL(2, \mathbb{R})$ depends on the choice of the linear subspace V (though this dependence can be removed by observing that any path in $SL(2, \mathbb{R})$ determines a path of Lagrangian subspaces in $\mathbb{R}^4$, see [**RS93**]). To obtain an index for paths $c\colon [0, \chi] \to SL(2, \mathbb{R})$ beginning at $c(0) = \mathrm{Id}$ which does not depend on such a choice we proceed as follows.

A path $c\colon [0, \chi] \to SL(2, \mathbb{R})$ defines a path α in the space of orientation preserving homeomorphisms of S^1 by $\alpha(s)(X) = c(s)X/\|c(s)X\|$ ($s \in [0, \chi], X \in S^1$). For each s and each X we have $\alpha(s)(-X) = -\alpha(s)(X)$. This implies that $|\operatorname{rot}(c, X) - \operatorname{rot}(c, Y)| < \pi$ for any two points $X, Y \in S^1$. Now if there is some $p \in \mathbb{Z}$ such that $\operatorname{rot}(c, X) \in (2p\pi, 2(p+1)\pi)$ for all $X \in S^1$ then we define $\mu(c) = 2p+1$. By continuity, otherwise there is some $X \in S^1$ and some $p \in \mathbb{Z}$ with $\operatorname{rot}(c, X) = 2p\pi$. By the above discussion, the number p is unique and we define $\mu(c) = 2p$. With this definition of a Maslov index for paths in $SL(2, \mathbb{R})$ beginning at the identity, the statement of the lemma is obvious.

The fundamental group of $SL(2, \mathbb{R})$ with basepoint the identity is infinitely cyclic and generated by the loop $t \to e^{2\pi i t}$ (viewed as a loop in $U(1) \subset SL(2, \mathbb{R})$). In particular, there is a natural group isomorphism $\rho\colon \pi_1(SL(2, \mathbb{R})) \to \mathbb{Z}$. We claim that the Maslov index defined in the previous paragraph has the following properties.

(1) $\mu(c)$ only depends on the homotopy class of c with fixed endpoints.
(2) If $\alpha \in \pi_1(SL(2, \mathbb{R}))$ and $c\colon [0, 1] \to SL(2, \mathbb{R})$ is any arc then

$$\mu(\alpha \cdot c) = 2\rho(\alpha) + \mu(c).$$

(3) $\mu(c^{-1}) = -\mu(c)$.
(4) For a constant invertible symmetric matrix S with $\|S\| < 2\pi$ and for $c(t) = \exp tJS$ ($t \in [0, 1]$) we have

$$\mu(c) = \frac{1}{2}\operatorname{signature}(S).$$

To see that these properties indeed hold true, note that the first and third statements are immediate from the definition. By definition, the Maslov index of the standard rotation $\alpha\colon t \to e^{2\pi i t} \in U(1)$ equals $\mu(\alpha) = 2 = 2\rho(\alpha)$ and hence the second statement follows from the definition and the first since the arc $\alpha \cdot c$ is homotopic with fixed endpoints to the concatentation of c with (a representative of) α. To see the fourth statement, observe that since S is symmetric by assumption, we have $\|S\| < 2\pi$ if and only if the absolute values of the eigenvalues of S are smaller than 2π.

Now by [**HWZ95**, Theorem 3.2], the above properties uniquely determine the Maslov index of paths in $SL(2,\mathbb{R})$ beginning at the identity and ending at a matrix which does not have one as an eigenvalue as used by Hofer, Wysocki and Zehnder ([**HWZ95**, Section 3], see also [**S99**, Chapter 2] for alternative definitions). Together this completes the proof of the lemma. □

To calculate the total rotation angle of a vector under the curve $\Phi\colon [0,T] \to SL(2,\mathbb{R})$ defined above we use complex coordinates and view the vector fields N, M as $\mathbb{C}^2$-valued functions on Σ. For a curve $\gamma\colon [0,b] \to \Sigma$ we abbreviate $N(t) = N(\gamma(t))$ and $M(t) = M(\gamma(t))$. Define a $U(2)$-valued curve $O\colon [0,\chi] \to U(2)$ by the requirement that for each $t \in [0,\chi]$, $O(t)$ is given with respect to the standard basis of $\mathbb{C}^2$ by the matrix

$$O(t) := (N(t), M(t)), \qquad \text{i.e. } O(t)\begin{pmatrix} a \\ b \end{pmatrix} = aN(t) + bM(t) \quad \text{for } a, b \in \mathbb{C}.$$

The image of the complex line $\{0\} \times \mathbb{C} \subset \mathbb{C}^2$ under the map $O(t)$ is just the complex line $L(\gamma(t))$. Therefore for each t, $\pi \circ O(t)$ is an $\mathbb{R}$-linear isomorphism of $\{0\} \times \mathbb{C}$ onto $\xi(\gamma(t))$.

From now on we use coordinates in $\mathbb{C}^2$. Without loss of generality we can assume that $\xi_{\gamma(0)} = L_{\gamma(0)}$ and hence we have $\pi M(0) = M(0) = M$. If we define a unitary 2×2 matrix $U(t)$ as

$$U(t) := \left(N(t), \frac{\pi^{-1} d\Psi_t M}{\|\pi^{-1} d\Psi_t M\|} \right), \tag{4.1}$$

then the turning angle of $O(t)^{-1}\pi^{-1} d\Psi_t M$ about zero (i.e. the rotation of M) is just the argument of $\det(U(t))$.

Define a unit vector field $\tilde{M}$ along γ by

$$\tilde{M}(t) = \frac{\pi^{-1} d\Psi_t M}{\|\pi^{-1} d\Psi_t M\|}.$$

The following lemma is the main technical tool for a calculation of the Maslov index of γ. For its formulation, recall that the *second fundamental form* of the hypersurface Σ in $\mathbb{C}^2$ is the symmetric bilinear form $\Pi\colon T\Sigma \times T\Sigma \to \mathbb{R}$ which is defined as follows. Let X, Y be vector fields on $\Sigma \subset \mathbb{R}^4$; then

$$\Pi(X,Y) = -\langle dY(X), N \rangle = \langle dN(X), Y \rangle.$$

The *shape operator* of Σ is the section A of the bundle $T^*\Sigma \otimes T\Sigma$ defined by $\Pi(X,Y) = \langle AX, Y \rangle$.

LEMMA 4.4. *For* $t_0 \in [0,T]$ *we have*

$$\begin{aligned}\frac{\partial}{\partial t}\det(U(t))|_{t=t_0} &= \frac{\partial}{\partial t}O(t)^{-1}\tilde{M}(t)|_{t=t_0}\\ &= i\phi(\gamma(t_0))(\Pi(JN(t_0),JN(t_0))+\Pi(\tilde{M}(t_0),\tilde{M}(t_0)))\det(U(t))\end{aligned}$$

with $\phi(\gamma(t_0)) = \|X(\gamma(t_0))\| = \|\gamma'(t_0)\| = \frac{2}{\langle p, N(t_0)\rangle}$.

PROOF. In the sequel we always view the second fundamental form Π of Σ as a bilinear form on a subspace of $\mathbb{R}^4$. Let $\pi_2\colon \mathbb{C}^2 \to \{0\}\times\mathbb{C}$ be the orthogonal projection. Using the simple fact that

$$O^{-1}\circ\pi^{-1} = \pi_2\circ O^{-1}$$

we deduce

$$\begin{aligned}\frac{\partial}{\partial t}O(t)^{-1}\tilde{M}(t) &= \frac{\partial}{\partial t}\frac{O(t)^{-1}\pi^{-1}d\Psi_t M}{\|\pi^{-1}d\Psi_t M\|}|_{t=t_0}\\ &= \pi_2\left(\frac{\partial}{\partial t}O(t)^{-1}|_{t=t_0}\right)\frac{d\Psi_{t_0}M}{\|\pi^{-1}d\Psi_{t_0}M\|}\\ &+ \pi_2 O(t_0)^{-1}\left(\frac{\partial}{\partial t}\frac{1}{\|\pi^{-1}d\Psi_t M\|}\Big|_{t=t_0}\right)d\Psi_{t_0}M\\ &+ \pi_2 O(t_0)^{-1}\frac{1}{\|\pi^{-1}d\Psi_{t_0}M\|}\left(\frac{\partial}{\partial t}d\Psi_t M|_{t=t_0}\right).\end{aligned}\tag{4.2}$$

The first term in our equation can be rewritten as

$$\begin{aligned}&\pi_2\left(\frac{\partial}{\partial t}O(t)^{-1}|_{t=t_0}\right)\frac{d\Psi_{t_0}M}{\|\pi^{-1}d\Psi_{t_0}M\|}\\ &= -\pi_2 O(t_0)^{-1}\left(\frac{\partial}{\partial t}O(t)|_{t=t_0}O(t_0)^{-1}\right)\frac{d\Psi_{t_0}M}{\|\pi^{-1}d\Psi_{t_0}M\|}\\ &= *.\end{aligned}$$

By definition, for every t the vectors $\{N(t),\tilde{M}(t)\}$ form a unitary basis of $\mathbb{C}^2$. Since $\{N(t),M(t)\}$ is also such a unitary basis, there is a smooth function $\psi\colon [0,\chi]\to\mathbb{R}$ such that $\tilde{M}(t) = e^{i\psi(t)}M(t)$ for all t. We now use the second fundamental form Π to calculate the differential of the matrix valued curve $O(t) = (N(t),M(t))$. For this recall the definition of the shape operator $A\colon T\Sigma\to T\Sigma$ of Σ and of the orthogonal projection $P\colon T\Sigma\to L$. Since $M(t) = \hat{M}N(t)$ and $\gamma'(t) = \phi(t)JN(t)$ we have

$$\frac{\partial}{\partial t}O(t) = \phi(t)(AJN(t),\hat{M}AJN(t)).$$

Thus with respect to the complex basis $(N(t),M(t))$ of $\mathbb{C}^2$ we have

$$\frac{\partial}{\partial t}O(t) = \phi(t)\begin{pmatrix} i\Pi(JN(t),JN(t)) & -\overline{PA(JN(t))}\\ PA(JN(t)) & -i\Pi(JN(t),JN(t))\end{pmatrix}$$

where we view $PA(\phi JN(t))$ as a complex multiple of $M(t)$.

By the definition of the function ψ we can write

$$\frac{O(t_0)^{-1}d\Psi_{t_0}M}{\|\pi^{-1}d\Psi_{t_0}M\|} = \begin{pmatrix} ic(t_0)\\ e^{i\psi(t_0)}\end{pmatrix}\quad\text{for some } c(t_0)\in\mathbb{R}.$$

Since $\tilde{M}(t) = e^{i\psi(t)}M(t)$ and $O^{-1}(t)M(t) = \binom{0}{1}, O^{-1}N(t) = \binom{1}{0}$ we deduce that

$$* = -\,\pi_2\phi(t_0)\begin{pmatrix} -c(t_0)\Pi(JN(t_0), JN(t_0)) - e^{i\psi(t_0)}\overline{PA(JN(t_0))} \\ ic(t_0)PA(JN(t_0)) - ie^{i\psi(t_0)}\Pi(JN(t_0), JN(t_0)) \end{pmatrix}.$$

Now let exp be the exponential map of the hypersurface Σ with respect to the Riemannian metric induced by the euclidean metric. We use this exponential map to compute the third term in (4.2).

$$\begin{aligned}
\pi_2 O(t_0)^{-1}\frac{\frac{\partial}{\partial t}d\Psi_t M|_{t=t_0}}{\|\pi^{-1}d\Psi_{t_0}M\|} &= \frac{\pi_2 O(t_0)^{-1}}{\|\pi^{-1}d\Psi_{t_0}M\|}\frac{\partial}{\partial t}\frac{\partial}{\partial s}\Psi_t(\exp_{\gamma(0)}(sM))|_{t=t_0,s=0} \\
&= \frac{\pi_2 O(t_0)^{-1}}{\|\pi^{-1}d\Psi_{t_0}M\|}\frac{\partial}{\partial s}(\phi(\Psi_{t_0}(\exp_{x_0}(sM))JN(\Psi_{t_0}(\exp_{\gamma(0)}(sM)))) \\
&= \pi_2 O^{-1}(t_0)\left(\frac{1}{\|\pi^{-1}d\Psi_{t_0}M\|}(\frac{\partial}{\partial s}\phi(\Psi_{t_0}(sM))JN(t_0)|_{s=0}) + \phi JA(d\Psi_{t_0}M)\right) \\
&= \pi_2 O^{-1}(t_0)\phi(t_0)JA(\tilde{M}(t_0) + c(t_0)JN) \\
&= \phi(t_0)e^{i\psi(t_0)}\big(i\Pi(\tilde{M},\tilde{M}) - \Pi(\tilde{M}, J\tilde{M}) + ic\Pi(JN,\tilde{M}) - c\Pi(JN, J\tilde{M})\big).
\end{aligned}$$

The tangent vector of a curve c in $\mathbb{C}$ with constant norm always has the form $\dot{c} = irc$ for some $r \in \mathbb{R}$, so we can neglect the radial parts of the above equations. Summing up the three terms in (4.2) yields

$$\frac{\partial}{\partial t}\det(U(t))|_{t=t_0} = i\phi(\Pi(JN, JN) + \Pi(\tilde{M},\tilde{M}))\det(U(t_0)).$$

□

The following corollary is immediate from Lemma 4.4 and the definition of the rotation of a vector with respect to an arc in $SL(2,\mathbb{R})$.

COROLLARY 4.5. *Let γ be a closed Reeb-orbit on Σ with period T. Then*

$$\operatorname{rot}(\Phi, M(0)) = \int_0^T |\gamma'|(\Pi(JN, JN) + \Pi(\tilde{M},\tilde{M}))dt$$

where $\tilde{M}(t) = \frac{\pi^{-1}d\Psi_t M(0)}{\|\pi^{-1}d\Psi_t M(0)\|}$.

Now we specialize again to the case that C is a compact strictly convex body in $\mathbb{C}^2$ with smooth boundary Σ which contains the origin in its interior. Recall that the *total curvature* of a smooth curve $\gamma\colon [0,t] \to \mathbb{C}^2$ parametrized by arc length is defined by

$$\kappa(\gamma) = \int_0^T \|\gamma''(t)\|dt.$$

The next corollary is immediate from Lemma 4.3 and lemma 4.4.

COROLLARY 4.6. *Let Σ be the boundary of a compact strictly convex body $C \subset \mathbb{C}^2$. If the principal curvatures $a \geq b \geq c$ of Σ satisfy the pointwise pinching condition $a \leq b+c$ then the total curvature of a periodic Reeb orbit of Maslov index 3 is smaller than 4π.*

PROOF. Let Σ be the boundary of a compact strictly convex body $C \subset \mathbb{C}^2$ with principal curvatures $a \geq b \geq c$ satisfying the pinching condition $a \leq b + c$. Let $\gamma\colon [0,T] \to \Sigma$ be a periodic Reeb orbit of Maslov index 3. We assume that γ

is parametrized by arc length on $[0,T]$. By Lemma 4.3, the rotation of the vector $M(\gamma(0))$ under the derivative of the Reeb flow is smaller than 4π.

Denote by $N(t)$ the normal field of the sphere restricted to the curve γ. Then $\gamma'(t) = JN(t)$ and therefore

$$\kappa(\gamma) = \int_0^T \|\frac{\partial}{\partial t}N(t)\|dt \le \int_0^T a(\gamma(t))dt \le \int_0^T b(\gamma(t)) + c(\gamma(t))dt < 4\pi$$

by Corollary 4.5. □

We use Corollary 4.6 to complete the proof of Theorem 1.3 from the introduction.

PROPOSITION 4.7. *Let Σ be the boundary of a compact strictly convex domain $C \subset \mathbb{C}^2$. If the principal curvatures $a \ge b \ge c$ of Σ satisfy the pointwise pinching condition $a \le b + c$ then a periodic Reeb orbit on Σ of Maslov index 3 bounds an embedded symplectic disc $f\colon (D, \partial D) \to (C, \gamma)$. In particular, γ has self-linking number -1.*

PROOF. Define the *crookedness* of a smooth closed curve $\gamma\colon S^1 \to \mathbb{R}^4$ to be the minimum of the numbers $m(\gamma, v)$ where $m(\gamma, v)$ is the number of minima of the function $t \to \langle \gamma(t), v\rangle, v \in S^3$. By a result of Milnor [**Mil50**], the crookedness of a curve of total curvature smaller than 4π equals one. Thus by Corollary 4.6, if γ is a periodic Reeb orbit on Σ of Maslov index 3 then there is some $v \in S^3$ such that the restriction to γ of the function $\phi\colon x \to \langle x, v\rangle$ assumes precisely one maximum and one minimum. We may moreover assume that these are the only critical points of the restriction of ϕ to γ and that they are non-degenerate (see [**Mil50**]).

Let a be the unique minimum of ϕ on γ. Assume that γ is parametrized in such a way that $\gamma(0) = a$. Let $\gamma_2\colon [0,\sigma] \to \Sigma$ be the parametrized subarc of γ issuing from $\gamma_2(0) = a$ which connects a to the unique maximum b of ϕ on γ. Let $\gamma_1\colon [0,\sigma] \to \Sigma$ be the parametrization of the second subarc of γ connecting a to b such that $\phi(\gamma_1(t)) = \phi(\gamma_2(t))$ for all $t \in [0,\sigma]$; this is possible by construction and by our choice of ϕ. Then the symplectic disc obtained from this parametrization by linear filling as in the proof of Lemma 4.1 is embedded. By Corollary 3.6, this implies that the self-linking number of γ equals -1. □

Acknowledgement: The authors thank the referee of an earlier version of this paper for pointing out a gap in the proof of Theorem 2 and for suggesting the statement of Theorem 1.

References

[Eke90] I. Ekeland, *Convexity methods in Hamiltonian mechanics*, volume 19 of *Ergebnisse der Mathematik und ihrer Grenzgebiete (3) [Results in Mathematics and Related Areas (3)]*. Springer-Verlag, Berlin, 1990.

[Eli92] Y. Eliashberg, *Contact 3-manifolds twenty years since J. Martinet's work*. Ann. Inst. Fourier (Grenoble), **42** (1992), 165–192.

[Eli93] Y. Eliashberg, *Legendrian and transversal knots in tight contact 3-manifolds.* In *Topological methods in modern mathematics* (Stony Brook, NY, 1991), 171–193, Publish or Perish, Houston, TX, 1993.

[GS99] R.Gompf and A. Stipsicz, *4-Manifolds and Kirby calculus*, volume 20 of *Graduate Studies in Mathematics.* AMS, Providence, RI, 1999.

[H07] S. Hainz, *Eine Riemannsche Betrachtung des Reeb-Flusses.* Bonner Mathematische Schriften, No. 382, 2007.

[HWZ95] H. Hofer, K. Wysocki, and E. Zehnder, *Properties of pseudo-holomorphic curves in symplectisations II: Embedding controls and algebraic invariants*, Geom. Func. Anal. **5** (1995), 270–328.

[HWZ96] H. Hofer, K. Wysocki, and E. Zehnder, *Unknotted periodic orbits for Reeb flows on the three-sphere*. Topol. Methods Nonlinear Anal., **7** (1996), no. 2, 219–244.

[HWZ98] H. Hofer, K. Wysocki, and E. Zehnder, *The dynamics on three-dimensional strictly convex energy surfaces*. Ann. of Math. (2), **148** (1998), no. 1, 197–289.

[HWZ99] H. Hofer, K. Wysocki, and E. Zehnder, *A characterization of the tight 3-sphere. II*. Comm. Pure Appl. Math., **52** (1999), no. 9, 1139–1177.

[McD91] D. McDuff, *The local behaviour of holomorphic curves in almost complex 4-manifolds*. J. Differential Geom. **34** (1991), no. 4, 143–164.

[Mil50] J. W. Milnor, *On the total curvature of knots*. Ann. of Math. (2), **52** (1950), 248–257.

[Rab79] P. Rabinowitz, *Periodic solutions of Hamiltonian systems on a prescribed energy surface*. J. Differential Equations **33** (1979), 336–352.

[RS93] J. Robbin and D. Salamon, *The Maslov index for paths*, Topology **32** (1993), 827–844.

[S99] D. Salamon, *Lectures on Floer homology*, in *Symplectic Geometry and Topology*, Y. Eliashberg and L. Traynor, eds. IAS/Park City Mathematics Series, Amer. Math. Soc. 1999.

[W79] A. Weinstein, *On the hypothesis of Rabinowitz' periodic orbit theorems*. J. Differential Equations **33** (1979), 353–358.

Mathematisches Institut der Universität Bonn,, Beringstrasse 1, D-53115 Bonn

Mathematisches Institut der Universität Bonn,, Beringstrasse 1, D-53115 Bonn
E-mail address: ursula@math.uni-bonn.de

Proceedings of Symposia in Pure Mathematics
Volume **82**, 2011

Twisted Alexander polynomials and fibered 3–manifolds

Stefan Friedl and Stefano Vidussi

ABSTRACT. In a series of papers the authors proved that twisted Alexander polynomials detect fibered 3-manifolds, and they showed that this implies that a closed 3–manifold N is fibered if and only if $S^1 \times N$ is symplectic. In this note we summarize some of the key ideas of the proofs. We also give new evidence to the conjecture that if M is a symplectic 4–manifold with a free S^1–action, then the orbit space is fibered.

1. Introduction

1.1. Definitions and previous results. A *manifold pair* is a pair (N, ϕ) where $N \neq S^1 \times D^2, N \neq S^1 \times S^2$ is an orientable connected 3–manifold with toroidal or empty boundary and $\phi \in H^1(N;\mathbb{Z}) = \mathrm{Hom}(\pi_1(N), \mathbb{Z})$ is non–trivial. We say that a manifold pair (N, ϕ) *fibers over* S^1 if there exists a fibration $p : N \to S^1$ such that the induced map $p_* : \pi_1(N) \to \pi_1(S^1) = \mathbb{Z}$ coincides with ϕ. Given a manifold pair (N, ϕ) the *Thurston norm* of ϕ is defined as

$$||\phi||_T = \min\{\chi_-(\Sigma) \,|\, \Sigma \subset N \text{ properly embedded surface dual to } \phi\}.$$

Here, given a compact surface Σ with connected components $\Sigma_1 \cup \cdots \cup \Sigma_k$, we define $\chi_-(\Sigma) = \sum_{i=1}^k \max\{-\chi(\Sigma_i), 0\}$. Thurston [**Th86**] showed that this defines a seminorm on $H^1(N;\mathbb{Z})$ which can be extended to a seminorm on $H^1(N;\mathbb{R})$.

Given a manifold pair (N, ϕ) and a homomorphism $\alpha : \pi_1(N) \to G$ to a finite group G we can consider the corresponding twisted Alexander polynomial $\Delta^\alpha_{N,\phi} \in \mathbb{Z}[t^{\pm 1}]$. This invariant was initially introduced by Lin [**Lin01**], Wada [**Wa94**] and Kirk–Livingston [**KL99**]. We will recall the definition in Section 2 and we refer to [**FV10**] for a survey of the theory of twisted Alexander polynomials.

We say that $\Delta^\alpha_{N,\phi} \in \mathbb{Z}[t^{\pm 1}]$ is *monic* if its top coefficient equals ± 1. Note that $\Delta^\alpha_{N,\phi}$ is palindromic, in particular if its top coefficient equals ± 1, then its bottom coefficient also equals ± 1. Given a polynomial $p(t) \in \mathbb{Z}[t^{\pm 1}]$ with $p = \sum_{i=k}^l a_i t^i, a_k \neq 0, a_l \neq 0$ we define $\deg(p) = l - k$.

We have the following theorem, that gives a characterization of fibered 3-manifolds in terms of twisted Alexander polynomials.

2010 *Mathematics Subject Classification.* Primary 57M27.

Key words and phrases. Twisted Alexander polynomials, fibered manifolds, symplectic manifolds.

The second author was partially supported by NSF grant #0906281.

THEOREM 1.1. *Let (N, ϕ) be a manifold pair. Then (N, ϕ) is fibered if and only if for any epimorphism $\alpha : \pi_1(N) \to G$ onto a finite group the twisted Alexander polynomial $\Delta^{\alpha}_{N,\phi} \in \mathbb{Z}[t^{\pm 1}]$ is monic and*

$$deg(\Delta^{\alpha}_{N,\phi}) = |G| \, \|\phi\|_T + (1 + b_3(N))\, div\, \phi_\alpha,$$

where ϕ_α denotes the restriction of $\phi : \pi_1(N) \to \mathbb{Z}$ to $Ker(\alpha)$, and where we denote by $div\, \phi_\alpha \in \mathbb{N}$ the divisibility of ϕ_α, i.e.

$$div\, \phi_\alpha = \max\{n \in \mathbb{N} \mid \phi_\alpha = n\psi \text{ for some } \psi : Ker(\alpha) \to \mathbb{Z}\}.$$

The 'only if' direction has been shown at various levels of generality by Cha [**Ch03**], Kitano and Morifuji [**KM05**], Goda, Kitano and Morifuji [**GKM05**], Pajitnov [**Pa07**], Kitayama [**Kiy07**], [**FK06**] and [**FV10**, Theorem 6.2]. We will also outline the proof in Section 2. The 'if' direction is the main result of [**FV08d**].

The main goal of this paper is to provide a summary of the proof, and to show a few ways that the approach can be generalized to prove new results.

Revisiting the proof of Theorem 1.1 will also show that in fact the following refinement of Theorem 1.1 holds:

THEOREM 1.2. *Let (N, ϕ) be a manifold pair such that $\pi_1(N)$ is residually finite solvable (we refer to Section 3.4 for the definition). Then (N, ϕ) is fibered if and only if for any epimorphism $\alpha : \pi_1(N) \to G$ onto a finite solvable group the twisted Alexander polynomial $\Delta^{\alpha}_{N,\phi} \in \mathbb{Z}[t^{\pm 1}]$ is monic and if the following equality holds*

$$deg(\Delta^{\alpha}_{N,\phi}) = |G| \, \|\phi\|_T + (1 + b_3(N))\, div\, \phi_\alpha.$$

Note that 3-manifolds with residually finite solvable fundamental group are fairly frequent. For example, as we will see in Theorem 3.4, any 3-manifold has a finite cover such that its fundamental group is residually finite solvable. Also note that the fundamental group of a fibered 3-manifold is always residually finite solvable.

1.2. Fibered manifolds and symplectic 4–manifolds. The main application of the "if" direction of Theorem 1.1 is in the proof of the following Theorem.

THEOREM 1.3. *Let N be a closed 3–manifold. Then $S^1 \times N$ is symplectic if and only if N is fibered.*

PROOF. *(outline)* We first consider the 'if' direction. This direction was first proved by Thurston [**Th76**], and we present here a proof that is well-known to the experts. Let $p : N \to S^1$ be a fibration. We write $\psi = p^*(dt)$ where dt is the canonical non-degenerate closed 1–form on $S^1 = \mathbb{R}/\mathbb{Z}$. By [**Ca69**] we can find a metric on N such that ψ is harmonic. Denote by $*\psi$ the dual closed 2–form. We now consider the 1–form ds on S^1 as a 1–form on $S^1 \times N$ by the pull back operation, similarly we consider ψ and $*\psi$ as forms on $S^1 \times N$. With this convention we now define:

$$\omega = ds \wedge \psi + *\psi.$$

Clearly ω is a closed 2–form on $S^1 \times N$. We furthermore calculate that

$$\omega \wedge \omega = 2ds \wedge \psi \wedge *\psi.$$

But since ψ is non-zero everywhere, it follows that $\psi \wedge *\psi$ is a 3-form on N which is non-zero everywhere. Hence $\omega \wedge \omega$ is a 4-form on $S^1 \times N$ which is non-zero everywhere. This shows that ω is a symplectic form on $S^1 \times N$.

The 'only if' direction follows by showing that if $S^1 \times N$ is symplectic, then there exists a class $\phi \in H^1(N, \mathbb{Z})$ determined by the symplectic form that satisfies the constraints described in Theorem 1.1. This is achieved building on an idea of Kronheimer in [**Kr99**]. Suppose that N is a closed 3-manifold such that $S^1 \times N$ admits a symplectic form ω. Without loss of generality we can assume that ω represents an integral class. Taubes proved in [**Ta94, Ta95**] that the Seiberg–Witten invariants of a symplectic 4–manifold satisfy very stringent constraints, that can be viewed as akin to a condition of "monicness." This, together with the relation between Seiberg–Witten invariants of $S^1 \times N$ and the Alexander polynomial of N, due to Meng and Taubes, translates in the condition that $\Delta_{N,\phi}$ is monic, where $\phi \in H^1(N, \mathbb{Z})$ is the Künneth component of $[\omega] \in H^2(S^1 \times N; \mathbb{Z}) \cong H^1(N; \mathbb{Z}) \oplus H^2(N; \mathbb{Z})$. Next, thanks to a theorem of Donaldson ([**Do96**]), there exists a symplectic surface dual to (a sufficiently large multiple of) the symplectic form. Such surface satisfies the usual adjunction formula for symplectic surfaces. This formula, played against Kronheimer's adjunction inequality for manifolds of type $S^1 \times N$, gives a constraint on the top degree of the Alexander polynomial $\Delta_{N,\phi}$ in terms of the Thurston norm, more precisely

$$\deg(\Delta_{N,\phi}) = \|\phi\|_T + 2\mathrm{div}\,\phi.$$

The constraints above hold for all finite covers of N, as all finite covers of $S^1 \times N$ are symplectic as well. The connection between the twisted Alexander polynomials of N and the ordinary Alexander polynomials of the finite covers of N entails at this point that for any epimorphism $\alpha : \pi_1(N) \to G$ to a finite group the twisted Alexander polynomial $\Delta^{\alpha}_{N,\phi}$ is monic and

$$\deg(\Delta^{\alpha}_{N,\phi}) = |G|\,\|\phi\|_T + 2\mathrm{div}\,\phi_\alpha.$$

(We refer to [**FV08b**] for the details of the argument). The theorem follows at this point from Theorem 1.1. □

We would like to mention that an alternative proof of the "only if" direction of Theorem 1.3 in the case that $b_1(N) = 1$, and under a technical condition in the general case, follows from combining the work of Kutluhan–Taubes [**KT09**], Kronheimer–Mrowka [**KM10**] and Ni [**Ni08**]. This proof requires a more sophisticated study of the Seiberg–Witten theory of $S^1 \times N$ in the symplectic case.

1.3. Non–fibered manifolds and vanishing twisted Alexander polynomials. It is natural to ask whether the conditions in Theorem 1.1 can be weakened. In particular, in light of some partial results discussed below, we propose the following conjecture.

CONJECTURE 1.4. *Let (N, ϕ) be a manifold pair. If (N, ϕ) is not fibered, then there exists an epimorphism $\alpha : \pi_1(N) \to G$ onto a finite group G such that $\Delta^{\alpha}_{N,\phi} = 0 \in \mathbb{Z}[t^{\pm 1}]$.*

Besides the interest *per se* in sharpening the results of Theorem 1.1 there are other reasons to investigate Conjecture 1.4. First, the proof of Theorem 1.3 would be significantly simplified, bypassing the use of Kronheimer's refined adjunction inequality: Taubes' nonvanishing result for Seiberg–Witten invariants of symplectic manifolds would suffice to carry the argument. But more importantly, Conjecture 1.4 would imply a result akin to Theorem 1.3 for all symplectic 4–manifolds that carry a free circle action. For those manifolds, in fact, a refined adjunction inequality in the spirit of [**Kr99**] does not seem available, and Taubes' constraints translate to a mere monicness of the twisted Alexander polynomials of the orbit space. We state these observations in the following form, referring to [**FV08a**] for details and for the extent to which the converse holds.

THEOREM 1.5. *Let M be a 4–manifold which carries a free circle action with orbit space N. If Conjecture 1.4 holds for N, then M admits a symplectic structure only if N fibers over the circle.*

We also refer to [**Ba01**] and [**Bo09**] for related work on the problem of determining which 4–manifolds with a free circle action admit a symplectic structure. We refer also to the work by Silver and Williams [**SW09a, SW09b**] and by Pajitnov [**Pa10**] for several interesting connections of Conjecture 1.4 to other problems in 3–dimensional topology.

Conjecture 1.4 can be proven to hold for various classes of manifolds. In order to describe them in detail, we must introduce some definitions.

Let π be a group and $\Gamma \subset \pi$ a subgroup. We say Γ is *separable* if for any $g \in \pi \setminus \Gamma$ there exists an epimorphism $\alpha : \pi \to G$ onto a finite group G such that $\alpha(g) \notin \alpha(\Gamma)$. Put differently, we can tell that g is not in Γ by going to a finite quotient. We say π is *locally extended residually finite (LERF)* if any finitely generated subgroup of π is separable.

The following theorem proves Conjecture 1.4 in various special cases:

THEOREM 1.6. *Let (N, ϕ) be a manifold pair. Suppose that $\Delta^{\alpha}_{N,\phi} \neq 0 \in \mathbb{Z}[t^{\pm 1}]$ for any epimorphism $\alpha : \pi_1(N) \to G$ onto a finite group G. Furthermore suppose that one of the following holds:*

1. *$N = S^3 \setminus \nu K$ and K is a genus one knot,*
2. *$||\phi||_T = 0$,*
3. *N is a graph manifold,*
4. *$\pi_1(N)$ is LERF.*

Then (N, ϕ) fibers over S^1.

We refer to [**FV07a**, Theorem 1.3] and [**FV08c**, Theorem 1, Proposition 4.6, Corollary 5.6] for details and proofs. Note that it is conjectured (cf. [**Th82**]) that $\pi_1(N)$ is LERF for any hyperbolic 3–manifold N.

In Section 4 we will give new conditions under which Conjecture 1.4 holds.

Acknowledgments. The first author would like to express his gratitude to the organizers of the Georgia International Topology Conference 2009 for the opportunity to speak and for organizing a most enjoyable and interesting meeting.

2. Twisted invariants of 3–manifolds

We recall the definition of twisted homology and cohomology and their basic properties. Let X be a topological space and let $\rho : \pi_1(X) \to \mathrm{GL}(n, R)$ be a representation. Denote by $\widetilde{X}$ the universal cover of X. Letting $\pi = \pi_1(X)$, we use the representation ρ to regard R^n as a left $\mathbb{Z}[\pi]$–module. The chain complex $C_*(\widetilde{X})$ is also a left $\mathbb{Z}[\pi]$–module via deck transformations. Using the natural involution $g \mapsto g^{-1}$ on the group ring $\mathbb{Z}[\pi]$, we can view $C_*(\widetilde{X})$ as a right $\mathbb{Z}[\pi]$–module and form the twisted homology groups

$$H_*^\rho(X;R^n) = H_*(C_*(\widetilde{X}) \otimes_{\mathbb{Z}[\pi]} R^n).$$

For most of the paper we will be interested in a particular type of representation. Let $\phi \in H^1(X;\mathbb{Z})$ and let $\alpha : \pi_1(X) \to \mathrm{GL}(n,\mathbb{Z})$ be a representation. We can now define a left $\mathbb{Z}[\pi_1(X)]$–module structure on $\mathbb{Z}^n \otimes_{\mathbb{Z}} \mathbb{Z}[t^{\pm 1}] =: \mathbb{Z}^n[t^{\pm 1}]$ as follows:

$$g \cdot (v \otimes p) := (\alpha(g) \cdot v) \otimes (\phi(g) \cdot p) = (\alpha(g) \cdot v) \otimes (t^{\phi(g)}p),$$

where $g \in \pi_1(X), v \otimes p \in \mathbb{Z}^n \otimes_{\mathbb{Z}} \mathbb{Z}[t^{\pm 1}] = \mathbb{Z}^n[t^{\pm 1}]$. Put differently, we get a representation $\alpha \otimes \phi : \pi_1(X) \to \mathrm{GL}(n, \mathbb{Z}[t^{\pm 1}])$.

We call the resulting twisted module $H_1^{\alpha\otimes\phi}(X;\mathbb{Z}^n[t^{\pm 1}])$ the *twisted Alexander module of* (X,ϕ,α). When ϕ and α are understood, then we will just write $H_*(X;\mathbb{Z}^n[t^{\pm 1}])$. Now suppose X has finitely many cells in all dimensions. Using that $\mathbb{Z}[t^{\pm 1}]$ is a Noetherian UFD it follows that$H_i^{\alpha\otimes\phi}(X;\mathbb{Z}^n[t^{\pm 1}])$ is a finitely generated module over $\mathbb{Z}[t^{\pm 1}]$. We now denote by $\Delta^\alpha_{X,\phi,i} \in \mathbb{Z}[t^{\pm 1}]$ the order of $H_1^{\alpha\otimes\phi}(X;\mathbb{Z}^n[t^{\pm 1}])$ and refer to it as the *twisted Alexander polynomial of* (X,ϕ,α). We refer to [**Tu01**] or [**FV10**, Section 2] for the precise definitions. Note that the twisted Alexander polynomials are well–defined up to multiplication by an element of the form $\pm t^k, k \in \mathbb{Z}$.

We adopt the convention that we drop α from the notation if α is the trivial representation to $\mathrm{GL}(1,\mathbb{Z})$. If $\alpha : \pi_1(N) \to G$ is a homomorphism to a finite group G, then we get the regular representation $\pi_1(N) \to G \to \mathrm{Aut}(\mathbb{Z}[G])$ where the second map is given by left multiplication. We can identify $\mathrm{Aut}(\mathbb{Z}[G])$ with $\mathrm{GL}(|G|,\mathbb{Z})$ and we obtain the corresponding twisted Alexander polynomial $\Delta^\alpha_{N,\phi}$.

As an example we give an outline of the proof of the 'only if' direction in Theorem 1.1.

LEMMA 2.1. *Let (N,ϕ) be a fibered manifold pair. Then for any epimorphism $\alpha : \pi_1(N) \to G$ onto a finite group the twisted Alexander polynomial $\Delta^\alpha_{N,\phi} \in \mathbb{Z}[t^{\pm 1}]$ is monic and the following equality holds*

$$deg(\Delta^\alpha_{N,\phi}) = |G|\, \|\phi\|_T + (1 + b_3(N))\, div\, \phi_\alpha.$$

PROOF. First note that exists a short exact Mayer–Vietoris sequence

$$0 \to H_1(\Sigma;\mathbb{Z}[G]) \otimes \mathbb{Z}[t^{\pm 1}] \xrightarrow{t\iota_+ - \iota_-} H_1(N\backslash\Sigma;\mathbb{Z}[G]) \otimes \mathbb{Z}[t^{\pm 1}] \to H_1(N;\mathbb{Z}[G][t^{\pm 1}]) \to 0,$$

where Σ is a fiber of (N,ϕ). Note that in the case of a knot complement and untwisted coefficients this is just the usual exact sequence relating the homology of a Seifert surface to the Alexander module of a knot (cf. e.g. [**Lic97**, Theorem 6.5]). We write $r = \operatorname{rank} H_1(\Sigma;\mathbb{Z}[G])$, where the rank is taken as a $\mathbb{Z}$–module. Picking a basis (over $\mathbb{Z}$) for $H_1(\Sigma;\mathbb{Z}[G])$ and $H_1(N \setminus \Sigma;\mathbb{Z}[G]) = H_1(\Sigma \times [0,1];\mathbb{Z}[G])$ we can represent $\iota_\pm$ by $r \times r$-matrices $A_\pm$. It follows from the definition of the Alexander polynomial that

$$\Delta^\alpha_{N,\phi} = \det(tA_+ - A_-) = t^r \det(A_+) + \cdots + \det(-A_-).$$

Since $\iota_\pm$ are homotopy equivalences it follows that $\det(A_\pm) = \pm 1$. In particular $\Delta^\alpha_{N,\phi}$ is monic and of degree r. It remains to determine r. First note that we have

$$\sum_{i=0}^{2}(-1)^i \operatorname{rank} H_i(\Sigma;\mathbb{Z}[G]) = |G| \cdot \sum_{i=0}^{2}(-1)^i \operatorname{rank} H_i(\Sigma;\mathbb{Z}) = |G| \cdot (-\chi(\Sigma)) = |G|\, \|\phi\|_T.$$

Furthermore recall that Σ is closed if and only if N is closed. The formula for r now follows from a direct calculation of the rank of $H_0(\Sigma;\mathbb{Z}[G])$ and from duality in the case that Σ is closed. □

3. Summary of the proof of Theorem 1.1

Our goal is to give an outline of the proof of Theorem 1.1 given in [**FV08d**] without descending into the many technical details required in a rigorous write up. We are acutely aware of the fact that the proof in [**FV08d**] hides the forest behind a wall of trees.

3.1. Step A: First observations. Let (N,ϕ) be a manifold pair and $k \in \mathbb{N}$. Note that (N,ϕ) fibers if and only if $(N,k\phi)$ fibers and note that $||k\phi||_T = k||\phi||_T$. It follows now easily that it suffices to prove Theorem 1.1 for primitive $\phi \in H^1(N;\mathbb{Z})$.

Theorem A. *Let (N,ϕ) be a manifold pair with $\phi \in H^1(N;\mathbb{Z})$ primitive. Assume that $\Delta_{N,\phi} \neq 0$. Then the following hold:*

(1) *There exists a connected Thurston norm minimizing surface Σ dual to ϕ.*
(2) *Any connected surface Σ dual to ϕ intersects any boundary torus, in particular Σ is closed if and only if N is closed.*
(3) *If $\Delta^{\alpha}_{N,\phi} \neq 0$ for any epimorphism $\alpha : \pi_1(N) \to G$ onto a finite group, then N is irreducible.*
(4) *The pair (N,ϕ) fibers over S^1 if and only if the maps $\iota_\pm : \pi_1(\Sigma) \to \pi_1(N \setminus \nu\Sigma)$ are isomorphisms.*

Proof. If $\Delta_{N,\phi} \neq 0$, then it follows from [**McM02**, Section 4 and Proposition 6.1] that there exists a connected Thurston norm minimizing surface Σ dual to ϕ.

Now let Σ be any connected surface dual to ϕ. Suppose that there exists a boundary torus T of N which Σ does not intersect. Then T lifts to the infinite cyclic cover $\hat{N}$ of N determined by $\phi : \pi_1(N) \to \mathbb{Z}$, in particular $\hat{N}$ contains infinitely many tori in its boundary. A standard argument now shows that $b_1(\hat{N}) = \infty$, but it is well-known (cf. [**Tu01**]) that $b_1(\hat{N}) = \deg\Delta_{N,\phi}$.

Statement (3) follows from an argument of McCarthy [**McC01**] (see also [**Bo09**] and [**FV08d**, Lemma 7.1]). Note that the proof of (3) relies on the fact that 3-manifold groups are residually finite, which is a consequence of the proof of the Geometrization Conjecture (cf. [**Th82**] and [**He87**]). The final statement is a consequence of Stallings' fibering theorem ([**St62**] and [**He76**]). □

Throughout this section Σ will always denote a connected Thurston norm minimizing surface dual to ϕ. We write $M = N \setminus \nu\Sigma$ and denote the two canonical inclusion maps of Σ into ∂M by $\iota_\pm$. Since $\Sigma \subset N$ is Thurston norm minimizing it follows from Dehn's lemma that the inclusion induced maps $\pi_1(\Sigma) \to \pi_1(N)$ and $\pi_1(M) \to \pi_1(N)$ are injective. In particular we can view $\pi_1(\Sigma)$ and $\pi_1(M)$ as subgroups of $\pi_1(N)$.

Given an epimorphism $\alpha : \pi_1(N) \to G$ onto a finite group G we say $\Delta^{\alpha}_{N,\phi}$ has *Property (M)* if $\Delta^{\alpha}_{N,\phi} \in \mathbb{Z}[t^{\pm 1}]$ is monic and if

$$\deg(\Delta^{\alpha}_{N,\phi}) = |G|\,\|\phi\|_T + (1 + b_3(N))\mathrm{div}\phi_\alpha$$

holds.

3.2. Step B: Extracting information from twisted Alexander polynomials. In view of Theorem A our strategy is now to translate the information coming from twisted Alexander polynomials into information on the maps $\iota_\pm : \pi_1(\Sigma) \to \pi_1(M)$. We start with considering the untwisted polynomial:

LEMMA 3.1. *If $\Delta_{N,\phi}$ has Property (M), then the maps $\iota_\pm : H_1(\Sigma;\mathbb{Z}) \to H_1(M;\mathbb{Z})$ are isomorphisms.*

This lemma is well-known in the case of the untwisted Alexander polynomial for knots. An early reference is given by [**CT63**] but see also [**Ni07**, Proposition 3.1] or [**GS08**].

PROOF. We translate the information on twisted Alexander polynomials into information on the maps $\iota_\pm : \pi_1(\Sigma) \to \pi_1(M)$ by considering, as in Lemma 2.1, the following long exact Mayer–Vietoris sequence:
(3.1)

$$\begin{array}{rcl} & \dots & \to H_2(N;\mathbb{Z}[t^{\pm1}]) \to \\ \to H_1(\Sigma;\mathbb{Z}) \otimes \mathbb{Z}[t^{\pm1}] & \xrightarrow{t\iota_+ - \iota_-} H_1(M;\mathbb{Z}) \otimes \mathbb{Z}[t^{\pm1}] & \to H_1(N;\mathbb{Z}[t^{\pm1}]) \to \\ \to H_0(\Sigma;\mathbb{Z}) \otimes \mathbb{Z}[t^{\pm1}] & \xrightarrow{t\iota_+ - \iota_-} H_0(M;\mathbb{Z}) \otimes \mathbb{Z}[t^{\pm1}] & \to H_0(N;\mathbb{Z}[t^{\pm1}]) \to 0. \end{array}$$

Now suppose that $\Delta_{N,\phi} \in \mathbb{Z}[t^{\pm1}]$ is monic and that

$$\deg\Delta_{N,\phi} = \|\phi\|_T + (1 + b_3(N))$$

holds. Note that this in particular implies that $H_1(N;\mathbb{Z}[t^{\pm1}])$ is $\mathbb{Z}[t^{\pm1}]$-torsion, since $H_0(\Sigma;\mathbb{Z}) \otimes \mathbb{Z}[t^{\pm1}]$ is a free $\mathbb{Z}[t^{\pm1}]$-module it follows immediately that the map $H_1(N;\mathbb{Z}[t^{\pm1}]) \to H_0(\Sigma;\mathbb{Z}) \otimes \mathbb{Z}[t^{\pm1}]$ is the trivial map. Now recall that Σ is connected and that Σ is closed if and only if N is closed. In our context this implies that $\|\phi\|_T + (1+b_3(N))$ equals twice the genus g of Σ. Using elementary arguments one can show that $H_1(M;\mathbb{Z})$ is a free abelian group of the same rank as $H_1(\Sigma;\mathbb{Z})$, namely $2g$. Picking bases for $H_1(\Sigma;\mathbb{Z})$ and $H_1(M;\mathbb{Z})$ we denote the corresponding $2g \times 2g$–matrices for $\iota_\pm$ by $A_\pm$. It now follows from the definition of the Alexander polynomial that

$$\Delta_{N,\phi} = \det(tA_+ - A_-) = \det(A_+)t^{2g} + \dots + \det(-A_-).$$

Recall that we assumed that $\Delta_{N,\phi}$ is monic and that $\deg\Delta_{N,\phi} = 2g$. Since $\Delta_{N,\phi}$ is palindromic it now follows that A_- and A_+ are invertible matrices, in particular the maps $\iota_\pm : H_1(\Sigma;\mathbb{Z}) \to H_1(M;\mathbb{Z})$ are isomorphisms. □

Clearly the conclusion of the claim is not enough to deduce that $\pi_1(\Sigma) \to \pi_1(M)$ is an isomorphism. In fact there exist many non-fibered knots whose Alexander polynomial has Property (M). We will therefore use the information coming from all twisted Alexander polynomials.

Using the idea of the proof the previous claim we can show the following (we refer to [**FV08d**, Theorem 3.2] for details). If $\alpha : \pi_1(N) \to G$ is an epimorphism onto a finite group such that $\Delta^\alpha_{N,\phi}$ has Property (M), then the maps $\iota_\pm : H_1(\Sigma;\mathbb{Z}[G]) \to H_1(M;\mathbb{Z}[G])$ are isomorphisms. In fact, considering the 'H_0-part' of the Mayer–Vietoris sequence (3.1) we see that the assumption $\Delta^\alpha_{N,\phi} \neq 0$ implies that the maps $\iota_\pm : H_0(\Sigma;\mathbb{Z}[G]) \to H_0(M;\mathbb{Z}[G])$ are isomorphisms. Using well-known properties of 0-th homology groups (cf. e.g. [**HS97**, Section VI]) this condition is equivalent to

$$\mathrm{Im}\{\pi_1(\Sigma) \xrightarrow{\iota_\pm} \pi_1(M) \xrightarrow{\alpha} G\} = \mathrm{Im}\{\pi_1(M) \xrightarrow{\alpha} G\}.$$

For future reference we now summarize the results of the above discussion in the following theorem.

Theorem B. *Let $\alpha : \pi_1(N) \to G$ be an epimorphism onto a finite group such that $\Delta^\alpha_{N,\phi} \neq 0$, then*

$$(3.2) \qquad Im\{\pi_1(\Sigma) \xrightarrow{\iota_\pm} \pi_1(M) \xrightarrow{\alpha} G\} = Im\{\pi_1(M) \xrightarrow{\alpha} G\}.$$

If furthermore $\Delta^\alpha_{N,\phi}$ has Property (M), then

$$(3.3) \qquad \iota_\pm : H_1(\Sigma; \mathbb{Z}[G]) \to H_1(M; \mathbb{Z}[G])$$

are isomorphisms

Our goal now is to show that the information we just obtained from twisted Alexander polynomials is in fact enough to deduce that $\iota_\pm : \pi_1(\Sigma) \to \pi_1(M)$ are isomorphisms.

3.3. Step C: Finite solvable quotients. First recall that in the untwisted case we obtained the following conclusion: if $\Delta_{N,\phi}$ has Property (M), then the maps $\iota_\pm : H_1(\Sigma; \mathbb{Z}) \to H_1(M; \mathbb{Z})$ are isomorphisms. Another way of saying this is that the maps $\iota_\pm : \pi_1(\Sigma) \to \pi_1(M)$ 'look like an isomorphism on the abelian level'. Our goal is now to show that if all twisted Alexander polynomials corresponding to finite solvable groups have Property (M), then the maps $\iota_\pm : \pi_1(\Sigma) \to \pi_1(M)$ 'look like isomorphisms on the finite solvable level'. More precisely, we will prove the following theorem.

Theorem C. *Let (N, ϕ) be a manifold pair such that for any epimorphism $\pi_1(N) \to S$ onto a finite solvable group the polynomial $\Delta^\alpha_{N,\phi}$ has Property (M). Then for any finite solvable group S the induced maps*

$$\iota^*_\pm : Hom(\pi_1(M), S) \to Hom(\pi_1(\Sigma), S)$$

are bijections.

The outline of the proof of Theorem C will require the remainder of this section. We will now need to introduce a couple of definitions. Given a solvable group S we denote by $\ell(S)$ its derived length, i.e. the length of the shortest decomposition into abelian groups. Note that $\ell(S) = 0$ if and only if $S = \{e\}$.

Given $n \in \mathbb{N} \cup \{0\}$ we denote by $\mathcal{S}(n)$ the statement that for any finite solvable group S with $\ell(S) \leq n$ the maps

$$\iota^*_\pm : \mathrm{Hom}(\pi_1(M), S) \to \mathrm{Hom}(\pi_1(\Sigma), S)$$

are bijections. It is a straightforward exercise to see that $\iota_\pm : H_1(\Sigma; \mathbb{Z}) \to H_1(M; \mathbb{Z})$ are isomorphisms if and only if $\mathcal{S}(1)$ holds.

Note that Theorem C says that $\mathcal{S}(n)$ holds for all n if all twisted Alexander polynomials corresponding to finite solvable groups have Property (M). We will show that this does indeed hold by induction on n. For the induction argument we use the following auxiliary statement: Given $n \in \mathbb{N} \cup \{0\}$ we denote by $\mathcal{H}(n)$ the statement that for any epimorphism $\beta : \pi_1(M) \to T$ where T is finite solvable with $\ell(T) \le n$ the maps

$$\iota_\pm : H_1(\Sigma; \mathbb{Z}[T]) \to H_1(M; \mathbb{Z}[T])$$

are isomorphisms.

PROPOSITION 3.2. [**FV08d**, Proposition 3.3] *If $\mathcal{H}(n)$ and $\mathcal{S}(n)$ hold, then $\mathcal{S}(n+1)$ holds as well.*

PROOF. Let G be a group and $\alpha : G \to S$ an epimorphism onto a solvable group of derived length n. Then we obtain the following short exact sequence

$$0 \to H_1(G; \mathbb{Z}[S]) \to G/[\mathrm{Ker}(\alpha), \mathrm{Ker}(\alpha)] \xrightarrow{\alpha} S \to 1.$$

In particular if we can control solvable quotients of derived length at most n and the corresponding first homology groups, then we can control solvable information on G up to length $n+1$. The proposition now follows from elaborating this principle. We refer to [**FV08d**, Section 3.3] for the full details. □

PROPOSITION 3.3. [**FV08d**, Proposition 3.4] *Assume that $\Delta^\alpha_{N,\phi}$ has Property (M) for any epimorphism $\alpha : \pi_1(N) \to S$ onto a finite solvable group S with $\ell(S) \le n+1$. If $\mathcal{S}(n)$ holds, then $\mathcal{H}(n)$ holds as well.*

PROOF. In this proof we find it convenient to introduce the notation $\pi = \pi_1(N)$, $A = \pi_1(\Sigma)$ and $B = \pi_1(M)$. Recall that we can view A and B as subgroups of π. In fact we can view π as an HNN extension of B by A, more precisely we have a canonical isomorphism

$$\pi = \langle B, t \,|\, t\iota_-(g)t^{-1} = \iota_+(g), g \in A\rangle.$$

As usual we will normally just write $\pi = \langle B, t\,|, ti_-(A)t^{-1} = \iota_+(A)\rangle$. Let $\beta : B \to T$ be an epimorphism where T is finite solvable with $\ell(T) \le n$. Given a finitely generated group C we now define

$$C(T) = \bigcap_{\gamma \in Hom(C,T)} \mathrm{Ker}(\gamma).$$

It is straightforward to see that $C/C(T)$ is a finite solvable group with $\ell(C/C(T)) \le n$ (see [**FV08d**, Lemma 3.6]). It is a consequence of $\mathcal{S}(n)$ that the homomorphisms

$$\iota_\pm : A/A(T) \to B/B(T) \tag{3.4}$$

are in fact isomorphisms (see [**FV08d**, Lemma 3.6]). In particular we can define an epimorphism

$$\pi = \langle B, t\,|, t\iota_-(A)t^{-1} = \iota_+(A)\rangle \to \langle B/B(T), t \,|\, t\iota_-(A/A(T))t^{-1} = \iota_+(A/A(T))\rangle.$$

It is a consequence of (3.4) that the above group is in fact a semidirect product, i.e. we have an isomorphism

$$\langle B/B(T), t \,|\, t\iota_-(A/A(T))t^{-1} = \iota_+(A/A(T))\rangle \cong \mathbb{Z} \ltimes B/B(T),$$

where $1 \in \mathbb{Z}$ acts on $B/B(T)$ via $\iota_- \circ \iota_+^{-1}$. Since $B/B(T)$ is finite this automorphism has finite order, say k, therefore there exists an epimorphism

$$\alpha : \pi \to \mathbb{Z} \ltimes B/B(T) \to \mathbb{Z}/k \ltimes B/B(T) =: S.$$

Note that S is a finite solvable group of length $n+1$. It now follows from our assumption that the twisted Alexander polynomial $\Delta^\alpha_{N,\phi}$ has Property (M). The information coming from Theorem B is not quite what we wanted, since we replaced

$\beta : B \to T$ by $\alpha : B \to S$. But since $\text{Ker}(\alpha) \subset \text{Ker}(\beta)$, the latter homomorphism contains in fact the information coming from β, using a few technical arguments we can now deduce that

$$\iota_{\pm} : H_1(A; \mathbb{Z}[T]) \to H_1(B; \mathbb{Z}[T])$$

is an isomorphism as well. We refer to [**FV08d**, Section 3.4] for the full details. □

Theorem C is now an immediate consequence of Propositions 3.2 and 3.3 and of the fact, observed above, that $\Delta_{N,\phi}$ having Property (M) implies that $\mathcal{S}(1)$ holds.

3.4. Step D: Residually finite solvable fundamental groups. The conclusion of Theorem C loosely says that the maps $\iota_{\pm}$ 'look like isomorphisms on the finite solvable level' if all $\Delta^{\alpha}_{N,\phi}$ have Property (M). With the methods from the previous section Theorem C is the maximum information on the map $\iota_{\pm} : \pi_1(\Sigma) \to \pi_1(M)$ we can obtain from twisted Alexander polynomials.

In order to analyze the content of the conclusion of Theorem C we need the following definition. Let $\mathcal{P}$ be a property of groups (e.g. finite, finite solvable), then we say that a group π is *residually* $\mathcal{P}$ if for any non-trivial $g \in \mathcal{P}$ there exists a homomorphism $\alpha : \pi \to G$ to a group G with Property $\mathcal{P}$ such that $\alpha(g)$ is non-trivial. For example it is well-known that surface groups are residually finite solvable, and that 3-manifold groups are residually finite (cf. [**Th82**] and [**He87**]).

On the other hand 3-manifold groups are in general not residually finite solvable. For example if K is a non-trivial knot with Alexander polynomial equal to one, then standard arguments show that any homomorphism $\pi_1(S^3 \setminus \nu K) \to S$ to a solvable group S necessarily factors through the abelianization $\pi_1(S^3 \setminus \nu K) \to \mathbb{Z}$. In particular $\pi_1(S^3 \setminus \nu K)$ is not residually finite solvable. One can use such a knot to construct a manifold pair (N, ϕ) where $\iota_{\pm} : \text{Hom}(\pi_1(\Sigma), S) \to \text{Hom}(\pi_1(M), S)$ is a bijection for any solvable S, but such that $\pi_1(M)$ is not residually solvable. In particular M is not a product.

This discussion shows that the conclusion of Theorem C is not strong enough to ensure that $\iota_{\pm} : \pi_1(\Sigma) \to \pi_1(M)$ are isomorphisms, the problem being that 3-manifold groups are in general not residually finite solvable.

Before we continue we need to introduce a few more notions. We say a group has a property *virtually*, if there exists a finite index subgroup which has this property. Also recall, that given a prime p a *p–group* is a group whose order is a power of p. If G is a group which is residually a p–group, then we will normally just say *G is residually p*.

We can now formulate the following recent theorem of Matthias Aschenbrenner and the first author.

THEOREM 3.4. [**AF10**] *Let N be a 3-manifold. Then for almost all primes p the group $\pi_1(N)$ is virtually residually p.*

Recall that p-groups are finite solvable, in particular Theorem 3.4 says that 3-manifold groups are virtually residually finite solvable.

If N is hyperbolic then the theorem is a consequence of the fact that linear groups are virtually residually p (cf. e.g. [**We73**, Theorem 4.7]). The proof of that fact is so short and elegant that we think it is worthwhile mentioning.

PROOF OF THEOREM 3.4 FOR HYPERBOLIC N. We write $\pi = \pi_1(N)$. Since we assume that N is hyperbolic we can assume that π is a subgroup of $SL(2,\mathbb{C})$. Since π is finitely generated there exists a finitely generated subring R of $\mathbb{C}$ such that $\pi \subset \mathrm{GL}(n,R)$. It is well–known that for almost all primes p there exists a maximal ideal $\mathfrak{m}$ of R with $\mathrm{char}(R/\mathfrak{m}) = p$ (see [**LS03**, p. 376f]).

Now let p be a prime for which there exists a maximal ideal $\mathfrak{m}$ of R with $\mathrm{char}(R/\mathfrak{m}) = p$. We will show that π is virtually residually p. Before we continue note that $R/\mathfrak{m}^k$ is a finite ring for any $k \geq 1$ and that $\bigcap_{k=1}^\infty \mathfrak{m}^k = \{0\}$ by the Krull Intersection Theorem. For $k \geq 1$ we let

$$\pi_k = \mathrm{Ker}\left(\pi \to \mathrm{GL}(n,R) \to \mathrm{GL}(n,R/\mathfrak{m}^k)\right).$$

Each π_k is a normal subgroup of π, of finite index, and clearly $\pi_{k+1} \subset \pi_k$ for every $k \geq 1$. Moreover $\bigcap_{k=1}^\infty \pi_k = \{1\}$ since $\bigcap_{k=1}^\infty \mathfrak{m}^k = \{0\}$.

We claim that π_1 is residually p. We will prove this by showing that π_1/π_k is a p-group for any k. This in turn follows from showing that any non–trivial element in π_k/π_{k+1} has order p. In order to show this pick $A \in \pi_k$. By definition we can write

$$A = \mathrm{id} + C \qquad \text{for some } n \times n\text{-matrix } C \text{ with entries in } \mathfrak{m}^k.$$

From $p \in \mathfrak{m}$ and $k \geq 1$ we get that

$$\begin{aligned} A^p = (\mathrm{id}+C)^p &= \mathrm{id} + pC + \tfrac{p(p-1)}{2}C^2 + \cdots + C^p \\ &= \mathrm{id} + (\text{some } n \times n\text{-matrix with entries in } \mathfrak{m}^{k+1}). \end{aligned}$$

Hence $A^p \in \pi_{k+1}$. □

The combination of Theorem C and 3.4 shows that proving Theorem 1.1 becomes much easier, if we can go to finite covers. Fortunately the following lemma tells us that we can indeed do so:

LEMMA 3.5. *Let $p : N' \to N$ be a finite cover and let $\phi' = p^*(\phi)$. Then the following hold:*

1. *(N,ϕ) fibers if and only if (N',ϕ') fibers,*
2. *if $\Delta^\alpha_{N,\phi}$ has Property (M) for any epimorphism α from $\pi_1(N)$ onto a finite group, then $\Delta^\alpha_{N,\phi}$ has Property (M) for any epimorphism α from $\pi_1(N')$ onto a finite group.*

PROOF. The first statement can for example be proved using Stallings' fibering theorem [**St62**]: Indeed, (N,ϕ) fibers if and only if $\mathrm{Ker}(\phi)$ is finitely generated and (N',ϕ') fibers if and only if $\mathrm{Ker}(\phi')$ is finitely generated. But $\mathrm{Ker}(\phi')$ is subgroup of $\mathrm{Ker}(\phi)$ of finite index. In particular if one is finitely generated, then so is the other.

The second statement is fundamentally just an application of Shapiro's lemma, which says that the homology of a finite cover of a space N is nothing but the twisted homology of N. Making this principle work in this context is a little delicate though, and we refer the reader to [**FV08d**, Lemma 7.6] for the details. □

The following is now an immediate corollary to Theorem 3.4 and Lemma 3.5.

Theorem D. *Suppose the conclusion of Theorem 1.1 holds for all 3-manifolds such that $\pi_1(N)$ is residually finite solvable, then Theorem 1.1 holds for all 3-manifolds.*

Note that the original proof of Theorem 1.1 (that appeared before the first author and M. Aschenbrenner completed the proof of Theorem 3.4) required in [**FV08d**, Section 6] a rather convoluted argument based on the study of residual properties of each piece of the JSJ decomposition of N. The result of Theorem 3.4 therefore greatly simplifies the argument.

3.5. Step E. Reformulation in terms of sutured manifolds and Agol's theorem. In our final step we find it convenient to switch to the language of sutured manifolds. A *sutured manifold* is a triple (M, Σ_-, Σ_+) where M is an oriented 3-manifold, $\Sigma_\pm$ are (possibly disconnected) disjoint oriented subsurfaces of ∂M with the following properties:

(1) the orientation of Σ_+ agrees with the orientation of ∂M,
(2) the orientation of Σ_- is the opposite orientation of ∂M,
(3) the closure of $\partial M \setminus \Sigma_- \cup \Sigma_+$ consists of a union of annuli $A_1, \dots, A_n$ such that for any i the boundary of A_i consists of a boundary curve of Σ_- and of a boundary curve of Σ_+. Furthermore the boundary curves have to be oriented the same way.

A sutured manifold (M, Σ_-, Σ_+) is called *taut* if M is irreducible and if $\Sigma_\pm$ are Thurston norm minimizing in their homology class in $H_2(M, \partial\Sigma_\pm; \mathbb{Z})$. We refer to [**Ju06**], [**Ga83**, Definition 2.6] or [**CC03**, p. 364] for more on sutured manifolds.

The following are the two most important types of examples for us:

(1) If Σ is a oriented surface, then $(\Sigma \times [-1, 1], -\Sigma \times -1, \Sigma \times 1)$ is a taut sutured manifold. We will refer to it as a *product sutured manifold.*
(2) Let N be an irreducible 3–manifold with empty or toroidal boundary. Let Σ be a Thurston norm minimizing surface which intersects all boundary tori of N. Denote by M the result of cutting N along Σ and denote by $\Sigma_\pm$ the two copies of Σ in M. Then (M, Σ_-, Σ_+) is a taut sutured manifold.

Theorem E. *Let (M, Σ_-, Σ_+) be a taut sutured manifold. Suppose that $\pi_1(M)$ is residually finite solvable and suppose that for any finite solvable group S the induced maps*

$$\iota_\pm^* : Hom(\pi_1(M), S) \to Hom(\pi_1(\Sigma_\pm), S)$$

are bijections. Then M is a product on $\Sigma_\pm$.

Theorem 1.1 is an immediate consequence of Theorems A, C, D and E, and Theorem 1.2 is an immediate consequence of Theorems A, C and E.

Note that the statement of Theorem E can be generalized to a question about groups in general: Let $\mathcal{P}$ be a property of groups, let $\varphi : A \to B$ be a homomorphism of finitely presented groups which are residually $\mathcal{P}$ such that for any group G with Property $\mathcal{P}$ the map $\mathrm{Hom}(B, G) \to \mathrm{Hom}(A, G)$ is a bijection. Does this imply that φ is an isomorphism? For $\mathcal{P} = \{\text{finite}\}$ this question goes back to Grothendieck [**Gr70**] and was answered in the negative by Bridson and Grunewald [**BG04**]. We refer to [**AHKS07**] for more on the case $\mathcal{P} = \{\text{finite solvable}\}$.

This excursion into group theory shows that in order to prove Theorem E we can not rely on a miracle in group theory, but we need a miracle which comes from our 3-dimensional setting. This miracle is provided by a stunning theorem of Agol [**Ag08**]. To explain it we need one more definition.

A group π is called *residually finite* $\mathbb{Q}$*-solvable* or *RFRS* if there exists a filtration of groups $\pi = \pi_0 \supset \pi_1 \supset \pi_2 \dots$ such that the following hold:

(1) $\cap_i \pi_i = \{1\}$,
(2) π_i is a normal, finite index subgroup of π for any i,
(3) for any i the map $\pi_i \to \pi_i/\pi_{i+1}$ factors through $\pi_i \to H_1(\pi_i;\mathbb{Z})$,
(4) for any i the map $\pi_i \to \pi_i/\pi_{i+1}$ factors through $\pi_i \to H_1(\pi_i;\mathbb{Z})/\text{torsion}$.

Note that conditions (1), (2) and (3) are equivalent to saying that π is residually finite solvable. But condition (4) means that the RFRS condition is considerably more restrictive. The notion of an RFRS group was introduced by Agol [**Ag08**], we refer to Agol's paper for more information on RFRS groups. For our context it is important to note that free groups and surface groups are RFRS. Indeed, it is well-known that these groups are residually finite solvable, in particular there exists a sequence π_i with Properties (1), (2) and (3). But the extra condition (4) is now always satisfied since the first homology of any finite index subgroup of a free group or a surface group is always torsion free.

Given a sutured manifold $M = (M, \Sigma_-, \Sigma_+)$ the double D_M is defined to be the double of M along Σ_- and Σ_+. Note that the annuli $\partial M \setminus (\Sigma_- \cup \Sigma_+)$ give rise to toroidal boundary components of D_M.

The following theorem is implicit in the proof of [**Ag08**, Theorem 6.1].

THEOREM 3.6 (Agol). *Let $M = (M, \Sigma_-, \Sigma_+)$ be a connected, taut sutured manifold which is not a product sutured manifold. Suppose that $\pi_1(M)$ is RFRS. Then there exists an epimorphism $\alpha : \pi_1(M) \to S$ onto a finite solvable group, such that the corresponding cover $\tilde{M} = (\tilde{M}, \tilde{\Sigma}_-, \tilde{\Sigma}_+)$ of $M = (M, \Sigma_-, \Sigma_+)$ has the property that the class $[\tilde{\Sigma}_-] \in H_2(D_{\tilde{M}}, \partial D_{\tilde{M}};\mathbb{Z})$ lies on the closure of the cone over a fibered face of the Thurston norm ball of $D_{\tilde{M}}$.*

Before we delve into the details of the proof of Theorem E, let us take a step back and think about what Theorem 3.6 does for us. Agol's theorem has as input information on finite solvable quotients of $\pi_1(M)$ and as output it gives us a strong topological conclusion. This is exactly the type of statement we want to make in Theorem E. It is now just a matter of time till bending and twisting turns Theorem 3.6 into a proof of Theorem E.

PROOF OF THEOREM E. Let $M = (M, \Sigma_-, \Sigma_+)$ be a taut sutured manifold such that $\pi_1(M)$ is residually finite solvable and such that for any finite solvable group S the induced maps

$$\iota^* : \mathrm{Hom}(\pi_1(M), S) \to \mathrm{Hom}(\pi_1(\Sigma_\pm), S) \tag{3.5}$$

are bijections. We have to show that M is a product sutured manifold.

Let us now suppose that M is not a product sutured manifold. We write $\Sigma = \Sigma_-$. Recall that we pointed out above that the surface group $\pi_1(\Sigma)$ is RFRS. By assumption $\pi_1(M)$ is residually finite solvable and by (3.5) the finite solvable quotients of $\pi_1(M)$ and $\pi_1(\Sigma)$ 'look the same'. It is now fairly elementary to show that $\pi_1(M)$ is also RFRS (see [**FV08d**, Section 4.2] for details).

We can thus apply Theorem 3.6 to the taut sutured manifold (M, Σ_-, Σ_+). In fact applying arguments similar to the ones used in Lemma 3.5 we can without loss of generality assume that already the class $[\Sigma_-] \in H_2(D_M, \partial D_M;\mathbb{Z})$ lies on the closure of the cone over a fibered face F of the Thurston norm ball of D_M. Moreover, as by hypothesis M is not a product, we can assume that $[\Sigma_-]$ lies in the cone over the boundary of F, as otherwise (N, ϕ) would fiber already. We refer to [**FV08d**, Lemma 4.3] for details.

Note that D_M has an obvious involution r given by 'reflection', i.e. interchanging the two copies of M. Also recall that (3.5) implies in particular that the inclusion induced maps $H_1(\Sigma_\pm;\mathbb{Z}) \to H_1(M;\mathbb{Z})$ are isomorphisms. This means that homologically M looks like a product, and hence homologically D_M looks like $S^1 \times \Sigma$. More precisely, there exists a canonical isomorphism $\mathbb{Z}\cdot t \oplus H_1(\Sigma_-;\mathbb{Z}) \xrightarrow{\cong} H_1(D_M;\mathbb{Z})$, where t is an oriented curve with $r(t) = -t$ which intersects each of Σ_- and Σ_+ once. Note that the map $r: H_1(D_M;\mathbb{Z}) \to H_1(D_M;\mathbb{Z})$ restricts to the identity on $H_1(\Sigma;\mathbb{Z})$ and sends t to $-t$.

Applying duality we now obtain a dual isomorphism $H_2(D_M, \partial D_M;\mathbb{Z}) = \mathbb{Z}\cdot[\Sigma] \oplus V$ where r acts as $-\mathrm{id}$ on V. Recall that $r([\Sigma]) = [\Sigma]$ and that $[\Sigma]$ sits on the boundary of the face F. It follows that $[\Sigma]$ also sits on the boundary of the face $r(F)$. Clearly $r(F)$ is also a fibered face, and since r acts as $-\mathrm{id}$ on V we see that F and $r(F)$ are distinct faces. Also note that by the convexity of the Thurston norm ball F and $r(F)$ can not sit on the same plane. Schematically we now have the situation presented in Figure 1.

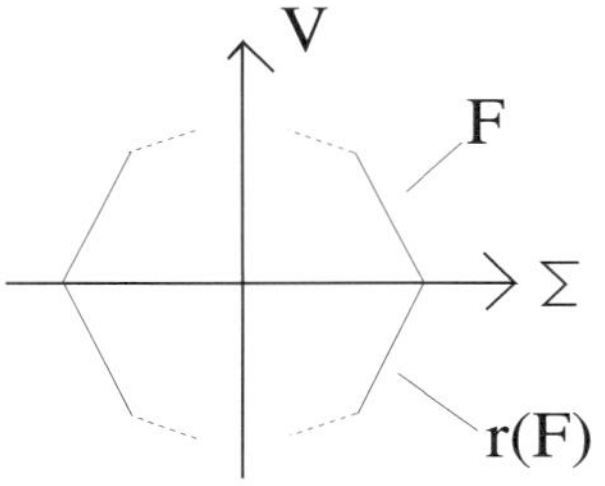

FIGURE 1. Thurston norm ball on $H_2(D_M, \partial D_M;\mathbb{R}) = \mathbb{R}\cdot\Sigma \oplus V$.

We now consider the information contained in (3.5) that comes from finite metabelian groups. We see that M looks like a product on the 'metabelian level,' and hence D_M looks like $S^1 \times \Sigma$ on the 'metabelian level.' Now recall that the multivariable Alexander polynomial of a 3-manifold is a metabelian invariant, we conclude that the multivariable Alexander polynomial of D_M equals the multivariable Alexander polynomial of $S^1 \times \Sigma$ which is well-known to be given by $(1-t)^{-\chi(\Sigma)}$, where $t \in H_1(S^1 \times M;\mathbb{Z}) = H_1(D_M;\mathbb{Z})$ is the same generator introduced above (we refer to [**FV08d**, Lemma 4.9] for details).

The norm ball dual to the Newton polygon of the multivariable Alexander polynomial is called the Alexander norm ball ([**McM02**]). Note that the Alexander norm ball of D_M is a convex subset of $\mathrm{Hom}(H_1(D_M;\mathbb{Z});\mathbb{R}) = H^2(D_M, \partial D_M;\mathbb{R})$. Since the Alexander polynomial is given by $(1-t)^{-\chi(\Sigma)}$ and since Σ is dual to t we see that the Alexander norm ball in our case is given by Figure 2. Note that for fibered classes the Thurston norm and the Alexander norm agree (see [**McM02**]). In particular the distinct fibered faces F and $r(F)$ of the Thurston norm ball have to lie on distinct faces of the Alexander norm ball. But the Alexander norm ball only has two faces, both preserved under reflection. This leads to a contradiction, which is schematically indicated by the mismatch of Figure 1 and Figure 2. This

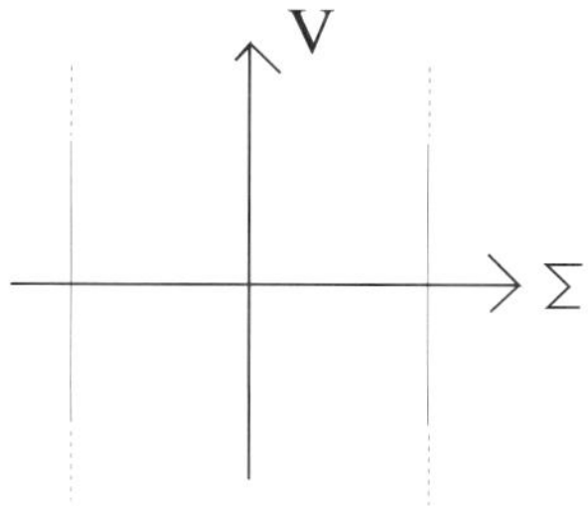

FIGURE 2. Alexander norm ball on $H_2(D_M, \partial D_M; \mathbb{R}) = \mathbb{R} \cdot \Sigma \oplus V$.

shows that the assumption that M is not a product leads to a contradiction. We refer to [**FV08d**, Section 4] for a formal and completely rigorous version of the above argument. □

4. Vanishing twisted Alexander polynomials for non–fibered manifolds

Throughout this section we use the notation from the previous section. In particular given a manifold pair (N, ϕ) where $\phi \in H^1(N, \mathbb{Z})$ is a primitive class with $\Delta_{N,\phi} \neq 0$, we will denote by Σ a connected Thurston norm minimizing surface dual to ϕ and we will write $M = N \setminus \nu\Sigma$. Furthermore we will denote the two natural inclusion maps of Σ into M by $\iota_\pm$. We recall the following theorem, whose proof we had outlined above:

Theorem B. *Let (N, ϕ) be a manifold pair and let $\alpha : \pi_1(N) \to G$ be an epimorphism onto a finite group such that $\Delta^\alpha_{N,\phi} \neq 0$, then*

$$Im\{\pi_1(\Sigma) \xrightarrow{\iota_\pm} \pi_1(M) \xrightarrow{\alpha} G\} = Im\{\pi_1(M) \xrightarrow{\alpha} G\}.$$

We will see in this section that Theorem B can be used in many situations to show that a non-fibered manifold pair has zero twisted Alexander polynomials.

THEOREM 4.1. [**FV08c**, Theorem 4.2] *Let (N, ϕ) be a non-fibered manifold pair such that ϕ is dual to a connected incompressible surface Σ. If $\pi_1(N)$ is LERF, then there exists an epimorphism $\alpha : \pi_1(N) \to G$ onto a finite group G such that $\Delta^\alpha_{N,\phi} = 0$.*

PROOF. We write $\Sigma = \Sigma_-$. By Theorem A we know that the monomorphism $\pi_1(\Sigma) \to \pi_1(M)$ is not an isomorphism, in particular the set $\pi_1(M) \setminus \pi_1(\Sigma)$ is nonempty. By the separability of $\pi_1(\Sigma) \subset \pi_1(N)$ we can now find for any $g \in \pi_1(M) \setminus \pi_1(\Sigma)$ an epimorphism $\alpha : \pi_1(N) \to G$ onto a finite group G such that $\alpha(g) \notin \alpha(\pi_1(\Sigma))$. In particular we have

$$\mathrm{Im}\{\pi_1(\Sigma) \xrightarrow{\iota} \pi_1(M) \xrightarrow{\alpha} G\} \subsetneq \mathrm{Im}\{\pi_1(M) \xrightarrow{\alpha} G\}.$$

The theorem now follows from Theorem B. □

It is an important open question whether fundamental groups of hyperbolic 3–manifolds are LERF (see [**Th82**, Question 15]), and various partial results of separability are known. In particular, Long and Niblo [**LN91**, Theorem 2] showed that the subgroup carried by an embedded torus is separable. It follows that the separability condition required in the proof of Theorem 4.1 is always satisfied if Σ is a torus. We now easily obtain the following.

THEOREM 4.2. [**FV08c**, Proposition 4.6] *Let N be a closed 3-manifold and $\phi \in H^1(N;\mathbb{Z})$ a non-trivial class with $||\phi||_T = 0$. Then (N,ϕ) is fibered if and only if for any epimorphism $\alpha : \pi_1(N) \to G$ onto a finite group G we have $\Delta^{\alpha}_{N,\phi} \neq 0$.*

Expanding on the ideas of Theorem 4.1 and 4.2 one can then continue to prove Theorem 1.6. We refer to [**FV08c**] for details.

Unfortunately not all 3-manifold groups are LERF (see [**NW01**]) and little is known even conjecturally about the separability properties of non-geometric 3-manifold groups. In the remainder of this section we will therefore give two examples of types of non-fibered manifold pairs where Theorem 4.1 cannot be applied, but where the weaker assumptions of Theorem B allow us to show that these pairs have twisted Alexander polynomials which are zero.

In order to prove our theorems we recall the following result of Long and Niblo [**LN91**]: Let Σ be an incompressible subsurface of the boundary of a 3–manifold M. Then $\pi_1(\Sigma) \subset \pi_1(M)$ is separable. This is often referred to as *peripheral subgroup separability.* We will exploit this result in two cases. The first is the case of the double of the complement of a nonfibered surface $\Sigma \subset N$. The second, perhaps of more conceptual breadth, is an application of 'virtual retractibility'.

We start with the first case. Let W be a 3–manifold with empty or toroidal boundary and let $\Sigma \subset W$ be an incompressible nonseparating connected properly embedded surface. Consider the manifold with boundary $M = W \setminus \nu\Sigma$. This manifold has two copies $\Sigma_\pm$ sitting in the boundary. Consider the double D_M of M along $\Sigma_- \cup \Sigma_+ \subset \partial M$. The images $\Sigma_\pm \subset D_M$ are nonseparating incompressible surfaces which are homologous in $H_2(D_M, \partial D_M;\mathbb{Z})$.

THEOREM 4.3. *Let D_M be defined as above, and let $\phi \in H^1(D_M,\mathbb{Z})$ be the primitive class Poincaré dual to $[\Sigma_\pm]$. If (D_M,ϕ) is a non-fibered pair, then there exists an epimorphism $\alpha : \pi_1(D_M) \to G$ onto a finite group G such that $\Delta^{\alpha}_{D_M,\phi} = 0$.*

PROOF. Suppose that (D_M,ϕ) is a non–fibered pair. First note that it is well-known that $\Sigma_+ \subset D_M$ is a fiber if and only if M is a product on Σ_+.

Note that we have a folding map $r : D_M \to M$ that is a retraction. In particular the induced map in homotopy $r_* : \pi_1(D_M) \to \pi_1(M)$ is an epimorphism, and has as right inverse the inclusion–induced map $i_* : \pi_1(M) \to \pi_1(D_M)$. Note that both r_* and i_* restrict to an isomorphism on the proper subgroups of the domain and image determined by a copy of $\pi_1(\Sigma_+)$. Consider now the proper subgroup $\pi_1(\Sigma_+) \subset \pi_1(M)$; by peripheral subgroup separability, there is an epimorphism $\beta : \pi_1(M) \to G$ to a finite group such that $\beta(\pi_1(\Sigma_+)) \subsetneq \beta(\pi_1(M))$. The surface $\Sigma_+ \subset D_M$ is an incompressible surface dual to ϕ; define $Z := D_M \setminus \nu\Sigma_+$. By the usual argument based on the incompressibility of Σ_+, $\pi_1(\Sigma_+)$ can be viewed as subgroup of $\pi_1(Z)$ and the latter is a subgroup of $\pi_1(D_M)$.

The inclusion–induced map $i_* : \pi_1(M) \to \pi_1(D_M)$ has image in $\pi_1(Z)$. It follows that if we let $\alpha = \beta \circ r_* : \pi_1(D_M) \to G$, then we have

$$\operatorname{Im}\{\pi_1(\Sigma_+) \xrightarrow{\iota} \pi_1(Z) \xrightarrow{\alpha} G\} \subsetneq \operatorname{Im}\{\pi_1(Z) \xrightarrow{\alpha} G\}.$$

It now follows from Theorem B that $\Delta^{\alpha}_{D_M,\phi} = 0$.

□

The second application of peripheral subgroup separability is in the context of *virtual retractions.* In [**LR08**] (see also [**LR05**]), Darren Long and Alan Reid define and explore the notion of *virtual retraction* of a group to one of its finitely generated subgroup, as well as various related properties. As the authors of [**LR08**] discuss, these notions are closely connected with subgroup separability properties of the group.

We start by giving the proper definitions, from [**LR08**], using a notation that adapts to the case we have in mind.

DEFINITION 4.4. Let π be a group and B a subgroup. Then a homomorphism $\theta : B \to G$ *extends over the finite index subgroup* $\hat{\pi} \subset \pi$ if $B \subset \hat{\pi}$ and if there exists a homomorphism $\Theta : \hat{\pi} \to G$ such that $\Theta|_B = \theta$.

DEFINITION 4.5. Let π be a group and B a subgroup. Then π *virtually retracts* onto B if the identity homomorphism $\theta = \mathrm{id}_B$ extends over some finite index subgroup of π.

Quite clearly, if π virtually retracts onto B, *any* homomorphism $\theta : B \to G$ extends over some finite index subgroup of π. A less trivial fact, observed in [**LR08**, Theorem 2.1], is that if π is LERF and if B is finitely generated, then any homomorphism θ onto a finite group extends over a finite index subgroup. In light of that the following theorem can be viewed as a generalization of Theorem 4.1.

THEOREM 4.6. *Let (N, ϕ) be a non-fibered manifold pair. Suppose that there exists a connected Thurston norm minimizing surface Σ dual to ϕ such that any homomorphism of $\pi_1(N \setminus \nu\Sigma)$ to a finite group extends to a finite index subgroup of $\pi_1(N)$. Then there exists an epimorphism $\alpha : \pi_1(N) \to G$ onto a finite group G such that $\Delta^{\alpha}_{N,\phi} = 0$.*

PROOF. We write $M := N \setminus \nu\Sigma$ and we write $\pi = \pi_1(N)$, $A = \pi_1(\Sigma)$ and $B = \pi_1(M)$. The manifold M has, as boundary, two copies of Σ; the incompressibility of Σ entails in particular the existence of inclusion–induced injective morphisms $\iota_\pm : A \hookrightarrow B \subset \pi$.

As (a copy of) Σ occurs as boundary component of M, the image under say i_+ of A in B (that we will denote by A as well) is separable by peripheral subgroup separability. This means that for any element $\gamma \in B \setminus A$ there exist an epimorphism $\theta : B \to H$ onto some finite group such that $\theta(\gamma) \notin \theta(A)$. Pick such an element γ. By assumption, θ extends to an epimorphism $\Theta : \hat{\pi} \to H$ where $\hat{\pi} \subset \pi$ is a finite index subgroup. The kernel $\ker \Theta \subset \hat{\pi}$ is a normal finite index subgroup of π. This subgroup may fail to be normal in π; define $\Gamma = \cap_{g \in \pi} g\,(\ker \Theta)\,g^{-1}$ to be its normal core in π, a normal finite index subgroup of both $\hat{\pi}$ and π. Denote by $G := \pi/\Gamma$, a finite group, and let $\alpha : \pi \to G$ be the quotient map. As $\hat{\pi}/\Gamma$ surjects on $\hat{\pi}/\ker \Theta$, it is not difficult to verify that the condition $\Theta(A) \subsetneq \Theta(B)$ entails that $\alpha(A) \subsetneq \alpha(B) \subset G$. The theorem is now an immediate consequence of Theorem B. □

Clearly, Theorem 4.6 applies when $\pi_1(N)$ virtually retracts to $\pi_1(M)$. Examples where this occurs are 3–manifolds whose fundamental group embeds into an all right hyperbolic Coxeter subgroup of $\mathrm{Isom}(\mathbb{H}^n)$ (see [**LR08**, Theorem 2.6]): these groups retract to any finitely generated geometrically finite subgroup, and the only finitely generated geometrically infinite subgroups are virtual fiber groups, hence excluded in the statement of Theorem 4.6.

References

[Ag08] I. Agol, *Criteria for virtual fibering*, J. Topol. **1** (2008), 269–284.

[AHKS07] G. Arzhantseva, P. de la Harpe, D. Kahrobaei and Z. Sunic, *The true prosoluble completion of a group: examples and open problems*, Geom. Dedicata 124 (2007), 5–26.

[AF10] M. Aschenbrenner and S. Friedl, *3–manifold groups are virtually residually p*, preprint, arXiv:1004.3619.

[Ba01] S. Baldridge, *Seiberg-Witten invariants of 4-manifolds with free circle actions*, Commun. Contemp. Math. **3** (2001), 341–353.

[Bo09] J. Bowden, *The topology of symplectic circle bundles*, Trans. Amer. Math. Soc., **361** (2009), no. 10, 5457-5468.

[BG04] M. Bridson and F. Grunewald, *Grothendieck's problems concerning profinite completions and representations of groups*, Ann. of Math. (2) 160 (2004), no. 1, 359–373.

[Ca69] E. Calabi, *An intrinsic characterization of harmonic one-forms*, Global Analysis, Papers in Honor of K. Kodaira, Univ. Tokyo Press, Tokyo, 1969, 101–117.

[CC03] A. Candel and L. Conlon, *Foliations. II*, Graduate Studies in Mathematics, **60**. American Mathematical Society, Providence, RI, 2003.

[Ch03] J. Cha, *Fibred knots and twisted Alexander invariants*, Trans. Amer. Math. Soc. **355** (2003), 4187–4200.

[CT63] R.H. Crowell and H.F. Trotter, *A class of pretzel knots*, Duke Math. J. **30** (1963), 373–377.

[Do96] S. Donaldson, *Symplectic submanifolds and almost-complex geometry*, J. Differential Geom., **44** (1996), 666–705.

[FK06] S. Friedl and T. Kim, *Thurston norm, fibered manifolds and twisted Alexander polynomials*, Topology, **45** (2006), 929–953.

[FV07a] S. Friedl and S. Vidussi, *Nontrivial Alexander polynomials of knots and links*, Bull. London Math. Soc., **39** (2007), 614–6222.

[FV08a] S. Friedl and S. Vidussi, *Symplectic 4-manifolds with a free circle action*, preprint, arXiv:0801.1513.

[FV08b] S. Friedl and S. Vidussi, *Twisted Alexander polynomials and symplectic structures*, Amer. J. Math. **130** (2008), no. 2, 455–484.

[FV08c] S. Friedl and S. Vidussi, *Symplectic $S^1 \times N^3$, surface subgroup separability, and vanishing Thurston norm*, J. Amer. Math. Soc. 21 (2008), 597–610.

[FV08d] S. Friedl and S. Vidussi, *Twisted Alexander polynomials detect fibered 3–manifolds*, preprint, arXiv:0805.1234.

[FV10] S. Friedl and S. Vidussi, *A survey of twisted Alexander polynomials*, The Mathematics of Knots: Theory and Application (Contributions in Mathematical and Computational Sciences), Springer, 2010, 45–94.

[Ga83] D. Gabai, *Foliations and the topology of 3–manifolds*, J. Differential Geom. **18** (1983), no. 3, 445–503.

[GKM05] H. Goda, T. Kitano and T. Morifuji, *Reidemeister Torsion, Twisted Alexander Polynomial and Fibred Knots*, Comment. Math. Helv. **80** (2005), no. 1, 51–61.

[GS08] H. Goda and T. Sakasai, *Homology cylinders in knot theory*, preprint, arXiv:0807.4034.

[Gr70] A. Grothendieck, *Représentations linéaires et compactification profinie des groupes discrets*, Manuscripta Math. **2** (1970), 375–396.

[He76] J. Hempel, 3-*Manifolds*, Ann. of Math. Studies, No. 86. Princeton University Press, Princeton, 1976.

[He87] J. Hempel, *Residual finiteness for* 3*-manifolds*, Combinatorial group theory and topology (Alta, Utah, 1984), 379–396, Ann. of Math. Stud., **111**, Princeton Univ. Press, Princeton, 1987.

[HS97] P. J. Hilton and U. Stammbach, *A Course in Homological Algebra*, second ed., Grad. Texts in Math. **4**, Springer, New York, 1997.

[Ju06] A. Juhasz, *Holomorphic disks and sutured manifolds*, Algebr. Geom. Topol. **6** (2006), 1429–1457.

[KL99] P. Kirk and C. Livingston, *Twisted Alexander invariants, Reidemeister torsion and Casson–Gordon invariants*, Topology **38** (1999), no. 3, 635–661.

[KM05] T. Kitano and T. Morifuji, *Divisibility of twisted Alexander polynomials and fibered knots*, Ann. Sc. Norm. Super. Pisa Cl. Sci. (5) **4** (2005), no. 1, 179–186.

[Kiy07] T. Kitayama, *Normalization of twisted Alexander invariants*, preprint, arXiv:0705.2371.

[Kr99] P. Kronheimer, *Minimal genus in* $S^1 \times M^3$, Invent. Math. **135** (1999), no. 1, 45–61.

[KM10] P. Kronheimer and T. Mrowka, *Knots, sutures and excision*, J. Differential Geom. **84** (2010), 301–364.

[KT09] Ç. Kutluhan and C. Taubes, *Seiberg-Witten Floer homology and symplectic forms on* $S^1 \times M^3$, Geom. Topol. **13** (2009), 493-525.

[Lic97] W. B. R. Lickorish, *An Introduction to Knot Theory*, Grad. Texts in Math. **175**, Springer, New York, 1997.

[Lin01] X. S. Lin, *Representations of knot groups and twisted Alexander polynomials*, Acta Math. Sin. (Engl. Ser.) **17** (2001), no. 3, 361–380.

[Lo87] D. Long, *Immersions and embeddings of totally geodesic surfaces*, Bull. London Math. Soc. **19** (1987), no. 5, 481–484.

[LN91] D. Long and G. Niblo, *Subgroup separability and 3-manifold groups*, Math. Z. **207** (1991), no. 2, 209–215.

[LR05] D. Long and A. W. Reid,, *Surface subgroups and subgroup separability in 3-manifold topology*, 25th Coloquio Brasileiro de Matematica, IMPA, Rio de Janeiro, 2005.

[LR08] D. Long and A. W. Reid, *Subgroup separability and virtual retractions of groups*, Topology, **47** (2008), 137–159.

[LS03] A. Lubotzky and D. Segal, *Subgroup growth*, Progr. Math., **212**, Birkhäuser Verlag, Basel, 2003.

[McC01] J. McCarthy, *On the asphericity of a symplectic* $M^3 \times S^1$, Proc. Amer. Math. Soc. **129** (2001), 257–264.

[McM02] C. T. McMullen, *The Alexander polynomial of a 3–manifold and the Thurston norm on cohomology*, Ann. Sci. Ecole Norm. Sup. (4) **35** (2002), no. 2, 153–171.

[Ni07] Y. Ni, *Knot Floer homology detects fibred knots*, Invent. Math. **170** (2007), no. 3, 577-608.

[Ni08] Y. Ni, *Addendum to: "Knots, sutures and excision"*, preprint, arXiv:0808.1327.

[NW01] G. A. Niblo and D. T. Wise, *Subgroup separability, knot groups and graph manifolds*, Proc. Amer. Math. Soc. **129** (2001), no. 3, 685–693.

[Pa07] A. Pajitnov, *Novikov homology, twisted Alexander polynomials, and Thurston cones*, St. Petersburg Math. J. **18** (2007), no. 5, 809–835.

[Pa10] A. Pajitnov, *On the tunnel number and the Morse-Novikov number of knots*, Algebr. Geom. Topol **10** (2010), 627–635.

[SW09a] D. Silver and S. Williams, *Nonfibered knots and representation shifts*, Algebraic Topology –Old and New, Banach Center Publications **85** (2009), 101–107.

[SW09b] D. Silver and S. Williams, *Twisted Alexander Polynomials and Representation Shifts*, Bull. London Math. Soc. **41** (2009), 535–540

[SW09c] D. Silver and S. Williams, *Alexander-Lin twisted polynomials*, preprint, arXiv:0908.1965.

[SW09d] D. Silver and S. Williams, *Dynamics of twisted Alexander invariants*, Topology Appl., **156** (2009), 2795–2811.

[St62] J. Stallings, *On fibering certain 3–manifolds*, Topology of 3–manifolds and related topics (Proc. The Univ. of Georgia Institute, 1961) Prentice-Hall, Englewood Cliffs, N.J., 1962, 95–100.

[Ta94] C. H. Taubes, *The Seiberg-Witten invariants and symplectic forms*, Math. Res. Lett. **1** (1994), 809–822.

[Ta95] C. H. Taubes, *More constraints on symplectic forms from Seiberg-Witten invariants*, Math. Res. Lett. **2** (1995), 9–13.

[Th76] W. P. Thurston, *Some simple examples of symplectic manifolds*, Proc. Amer. Math. Soc. **55** (1976), no. 2, 467–468.

[Th82] W. P. Thurston, *Three dimensional manifolds, Kleinian groups and hyperbolic geometry*, Bull. Amer. Math. Soc. **6** (1982), 357–381.

[Th86] W. P. Thurston, *A norm for the homology of 3–manifolds*, Mem. Amer. Math. Soc. **339** (1986), 99–130.

[Tu01] V. Turaev, *Introduction to combinatorial torsions*, Birkhäuser, Basel, 2001.

[Wa94] M. Wada, *Twisted Alexander polynomial for finitely presentable groups*, Topology **33** (1994), no. 2, 241–256.

[We73] B. A. F. Wehrfritz, *Infinite linear groups*, Springer, Berlin, 1973.

Mathematisches Institut, Universität zu Köln, Köln, Germany
E-mail address: stfriedl@gmail.com

Department of Mathematics, University of California, Riverside, CA 92521, USA
E-mail address: svidussi@math.ucr.edu

Proceedings of Symposia in Pure Mathematics
Volume **82**, 2011

Displacing Lagrangian toric fibers via probes

Dusa McDuff

ABSTRACT. This note studies the geometric structure of monotone moment polytopes (the duals of smooth Fano polytopes) using probes. The latter are line segments that enter the polytope at an interior point of a facet and whose direction is integrally transverse to this facet. A point inside the polytope is displaceable by a probe if it lies less than half way along it. Using a construction due to Fukaya–Oh–Ohta–Ono, we show that every rational polytope has a central point that is not displaceable by probes. In the monotone case, this central point is its unique interior integral point, and we show that every other point is displaceable by probes if and only if the polytope satisfies the star Ewald condition. (This is a strong version of the Ewald conjecture concerning the integral symmetric points in the polytope.) Further, in dimensions up to and including three every monotone polytope is star Ewald. These results are closely related to the Fukaya–Oh–Ohta–Ono calculations of the Floer homology of the Lagrangian fibers of a toric symplectic manifold, and have applications to questions introduced by Entov–Polterovich about the displaceability of these fibers.

1. Introduction

Symplectic toric manifolds form a nice family of examples in which to test various ideas; they can be described by simple diagrams and many of their invariants can be explicitly calculated. In this paper we discuss some geometric problems that arise when studying their natural family of Lagrangian submanifolds. Before describing our results, we shall illustrate the main definitions by some examples.

A symplectic toric manifold is a $2n$-dimensional symplectic manifold with an action of the n-torus T^n that is Hamiltonian, i.e. the action of the ith component circle is induced by a function $H_i : M \to \mathbb{R}$. These functions H_i fit together to give the moment map

$$\Phi : M \to \mathbb{R}^n, \quad \Phi(x) = \big(H_1(x), \dots, H_n(x)\big),$$

whose image is called the *moment polytope*. The first examples are:

2010 *Mathematics Subject Classification.* 53D12, 52B20, 14M25.

Key words and phrases. symplectic toric manifold, Lagrangian fiber, reflexive polytope, Fano polytope, Ewald conjecture, Floer homology, monotone polytope.

The author was partially supported by NSF grants DMS 0604769 and DMS 0905191.

• S^2 with its standard area form (normalized to have area 2) and with the S^1 action that rotates about the north-south axis; the corresponding Hamiltonian is the height function, and its moment polytope is the 1-simplex $[-1, 1]$; see Fig. 1.1.

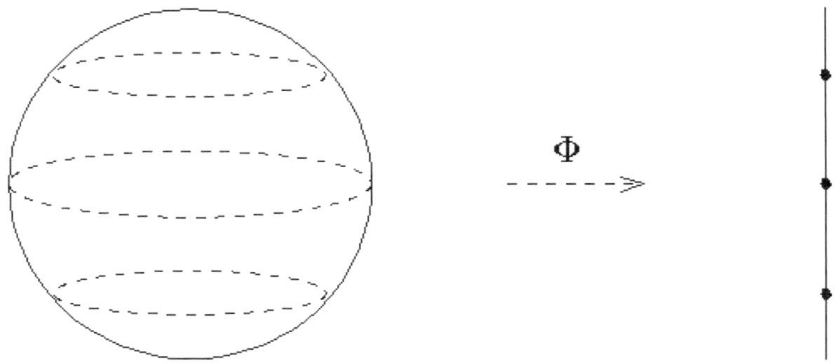

FIGURE 1.1. The S^1 action on S^2 with some of its orbits. Clearly each of them except for the equator can be displaced by an area preserving isotopy.

• $\mathbb{R}^{2n} = \mathbb{C}^n$ with the usual symplectic form, normalized as

$$\omega := \sum \frac{i}{2\pi} dz_j \wedge d\overline{z}_j = \frac{1}{\pi} \sum_{k=1}^{n} dx_{2k-1} \wedge dx_{2k},$$

and T^n action $(z_1, \dots, z_n) \mapsto (e^{2\pi i t_1} z_1, \dots, e^{2\pi i t_n} z_n)$ for $t_j \in S^1 \equiv \mathbb{R}/\mathbb{Z}$. Then the moment map is

$$\Phi_0 : (z_1, \dots, z_n) \mapsto \Big(|z_1|^2, \dots, |z_n|^2\Big) \in \mathbb{R}^n, \tag{1.1}$$

with image the first quadrant $\mathbb{R}^n_+$ in $\mathbb{R}^n$. Note that the inverse image under Φ of the line segment $(x, c_2, \dots, c_n), 0 \le x \le a$, where $c_j > 0$ for $j > 1$, is the product $D^2(a) \times T^{n-1}$, where $D^2(a)$ is a disc of ω-area a in the first factor $\mathbb{C}$ and the torus is a product of the circles $|z_j| = \sqrt{c_j}$ in the other copies of $\mathbb{C}$.

• the projective plane $\mathbb{C}P^2$ with the Fubini–Study form and T^2 action $[z_0; z_1; z_2] \mapsto [z_0; \lambda_1 z_2; \lambda_2 z_2]$. The moment map is

$$[z_0; z_1; z_2] \mapsto \Big(\frac{|z_1|^2}{\sum_i |z_i|^2}, \frac{|z_2|^2}{\sum_i |z_i|^2}\Big),$$

with image the standard triangle $\{x_1, x_2 \ge 0, x_1 + x_2 \le 1\}$ in $\mathbb{R}^2$.

More moment polytopes are illustrated in Figures 4.2 and 4.5 below. Good references are Audin [**Au**] and Karshon–Kessler–Pinsonnault [**KKP**]. Notice that the moment map simply quotients out by the T^n action.

One important fact here is that the moment polytope $\Phi(M)$ is a convex polytope Δ, satisfying certain integrality conditions at each vertex. (We give a precise statement in Theorem 2.2 below.) Another is that the symplectic form on M is determined by the polytope Δ; indeed every point in (M, ω) has a Darboux chart that is equivariantly symplectomorphic to a set of the form $\Phi_0^{-1}(V)$, where V is a neighborhood of some point in the first quadrant $\mathbb{R}^n_+$. Thus locally the action looks like that of T^n in $\mathbb{C}^n$. In particular the (regular) orbits of T^n are Lagrangian.[1] Hence the inverse image of each interior point $u \in \operatorname{int}\Delta$ is a smooth Lagrangian manifold L_u, that is $\omega|_{L_u} = 0$.

This note addresses the question of which of these toric fibers L_u are *displaceable by a Hamiltonian isotopy*, i.e. are such that there is a family of functions $H_t : M \to \mathbb{R}, t \in [0,1]$, whose associated flow $\phi_t, t \in [0,1]$, has the property that $\phi_1(L_u) \cap L_u = \emptyset$.

This question was first considered (from different points of view) in [**BEP**] and [**Ch**]. Biran, Entov and Polterovich showed in [**BEP**] that if the quantum homology $QH_*(M)$ (taken with appropriate coefficients) has a field summand then at least one of the fibers L_u is nondisplaceable. In later work (cf. [**EP06**, Theorem 2.1]), Entov–Polterovich managed to dispense with the condition on quantum homology. Even more recently, they showed in [**EP09**, Theorem 1.9] by a dynamical argument that if in addition (M, ω) is *monotone*, i.e. $[\omega]$ is a positive multiple of the first Chern class $c_1(M)$, then the fiber over the so-called *special point* u_0 of Δ is nondisplaceable.[2] However, for general polytopes their argument gives no information about which fibers might be nondisplaceable.

Cho [**Ch**], and later Fukaya, Oh, Ohta and Ono [**FOOO1, FOOO2**], took a more constructive approach to this problem. The upshot of this work is that for any toric manifold, one can define Floer homology groups $HF^*(L_u, \chi)$ (depending on various deformation parameters χ) that vanish whenever L_u is displaceable. Moreover, in [**FOOO1**, §9], the authors construct a point v_0 for which this Floer homology does not vanish for suitable χ, at least in the case when $[\omega]$ is a rational cohomology class. They show in [**FOOO1**, Thm. 1.5] that even in the nonrational case the corresponding fiber L_{v_0} cannot be displaced. They also show in the monotone case that v_0 coincides with the special point u_0 and that $HF^*(L_u, \chi) = 0$ for all other u; cf. [**FOOO1**, Thm. 7.11].

One of the main motivating questions for the current study was raised by Entov and Polterovich, who ask in [**EP09**] whether the special fiber L_{u_0} is *stem*, that is, whether all *other* fibers are displaceable. The results stated above show that from the Floer theory point of view this holds.

In this paper we develop a geometric way to displace toric fibers using probes. Our method is based on the geometry of the moment polytope Δ and makes sense for general rational polytopes. Using it, we show:

THEOREM 1.1. *If (M, ω) is a monotone toric symplectic manifold of (real) dimension ≤ 6 then the special fiber L_{u_0} is a stem.*

This is an immediate consequence of Proposition 4.7.

In our approach, this question of which toric fibers can be displaced is closely related to the well known Ewald conjecture in [**Ew**] about the structure of monotone polytopes Δ, namely that the set

$$\mathcal{S}(\Delta) = \{v \in \Delta \cap \mathbb{Z}^n \ : \ -v \in \Delta\}$$

[1] One way to prove this is to note that the functions H_i are in involution; the Poisson brackets $\{H_i, H_j\}$ vanish because T^n is abelian.

[2] If ω is normalized so that $[\omega] = c_1(M)$, then the smooth moment polytope Δ is dual to an integral Fano polytope P and u_0 is its unique interior integral point which is usually placed at $\{0\}$. We call such polytopes Δ *monotone*; cf. Definition 3.1.

of symmetric integral points in Δ contains an integral basis for $\mathbb{R}^n$.[3] By work of Øbro [**Ob**, Ch. 4.1], this is now known to hold in dimensions ≤ 8. However, in general it is not even known if $\mathcal{S}(\Delta)$ must be nonempty.

In Definition 3.5 we formulate a stronger, but still purely combinatorial, version of the Ewald property (called the *star Ewald* property) and prove the following result.

THEOREM 1.2. *A monotone polytope Δ satisfies the star Ewald condition if and only if every point in* $\operatorname{int}\Delta \smallsetminus \{0\}$ *can be displaced by a probe.*

COROLLARY 1.3. *If Δ is a monotone polytope in $\mathbb{R}^n$ for which all points except for u_0 are displaceable by probes, then $\mathcal{S}(\Delta)$ spans $\mathbb{R}^n$.*

We show in Proposition 4.7 that every 3-dimensional monotone polytope satisfies the star Ewald condition. The proof of this result is fairly geometric, and does not appear to generalize easily to higher dimensions. Therefore, before attempting such a generalization, it would seem sensible to carry out a computer check of the star Ewald condition using Øbro's methods.[4]

We then analyze the star Ewald condition for monotone polytopes that are bundles. (Definitions are given in §5.) By Lemma 5.2 the fiber and base of such a bundle must be monotone. Although it seems likely that the total space is star Ewald whenever the fiber and base are, we could only prove the following special case.

PROPOSITION 1.4. *Suppose that the monotone polytope Δ is a bundle over the k-simplex Δ^k, whose fiber $\widetilde{\Delta}$ satisfies the star Ewald condition. Then Δ satisfies the star Ewald condition.*

Using this, we show that L_{u_0} is a stem in various other cases, in particular for the 8-dimensional monotone manifold found by Ostrover–Tyomkin [**OT**] that does not have semisimple quantum homology.

During the course of the proof we show that if the polytope $\widehat{\Delta}$ is star Ewald, then the total space of every bundle over $\widehat{\Delta}$ with star Ewald fiber is itself star Ewald if and only if this is true for bundles with fiber the one-simplex Δ_1; see Proposition 5.3.

Finally, we discuss the notion of stable displaceability in §2.3. This notion was introduced by Entov–Polterovich in [**EP09**] as an attempt to generalize the notion of displaceability. However, we show in Proposition 2.10 that in many cases stably displaceable fibers are actually displaceable by probes.

Our arguments rely on the Fukaya–Oh–Ohta–Ono notion of the *central point* v_0. We explain this in §2.2, and then in Lemma 2.7 give a direct combinatorial proof of the following fact.

PROPOSITION 1.5. *For every rational polytope the point v_0 is not displaceable by probes.*

In some cases, it is easy to check that probes displace all points $u \in \operatorname{int}\Delta$ for which $HF^*(L_u, \chi) = 0$. For example, the results of Fukaya–Oh–Ohta–Ono concerning one point blow ups of $\mathbb{C}P^2$ and $\mathbb{C}P^1 \times \mathbb{C}P^1$ become very clear from this

[3] As is customary in this subject, Ewald works with the dual polytope P that is constructed from the fan. Hence his formulation looks very different from ours, but is equivalent.

[4] A. Paffenholz has recently made such a computer search, finding that the first dimension in which counterexamples occur is 6.

perspective: see Example 4.3 and Figure 4.2. However, by Lemma 4.1 this is no longer true for general Hirzebruch surfaces. Further, the Floer-theoretic nondisplaceable set

$$\mathcal{ND}_{HF} := \{u \in \operatorname{int} \Delta : HF^*(L_u, \chi) \neq 0\}$$

has dimension at most $n-1$, while we prove the following result in §4.1.

PROPOSITION 1.6. *There are 2-dimensional moment polytopes with a nonempty open set consisting of fibers that are not displaceable by probes.*

It is not at all clear whether these fibers really are nondisplaceable, or whether one just needs to find more elaborate ways of displacing them.

Varying the symplectic form

In general, the set of nondisplaceable fibers varies as one varies the toric symplectic form. In terms of the moment polytope this amounts to changing the support constants κ_i of the facets $\{x : \langle x, \eta_i\rangle \le \kappa_i\}$ without changing the normal vectors η_i. For each Δ we denote by $\Delta(\kappa)$ the polytope with support constants κ and normals equal to those of Δ, and by $\mathcal{C}_\Delta$ the set of $\kappa = (\kappa_i)$ for which $\Delta(\kappa)$ is analogous to Δ, i.e. is such that a set of facets has nonempty intersection in $\Delta(\kappa)$ if and only if it does in Δ. Let us say that $\Delta(\kappa)$ is *accessible* if all its points except for v_0 are displaceable by probes. Then we may ask:

- For which Δ is there some $\kappa \in \mathcal{C}_\Delta$ such that $\Delta(\kappa)$ is accessible?
- For which Δ is $\Delta(\kappa)$ accessible for all $\kappa \in \mathcal{C}_\Delta$?

If Δ is a product of simplices, it is obvious that $\Delta(\kappa)$ is always accessible. However in dimension 2 some trapezoids (the even ones) are also accessible for all κ; see Corollary 4.2. It is not clear what happens in higher dimensions.

Acknowledgements. This paper grew out of an attempt with Leonid Polterovich and Misha Entov to understand the displaceability of fibers of 2-dimensional monotone polytopes, and I wish to thank them for useful discussions. I also am very grateful to Benjamin Nill for his many penetrating comments on earlier drafts of this note and, in particular, for sharpening the original version of Lemma 3.7. Discussions with Fukaya, Ohta and Ono and with Chris Woodward helped to clarify some of the examples in §4.1. Finally, I would like to thank the referee for reading the manuscript so carefully and pointing out many small inaccuracies.

2. The method of probes

2.1. Basic notions. A line is called rational if its direction vector is rational. The *affine distance* $d_{\mathit{aff}}(x,y)$ between two points x, y on a rational line L is the ratio of their Euclidean distance $d_E(x,y)$ to the minimum Euclidean distance from 0 to an integral point on the line through 0 parallel to L. Equivalently, if ϕ is any integral affine transformation of $\mathbb{R}^n$ that takes x, y to the x_1 axis, then $d_{\mathit{aff}}(x,y) = d_E(\phi x, \phi y)$.

An affine hyperplane A is called *rational* if it has a primitive integral normal vector η, i.e. if it is given by an equation of the form $\langle x, \eta\rangle = \kappa$ where $\kappa \in \mathbb{R}$ and η is primitive and integral. The *affine distance* $d_\lambda(x, A)$ from a point x to a (rational) affine hyperplane A in the (rational) direction λ is defined to be

$$d_\lambda(x, A) := d_{\mathit{aff}}(x, y) \tag{2.1}$$

where $y \in A$ lies on the ray $x + a\lambda, a \in \mathbb{R}^+$. (If this ray does not meet A, we set $d_\lambda(x, A) = \infty$.) If the direction λ is not specified, we take it to be η. We shall say that an integral vector λ is *integrally transverse to* A if it can be completed to an integral basis by vectors parallel to A. Equivalently, we need $|\langle\lambda, \eta\rangle| = 1$ where η is the normal as above.

The next lemma shows that the affine distance of x from A is *maximal* along these affine transverse directions. If $A = \{\langle x, \eta\rangle = \kappa\}$, we define $\ell_A(x) := \kappa - \langle x, \eta\rangle$.

LEMMA 2.1. *Let A be the hyperplane $\ell_A(x) := \kappa - \langle x, \eta\rangle = 0$, where η is a primitive integral vector. Then for any rational points $u \notin A$ and $y \in A$*

$$d_{aff}(u, y) \leq |\ell_A(u)|,$$

with equality if and only if the primitive integral vector in the direction $y - u$ is integrally transverse to F.

PROOF. This is obvious if one chooses coordinates so that $A = \{x_1 = 0\}$. □

A (convex, bounded) polytope $\Delta \subset \mathbb{R}^n$ is called *rational* if each of its facets $F_i, i = 1, \dots, N$, is rational. Thus there are primitive integral vectors η_i (the outward normals) and constants $\kappa_i \in \mathbb{R}$ so that

$$\Delta = \big\{x \in \mathbb{R}^n \mid \langle\eta_i, x\rangle \leq \kappa_i, i = 1, \dots, N\big\}. \tag{2.2}$$

We denote by

$$\ell_i : \Delta \to \mathbb{R}, \quad x \mapsto \kappa_i - \langle\eta_i, x\rangle \tag{2.3}$$

the affine distance from $x \in \Delta$ to the facet F_i. Further, Δ is *simple* if exactly n facets meet at each vertex, and is *integral* if its set $\mathcal{V}(\Delta)$ of vertices are integral. (Integral polytopes are also known as *lattice* polytopes.) A simple, rational polytope is *smooth* if for each vertex $v \in \mathcal{V}(\Delta)$ the normals $\eta_i, i \in I_v$, of the facets meeting at v form a basis for the integral lattice $\mathbb{Z}^n$. This is equivalent to requiring that for each vertex v the n primitive integral vectors $e_i(v)$ pointing along the edges from v form a lattice basis.

Delzant proved the following foundational theorem in [**D**].

THEOREM 2.2. *There is a bijective correspondence between smooth polytopes in $\mathbb{R}^n$ (up to integral affine equivalence) and toric symplectic $2n$-manifolds (up to equivariant symplectomorphism).*

DEFINITION 2.3. Let w be a point of some facet F of a rational polytope Δ and $\lambda \in \mathbb{Z}^n$ be integrally transverse to F. The **probe** $p_{F,\lambda}(w) = p_\lambda(w)$ with direction $\lambda \in \mathbb{Z}^n$ and initial point $w \in F$ is the half open line segment consisting of w together with the points in $\operatorname{int}\Delta$ that lie on the ray from w in direction λ.

In the next lemma we can use any notion of length along a line, though the affine distance is the most natural.

LEMMA 2.4. *Let Δ be a smooth moment polytope. Suppose that a point $u \in \operatorname{int}\Delta$ lies on the probe $p_{F,\lambda}(w)$. Then if w lies in the interior of F and u is less than halfway along $p_{F,\lambda}(w)$, the fiber L_u is displaceable.*

PROOF. Let $\Phi : M \to \Delta$ be the moment map of the toric manifold corresponding to Δ, and consider $\Phi^{-1}(p)$ where $p := p_{F,\lambda}(w)$. We may choose coordinates on

$\mathbb{R}^n \supset \Delta$ so that $F = \{x_1 = 0\}$ and $\lambda = (1, 0, \ldots, 0)$. Formula (1.1) implies that there is a corresponding Darboux chart on M with coordinates $z_1, \ldots, z_n$ such that

$$\Phi^{-1}(p) = \{z : |z_1| \leq a, |z_i| = \text{ const}\},$$

where a is the affine length of the probe p. Hence there is a diffeomorphism from $\Phi^{-1}(p)$ to $D^2(a) \times T^{n-1}$ that takes the restriction of the symplectic form to $\mathrm{pr}^*(dx \wedge dy)$ where $\mathrm{pr} : \mathbb{R}^2 \times T^{n-1} \to \mathbb{R}^2$ is the projection and $D^2(a)$ is the disc with center 0 and area a. Further this diffeomorphism takes L_u to $\partial D^2(b) \times T^{n-1}$ where $b = d_{\mathit{aff}}(w, u)$. But when $b < a/2$ one can displace the circle $\partial D^2(b)$ in $D^2(a)$ by a compactly supported area preserving isotopy. Therefore L_u can be displaced inside $\Phi^{-1}(p)$ by an isotopy that preserves the restriction of ω. But this extends to a Hamiltonian isotopy of M that displaces L_u. □

DEFINITION 2.5. Let Δ be any rational polytope and $u \in \operatorname{int} \Delta$. If there is a probe $p_{F,\lambda}(w)$ through u that satisfies the conditions in Lemma 2.4 we say that u is **displaceable by the probe** $p_{F,\lambda}(w)$.

2.2. The point v_0. In [**FOOO1**], Fukaya, Oh, Ohta and Ono construct a point v_0 in Δ by the following procedure. For $u \in \Delta$, let $s_1(u) := \inf\{\ell_i(u) : 1 \leq i \leq N\}$ where $\ell_i(u)$ is as in equation (2.3). Let $P_0 := \Delta$ and $I_0 := \{1, \ldots, N\}$ and define

$$\begin{aligned} S_1 : &= \sup\{s_1(u) : u \in P_0\}, \\ P_1 : &= \{u \in P_0 : s_1(u) = S_1\}, \\ I_1 : &= \{i \in I_0 : \ell_i(u) = S_1 \text{ for all } u \in P_1\}. \end{aligned}$$

Then $\Delta \smallsetminus P_1 = \{u \in \Delta : \exists j \in I_0,\ \ell_j(u) < S_1\}$, and $P_1 = \{u \in \Delta : \ell_j(u) \geq S_1\ \forall j \in I_0\}$. It follows from the definition of S_1 that P_1 is nonempty and it is easy to check that it is convex. If the plane $\ell_i = S_1$ intersects $\operatorname{int} P_1$ but does not contain it, there will be points $u \in P_1$ with $\ell_i(u) < S_1$ which is impossible. Therefore for each $i \in I_0$, the function $\ell_i(u)$ is either equal to S^1 on P_1 or strictly greater than S_1 on P_1. In other words,

$$I_1 = \{i : \ell_i(u) = S_1 \text{ for some } u \in \operatorname{int} P_1\}.$$

It follows easily that $|I_1| \geq 2$ (since if $I_1 = \{i\}$ one can increase $s_1(u)$ by moving off P_1 along the direction $-\eta_i$.) Hence $\dim P_1 < n$.

As an example, observe that if Δ is a rectangle with sides of lengths $a < b$, then P_1 is a line segment of length $b - a$. Further, in the monotone case, we show at the beginning of §3 that one can choose coordinates so that Δ contains 0 and is given by equations of the form (2.2) with all $\kappa_i = 1$. Then, $\ell_i(0) = 1$ for all i. Moreover, any other point y of $\operatorname{int} \Delta$ lies on a ray from 0 that exits through some facet F_j. Hence $\ell_j(y) < 1$. Thus $S_1 = 1$ and $P_1 = \{0\}$.

Inductively, if $\dim P_k > 0$, define $s_{k+1} : P_k \to \mathbb{R}$, S_{k+1}, P_{k+1} and I_{k+1} by setting

$$\begin{aligned} s_{k+1}(u) : &= \begin{cases} \inf\{\ell_i(u) : \ell_i(u) > S_k\} & \text{if } u \in \operatorname{int} P_k, \\ S_k & \text{if } u \in \partial P_k, \end{cases}, \\ S_{k+1} : &= \sup\{s_{k+1}(u) : u \in P_k\}, \\ P_{k+1} : &= \{u \in P_k : s_{k+1}(u) = S_{k+1}\}, \\ I_{k+1} : &= \{i : \ell_i(u) = S_{k+1}, \text{ for all } u \in P_{k+1}\}. \end{aligned}$$

Arguing much as above, they show in [**FOOO1**, Proposition 9.1] that s_{k+1} is a continuous, convex piecewise affine function, that P_{k+1} is a nonempty convex polytope lying in $\operatorname{int} P_k$, and that

$$I_{k+1} = \{i : \ell_i(u) = S_{k+1} \text{ for some } u \in \operatorname{int} P_{k+1}\}.$$

It is not hard to see that $\dim P_{k+1} < \dim P_k$ unless the functions $\ell_i, i \in I_{k+1}$, are constant on P_k. Because Δ is bounded, at least one of the functions ℓ_j for $j \notin \cup_{r\le k} I_r$ must be nonconstant on P_k when $\dim P_k > \{0\}$. Therefore, after a finite number s of steps one must have $\dim P_{k+s} < \dim P_k$. Hence there is $K \le N$ such that P_K is a point; call it v_0. By [**FOOO1**, Theorem 1.4], $HF^*(L_{v_0}, \chi) \ne 0$ for suitable χ when Δ is rational. Observe also that

$$S_1 < S_2 < \cdots < S_K, \text{ and } I_k = \{i : \ell_i(v_0) = S_k\}. \tag{2.4}$$

Further, $\ell_j(v_0) > S_K$ for all $j \notin$ some I_k. Finally observe that if $V(P_k)$ denotes the plane spanned by the vectors lying in P_k for some $k \ge 0$, the fact that P_{k+1} lies in the interior of P_k implies that

> (*) *if* $\dim V(P_{k+1}) < \dim V(P_k)$, *the normals* $\eta_i, i \in I_{k+1}$, *project to vectors* $\overline{\eta}_i$ *in* $V(P_k)/V(P_{k+1})$ *whose nonnegative combinations* $\sum q_i \overline{\eta}_i, q_i \ge 0$, *span this quotient space.*

(Really one should think of the normals η_i as lying in the dual space to $\mathbb{R}^n$ so that this projection is obtained by restricting the linear functional $\langle \eta_i, \cdot \rangle$.)

REMARK 2.6. It is claimed in early versions of [**FOOO1**, Prop. 9.1] that $\dim P_{k+1} < \dim P_k$ for all k. But this need not be the case. For example, suppose that Δ is the product of Δ' with a long interval, where Δ' is a square with one corner blown up a little bit as in Figure 2.1. Then I_1 consists of the labels of the four facets of the square, I_2 contains just the label of the exceptional divisor, while I_3 contains the two facets at the ends of the long tube. Correspondingly, P_1 is an interval, P_2 is a subinterval of P_1 and $P_3 = \{v_0\}$ is a point.

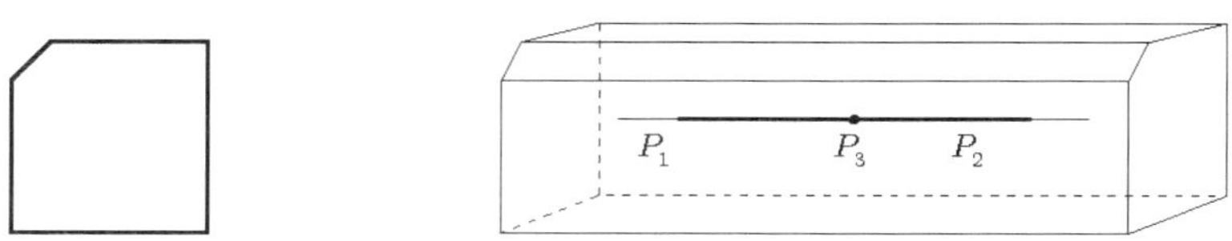

FIGURE 2.1. The construction of v_0 for the polytope considered in Remark 2.6.

LEMMA 2.7. *For every rational polytope* Δ *the point* v_0 *is not displaceable by a probe.*

PROOF. Suppose that v_0 is displaced by a probe $p = p_\lambda(w_1)$ that enters Δ at the point $w_1 \in \operatorname{int} F_1$ and exits Δ through $w_2 \in F_2$. Then

$$\ell_1(v_0) = d_{aff}(w_1, v_0) < d_{aff}(v_0, w_2) \le \ell_2(v_0),$$

where the last inequality follows from Lemma 2.1.

Recall from equation (2.1) that $d_{aff}(v_0, w_2)$ is just the affine distance $d_\lambda(v_0, F_2)$ of v_0 from F_2 in direction λ. Because the ray from v_0 to w_2 in direction λ goes through no facets of Δ until it meets F_2 (and perhaps some other facets as well) at w_2, we have

$$\ell_1(v_0) < d_\lambda(v_0, F_2) \le d_\lambda(v_0, F_i), \quad \text{for all } i.$$

A similar argument applied to the ray from v_0 in direction $-\lambda$ gives

$$\ell_1(v_0) < d_{-\lambda}(v_0, F_i) \quad \text{for all } i \neq 1.$$

(Here we use the fact that $w_1 \in \operatorname{int} F_1$ so that the ray meets F_1 before all other facets.) But if $d_\lambda(v_0, F_i) < \infty$ then $d_\lambda(v_0, F_i) \le \ell_i(v_0)$ by Lemma 2.1. Therefore, for all facets $F_i, i \neq 1$, that are not parallel to λ, we have

$$\ell_i(v_0) > \ell_1(v_0). \tag{2.5}$$

Now observe that because P_K is a single point $\{v_0\}$ the vectors $\eta_i, i \in I_k, 1 \le k \le K$, span $\mathbb{R}^n$. Therefore there is some $k \le K$ such that the $F_i, i \in I_k$, are not all parallel to λ. Let r be the minimum such k, and let $j \in I_r$ be such that F_j is not parallel to λ. If $j \neq 1$ then $S_r = \ell_j(v_0) > \ell_1(v_0)$ by equation(2.5). Hence equation (2.4) implies that $1 \in I_k$ for some $k < r$, which is impossible since λ is not parallel to F_1. On the other hand, if $j = 1$ the same reasoning shows that all other elements of I_r correspond to facets that are parallel to λ. Since by hypothesis the same is true for the elements of I_{r-1}. Therefore $\lambda \in V(P_k)$ for $k < r$ but $\lambda \notin V(P_r)$, so that λ has nonzero image in $V(P_{r-1})/V(P_r)$. But because there is only one $i \in I_r$ for which ℓ_i varies along λ this contradicts (*). □

See Example 4.3 for an example that illustrates how the point v_0 varies as the facets of Δ are moved.

REMARK 2.8. Later we need a slight generalization of this argument in which the set of functions $\ell_i, 1 \le i \le N$, that determine the facets of Δ are augmented by some other nonconstant linear functions $\ell'_j = \kappa'_j - \langle \cdot, \eta'_j \rangle, j \in J$, that are strictly positive on Δ. Thus the hyperplanes A'_j on which these functions vanish do not intersect Δ, so that the functions ℓ'_j correspond to ghost (or empty) facets of Δ. But then, for all $v \in \Delta$, $i \in \{1, \dots, N\}$ and $j \in J$, we have

$$\ell_i(v) = d_{\eta_i}(v, F_i) < d_{\eta_i}(v, A'_j) \le d_{\eta'_j}(v, A'_j) = \ell'_j(v).$$

Therefore the maximin procedure that constructs v_0 is unaffected by the presence of the ℓ'_j. Also, the proof of Lemma 2.7 goes through as before.

2.3. Stable Displaceability. We end this section with a brief digression about stably displaceable fibers. The following definitions are taken from Entov–Polterovich [**EP09**].

DEFINITION 2.9. A point $u \in \operatorname{int} \Delta$ of a smooth moment polytope is said to be **stably displaceable** if $L_u \times S^1$ is displaceable in $M_\Delta \times T^*S^1$ where S^1 is identified with the zero section. Moreover L_{u_1} (or simply u_1) is called a **stable stem** if all points in $\operatorname{int} \Delta \setminus u_1$ are *stably displaceable*

Theorem 2.1 of [**EP09**] states that L_u is stably displaceable if there is an integral vector $H \in \mathfrak{t}$ such that the corresponding circle action Λ_H satisfies the following conditions:

- the normalized Hamiltonian function K_H that generates Λ_H does not vanish on L_u;

- Λ_H is *compressible*, that is, when considered as a loop in the group $\mathrm{Ham}(M_\Delta, \omega)$ of Hamiltonian symplectomorphisms, some multiple of the circle Λ_H forms a contractible loop in $\mathrm{Ham}(M_\Delta, \omega)$.

It is easy to check that $K_H : M_\Delta \to \mathbb{R}$ has the form

$$K_H(x) = \langle H, \Phi(x) - c_\Delta \rangle,$$

where c_Δ is the center of gravity of Δ. The paper [**MT1**] makes a detailed study of those H for which Λ_H is compressible. This condition implies that the quantity $\langle H, c_\Delta \rangle$ depends linearly on the positions of the facets of Δ, and so the corresponding H are called *mass linear functions* on Δ.

There are two cases, according to whether the circle Λ_{kH} contracts in $\mathrm{Isom}\,(M)$ or only in $\mathrm{Ham}(M)$, where $\mathrm{Isom}\,(M)$ is the group of isometries of the canonical Kähler metric on $M := M_\Delta$ obtained by thinking of it as a (nondegenerate)[5] symplectic quotient $\mathbb{C}^N//T'$. In the first case H is called inessential, while in the second H is essential. The inessential case can be completely understood. The following argument uses the definitions and notation of [**MT1**] without explanation.[6]

PROPOSITION 2.10. *The fiber L_u is stably displaceable by an inessential H if and only if it may be displaced by a probe $p_{F,\lambda}(x)$ whose direction vector λ is parallel to all but two of the facets of Δ, namely the entering and exiting facets of the probe.*

PROOF. Suppose first that L_u is displaceable by a probe $p_{F,\lambda}(w)$ with the given property. Then, by [**MT1**, Lemma 3.4], the entering and exiting facets $F := F_1$ and F_2 of the probe are equivalent and there is an affine reflection of Δ that interchanges them. (Cf. [**MT1**, Definition 1.12].) Moreover, λ must be integrally transverse to the exiting facet F_2. Hence the hyperplane that is fixed by this symmetry contains the midpoint of the probe as well as the center of gravity c_Δ. Hence, if $H = \eta_1 - \eta_2$, K_H does not vanish on L_u. Moreover Λ_H is compressible by [**MT1**, Corollary 1.28]. Thus u is stably displaceable by an inessential H.

Conversely, if $u \in \mathrm{int}\,\Delta$ is stably displaceable, there is an inessential H such that $K_H(L_u) := \langle H, u - c_\Delta \rangle \neq 0$. Then [**MT1**, Corollary 1.28] implies that $H = \sum \beta_i \eta_i$ where $\sum_{i \in I} \beta_i = 0$ for each equivalence class of facets I. But each such H is a linear combination of (inessential) vectors H_α of the form $\eta_{\alpha_2} - \eta_{\alpha_1}$ where α_1, α_2 are equivalent. Therefore there is some pair α such that $\langle H_\alpha, u - c_\Delta \rangle < 0$. Let p be the probe from F_{α_1} through u in direction $\lambda = H_\alpha$. (Observe that λ does point into Δ since the η_i are outward normals.) Then the probe must start at some point in $\mathrm{int}\, F_{\alpha_1}$ since it is parallel to all facets that meet F_{α_1} and $u \in \mathrm{int}\,\Delta$. Moreover, because there is an affine symmetry that interchanges the facets $F_{\alpha_1}, F_{\alpha_2}$ while fixing the others, c_Δ must lie half way along this probe. Hence, because $\langle H_\alpha, u \rangle < \langle H_\alpha, c_\Delta \rangle$ this probe displaces u. □

The geometric picture for fibers stably displaceable by an essential mass linear H is much less clear. We show in [**MT1**] that there are no monotone polytopes in dimensions ≤ 3 with essential H. In fact, [**MT1**, Theorem 1.4] states that there is exactly one family $Y_a(\kappa)$ of 3-dimensional polytopes with essential H. They

[5]Here N is the number of facets of Δ, i.e. there are no "ghost" (or empty) facets. With this assumption the Kähler structure is unique.

[6]The above definition of inessential is equivalent to the one of [**MT1**, Definition 1.14] by [**MT1**, Corollary 1.28].

correspond to nontrivial bundles over S^2 with fiber $\mathbb{C}P^2$, and always have a symplectically embedded 2-sphere which is a section of the bundle and lies in a class A with $c_1(A) < 0$. Hence they cannot be monotone. (In [**MT1**, Example 1.1], this section is represented by the shortest vertical edge, which has Chern class $2-a_1-a_2$ where $a_1, a_2 \geq 1$ and $a_1 \neq a_2$.)

It is not clear whether there are higher dimensional monotone polytopes with essential H. In particular, at the moment there are no examples of monotone polytopes for which u_0 is known to be a stable stem but not known to be a stem.

3. Monotone polytopes

There are several possible definitions of a monotone (moment) polytope. We have chosen to use one that is very closely connected to the geometry of Δ.

DEFINITION 3.1. We shall call a simple smooth polytope Δ **monotone** if:

- Δ is an integral (or lattice) polytope in $\mathbb{R}^n$ with a unique interior integral point u_0,
- Δ satisfies the **vertex-Fano** condition: for each vertex v_j we have

$$v_j + \sum_i e_{ij} = u_0,$$

where $e_{ij}, 1 \leq i \leq n$, are the primitive integral vectors from v_j pointing along the edges of Δ.

It follows that for every vertex v one can choose coordinates for which $u_0 = (0, \dots, 0)$, $v = (-1, \dots, -1)$ and the facets through v are $\{x_i = -1\}$, $i = 1, \dots, n$. In particular $\ell_j(u_0) = 1$ for all facets F_j. Thus if we translate Δ so that $u_0 = \{0\}$ the structure constants κ_i in the formula (2.2) are all equal to 1.

REMARK 3.2. (i) An equivalent formulation is that Δ is a simple smooth lattice polytope with $\{0\}$ in its interior and such that the structure constants κ_i are all equal to 1. To see this, note that if v is a vertex and e_i are the primitive integral vectors along the edges from v then the lattice points in Δ may all be written as $v + m_i e_i$ for some non-negative integers $m_i \geq 0$. Thus 0 has such an expression, and in this case the m_i are just the structural constants. Thus our definition is equivalent to the usual definition of Fano for the dual polytope Δ^* (the simplicial polytope determined by the fan of Δ).

(ii) Although it is customary to assume that the point u_0 is the *unique* interior integral point, it is not necessary to do this. For if we assume only that $u_0 \in \operatorname{int} \Delta$ and that the vertex-Fano condition is satisfied by every vertex we may conclude as above that $\ell_j(u_0) = 1$ for all facets F_j. Therefore there cannot be another integral interior point u_1. For in this case, we must have $d_{\mathit{aff}}(u_0, y) > 1$ where $y \in F$ is the point where the ray from u_0 through u_1 exits Δ. But by Lemma 2.1 we must also have $d_{\mathit{aff}}(u_0, y) \leq |\ell_F(u_0)| = 1$, a contradiction.

It is well known that the monotone condition for moment polytopes is equivalent to the condition that the corresponding symplectic toric manifold $(M_\Delta, \omega_\Delta)$ is monotone in the sense that $c_1 := c_1(M) = [\omega_\Delta]$. A proof is given in [**EP09**, Proposition 1.8]. We include another for completeness. In the statement below we denote the moment map by $\Phi : M \to \Delta$. Recall also that by construction the affine length of an edge ϵ of Δ is just $\int_{\Phi^{-1}(\epsilon)} \omega_\Delta$.

LEMMA 3.3. *Let Δ be a smooth integral moment polytope with an interior integral point u_0. Then Δ is monotone if and only if the affine length of each edge ϵ of Δ equals $c_1(\Phi^{-1}(\epsilon))$.*

PROOF. Suppose that $v_j + \sum_i e_{ij} = u_0$ for all vertices. Suppose that $\epsilon =: ae_{01}$ is the edge between the vertices w_0 and w_1, and assume that the other edges ϵ_{0i} starting at w_0 and in the directions e_{0i} end at the points $w_i, 2 \le i \le n$. Move Δ by an integral linear transformation so that $w_0 = (0,\dots,0)$ and so that e_{0i} points along the ith coordinate direction, for $i = 1,\dots,n$. Then $w_1 = (a,0,\dots,0)$ and we need to check that $c_1(\Phi^{-1}(\epsilon_{01})) = a$. Note that in this coordinate system $u_0 = (1,\dots,1)$.

Consider the vertices $y_1 = w_0, y_2,\dots,y_n$ connected to w_1. There is one such vertex $y_j = w_1 + m_j e'_{1j}$ in each of the 2-faces $f_{01j} = \mathrm{span}(e_{01}, e_{0j})$, $j > 1$, containing e_{01}. (Here e'_{1j} is a primitive integral vector pointing from w_1 to y_j.) Therefore $e'_{1j} = (b_j, 0,\dots,0,1,0,\dots,0)$, where the 1 appears as the jth component.[7] Therefore the identity $(1,\dots,1) = (a,0,\dots,0) + (-1,0,\dots,0) + \sum_j e'_{1j}$ implies that

$$1 = a - 1 + \sum_{j\ge 2} b_j.$$

Now consider the S^1 action on M_Δ with Hamiltonian given by $pr_1 \circ \Phi$, where pr_1 denotes projection to the first coordinate. The weights of this action at $\Phi^{-1}(w_1)$ are $(-1, b_2,\dots,b_n)$ with sum $m_1 = -1 + \sum b_j$, while its weights at $\Phi^{-1}(w_0)$ are $(1,0\dots,0)$ with sum $m_0 = 1$. Therefore

$$c_1\big(\Phi^{-1}(e_{01})\big) = m_0 - m_1 = 1 - (-1 + \sum_j b_j) = a,$$

as required. The proof of the converse is similar. $\square$

In the next lemma we denote by $\mathcal{S} := \mathcal{S}(\Delta) := \Delta \cap (-\Delta) \cap \big(\mathbb{Z}^n \setminus \{0\}\big)$ the set of nonzero symmetric integral points of Δ, where we assume that $u_0 = \{0\}$.

LEMMA 3.4. *Let Δ be a monotone polytope. If U is a sufficiently small neighborhood of $u_0 = \{0\}$, then the set of direction vectors of the probes that displace some point in U is precisely $\mathcal{S}$.*

PROOF. Given U, let $\Lambda(U)$ be the set of direction vectors of probes $p_{F,\lambda}(w)$ that displace some point y in U.

We first claim that $\mathcal{S} \subset \Lambda(U)$ for all U. To see this, observe first that if $\lambda \in \mathcal{S}$ then λ (considered as a direction vector) is integrally transverse to every facet F containing the point $-\lambda$. (This holds because we may choose coordinates so that $u_0 = \{0\}$ and $F = \{x_1 = -1\}$.) Therefore for each such pair λ, F there is a probe $p_{F,\lambda}(-\lambda)$. This exits Δ at the point λ and has midpoint at $\{0\}$. If $-\lambda \in \mathrm{int}\, F$ this probe displaces all points less than half way along it. Moreover, if $\lambda \notin \mathrm{int}\, F$, then because U is open any probe $p_{F,\lambda}(w)$ starting at a point $w \in \mathrm{int}\, F$ sufficiently close to $-\lambda$ will displace some points of U. Hence $\mathcal{S} \subset \Lambda(U)$ as claimed.

We next claim that if $\lambda \in \Lambda(U)$ and U is sufficiently small then $\pm\lambda \in \Delta$. Since $\lambda \in \mathbb{Z}^n$ this means that $\lambda \in \mathcal{S}$, which will complete the proof.

---

[7]The jth component must be 1 because the $e'_{1j}, j > 1$, together with $-e_{01}$ form a lattice basis.

To prove the claim, consider a probe $p_{F,\lambda}(w)$ that displaces some $y \in U$, and choose coordinates $x_1, \dots, x_n$ on $\mathbb{R}^n$ so that $F \subset \{x_n = -1\}$. Then the direction λ is an integral vector with last coordinate $= 1$. Therefore $-\lambda$ is an integral point in the plane $\{x_n = -1\}$. To arrange that $-\lambda \in F$, assume that in this coordinate system U is contained in the Euclidean ball about $\{0\}$ with radius ε. Then if $y = (y_1, \dots, y_n) \in U$ is displaced by $p_{F,\lambda}(w)$ we must have $y - (1 - y_n)\lambda = w \in F$. Therefore the Euclidean distance of $-(1 - y_n)\lambda$ to F is at most ε. Since $|y_n| < \varepsilon$ and $-\lambda$ is integral, this implies that $-\lambda \in F$ if ε is sufficiently small. Similarly, because y is less than half way along the probe, $y + (1 - y_n)\lambda \in \Delta$. Therefore the Euclidean distance of $(1 - y_n)\lambda$ to Δ is at most ε, and so, by the integrality of λ we may assume that ε is so small that $\lambda \in \Delta$ also.

The permissible size of ε here depends only on the image of Δ in our chosen coordinate system. But we need make at most one such choice of coordinate system for each facet. Hence we may choose $\varepsilon > 0$ so small that the above argument works for all $\lambda \in \Lambda(U)$. □

3.1. Probes and the Ewald conjecture. The (dual version of the) **Ewald conjecture** of [**Ew**] claims that if Δ is a monotone polytope then the set $\mathcal{S}(\Delta)$ of integral symmetric points contains an integral basis of $\mathbb{R}^n$. Essentially nothing is known about the truth of this conjecture in general; for example, it is even not known whether $\mathcal{S}(\Delta)$ is nonempty. However, the conjecture has been checked by Øbro [**Ob**] in dimensions ≤ 8. Moreover, Øbro observes that in these dimensions a stronger form of the Ewald conjecture holds. Namely *in dimensions* ≤ 8 *for every facet* F, $\mathcal{S}(\Delta) \cap F$ *contains an integral basis for* $\mathbb{R}^n$.

To prove displaceability by probes one needs a slightly different condition. Given a face $f = \cap_{i \in I} F_i$ we shall denote by $\mathrm{Star}(f)$ the union $\cup_{i \in I} F_i$ of the facets containing f and by $\mathrm{star}(f)$ the union $\cup_{i,j \in I, i \neq j} F_i \cap F_j$ of the codimension 2 faces containing f. Further we define the deleted star $\mathrm{Star}^*(f)$ as:

$$\mathrm{Star}^*(f) := \mathrm{Star}(f) \smallsetminus \mathrm{star}(f) = \bigcup_{i \in I} F_i \smallsetminus \bigcup_{i \neq j, i,j \in I} F_i \cap F_j.$$

In particular, $\mathrm{Star}(F) = F = \mathrm{Star}^*(F)$ for any facet F.

Definition 3.5. Let Δ be any smooth polytope with $\{0\}$ in its interior. We will say that Δ satisfies the **weak Ewald condition** if $\mathcal{S}(\Delta)$ contains an integral basis of $\mathbb{R}^n$, and that it satisfies the **strong Ewald condition** if $\mathcal{S}(\Delta) \cap F$ contains an integral basis of $\mathbb{R}^n$ for every facet F. A face f satisfies the **star Ewald condition** if there is some element $\lambda \in \mathcal{S}(\Delta)$ with $\lambda \in \mathrm{Star}^*(f)$ but $-\lambda \notin \mathrm{Star}(f)$. Further, Δ satisfies the star Ewald condition (or, more succinctly, is **star Ewald**) if all its faces have this property.

Remark 3.6. (i) Because λ and $-\lambda$ cannot lie in the same facet F, the star Ewald condition is satisfied by any facet F for which $\mathcal{S} \cap F \neq \emptyset$.

(ii) If Δ is monotone then, because it has a unique interior integral point u_0, we must have $u_0 = \{0\}$ in the above definition.

(iii) The star Ewald condition makes sense for any (not necessarily smooth) polytope containing $\{0\}$ in its interior, and in particular for reflexive polytopes. These are integral polytopes such that $\{0\}$ has affine distance 1 from all facets. Thus, as in the monotone case, the special point $v_0 = \{0\} = P_1$ is reached at the first step of the maximin construction in §2. However, we shall not work in this generality because we are interested in the question of when there is a unique nondisplaceable point, and the examples in Remark 4.6 suggest that this happens only in the smooth case.

The relationships between the strong Ewald and star Ewald conditions are not completely clear. However, as we see in the next lemma, the star Ewald condition does imply the weak Ewald condition for monotone polytopes.[8]

LEMMA 3.7. *If a monotone polytope Δ has a vertex v such that every face containing v is star Ewald, then Δ satisfies the weak Ewald condition.*

PROOF. Choose coordinates for Δ so that $v = (-1, \dots. - 1)$ and the facets through v are $\{x_i = -1\}$. Then Δ lies in the quadrant $\{x_i \geq -1\}$, so that the coordinates of any point $\lambda \in \mathcal{S}(\Delta)$ must lie in $\{0, \pm 1\}$. By the star Ewald condition for the 0-dimensional face v (and renumbering the coordinates if necessary) we may assume that there exists some $\lambda \in \mathcal{S}(\Delta)$ with $\lambda_1 = -1, \lambda_2 = ... = \lambda_n = 0$. Now consider $f = \{x_2 = ... = x_n = -1\}$. Again, by the star Ewald condition for f (and renumbering if necessary) we find that there is $\lambda' \in \mathcal{S}(\Delta)$ with $\lambda'_2 = -1, \lambda'_3 = ... = \lambda'_n = 0$. Proceeding in this way we get n lattice points in $\mathcal{S}(\Delta)$ forming a lattice basis. □

Here is another easy result.

LEMMA 3.8. *If a facet F of a monotone polytope Δ contains a lattice basis consisting of points in $\mathcal{S}(\Delta)$, then each of its codimension 2 faces satisfies the star Ewald condition.*

PROOF. With coordinates as in the previous lemma, it suffices to consider a face $f = F_1 \cap F_2 = \{x_1 = x_2 = -1\}$ such that $\mathcal{S}(\Delta) \cap F_1$ contains a lattice basis. Since $\text{Star} f = F_1 \cup F_2$ and $\text{star} f = f$, we need to show that there is a symmetric point v in $F_1 \setminus f$ with $-v \notin F_1 \cup F_2$.

By assumption the points in $\mathcal{S}(\Delta) \cap F_1$ form a lattice basis. If some point in this set has the form $v_1 = (-1, 0, y_3 \dots, y_n)$ then we are done. Otherwise, there is a lattice basis consisting of points $v_1, \dots, v_n$ that all have second coordinate $y_2 = \pm 1$. The points $v_1, v_j \pm v_1, j \geq 2$, also form a lattice basis, and we may choose the signs so that the first coordinate of each $v_j \pm v_1, j \geq 2$, is zero. But then the second coordinates of these points are always multiples of 2, which is impossible, since they form a matrix of determinant ± 1. □

We now prove Theorem 1.2, which states that for monotone polytopes Δ the star Ewald condition is equivalent to the property that every point in $\text{int}\,\Delta \setminus \{0\}$ can be displaced by a probe.

PROOF OF THEOREM 1.2. For each point $x \in \Delta$ and disjoint face f denote by $C(f, x)$ the (relative) interior of the cone with vertex x and base f. Thus

$$C(f, x) = \{rx + (1 - r)y : r \in (0, 1), y \in \text{int} f\}.$$

(Here, the relative interior $\text{int} f$ is assumed to have the same dimension as f. In particular, for every vertex v, we have $v = \text{int}\, v$.) Thus

$$\text{int}\,\Delta \setminus \{0\} = \cup_f C(f, \{0\}).$$

[8]I am indebted to Benjamin Nill for sharpening my original claim.

If $p_\lambda(-\lambda)$ is a probe through $\{0\}$ starting at $-\lambda \in F \cap \mathcal{S}$, then $\lambda \in \Delta \smallsetminus F$ so that by convexity $C(F, \lambda) \subset \operatorname{int} \Delta$. It is then easy to check that all points in $C(F, \{0\})$ are displaceable by the probes $p_\lambda(x), x \in \operatorname{int} F$, in this direction λ.

More generally, for each face f of Δ, the points in $C(f, \{0\})$ are displaced by probes in the direction $\lambda \in \mathcal{S}$ if $-\lambda \in \operatorname{Star}^*(f)$ while $\lambda \notin \operatorname{Star}(f)$. For if $-\lambda \in F \cap \operatorname{Star}^*(f)$ then

$$W := C(f, -\lambda) \subset \operatorname{int} F, \quad \text{and} \quad C\big(W, \lambda\big) \subset \operatorname{int} \Delta.$$

(Here, by slight abuse of notation, we allow the base of our cone to be a subset of a face rather than the face itself.) Therefore we may displace the points in $C(f, \{0\})$ by the probes $p_\lambda(w)$ where $w \in W = C(f, -\lambda) \subset \operatorname{int} F$.

Conversely, let f be a face such that every point in $C(f, \{0\})$ can be displaced by a probe. We will show that f satisfies the star Ewald condition. To this end choose coordinates on Δ so that Δ has facets $F_i := \{x_i = -1\}, 1 \leq i \leq n$, where $f = \cap_{1 \leq i \leq d} F_i$.

If $f = F_1$, then for $t > 0$ consider the slice $\Delta_t := \Delta \cap \{x_1 = -t\}$. Because Δ is integral, there are no vertices in the slice $\{0 > x_1 > -1\}$. Therefore Δ_t is a smooth polytope for $0 < t < 1$ with facets $F_j \cap \Delta_t, j \in J$, where $J = \{j : 1 < j \leq N, F_j \cap F_1 \neq \emptyset\}$. Every probe in Δ_t is a probe in Δ. Therefore, by Lemma 2.7 there is a point $v_t \in \Delta_t$ that cannot be displaced by any probe in Δ_t. Hence the direction vector $\lambda = (\lambda_1, \ldots, \lambda_n)$ of any probe that displaces v_t must have $\lambda_1 \neq 0$. Now observe that because $\{0\}$ is the unique point with $\ell_i = 1$ for all i, the construction of the special point in Section 2.2 implies that $v_t \to \{0\}$ as $t \to 0$. Therefore v_t is in the neighborhood U of Lemma 3.4 for sufficiently small t so that $\lambda \in \mathcal{S}(\Delta)$. If $\lambda_1 > 0$ then the probe must originate from a point in F_1. Letting $t \to 0$ we see that $-\lambda \in F_1$. On the other hand, if $\lambda_1 < 0$ a similar argument shows that $\lambda \in F_1$. Thus in both cases $\mathcal{S} \cap F_1 \neq 0$, as required by the star condition for $f = F_1$.

Now suppose that $\dim f = n - d < n - 1$ and let $\Lambda(f)$ be the set of directions λ of probes $p_\lambda(w)$ that displace points of $C(f, \{0\})$ arbitrarily close to $\{0\}$. For each $\lambda \in \Lambda(f)$ denote

$$U_\lambda(f) = \{y \in C(f, \{0\}) \,:\, y \text{ is displaced by a probe in direction } \lambda\}.$$

Then $\{0\}$ is in the closure of each $U_\lambda(f)$, and

$$\bigcup_{\lambda \in \Lambda(f)} U_\lambda(f)$$

contains all points in $C(f, \{0\})$ sufficiently close to $\{0\}$.

Now, for each facet F containing f consider the set

$$W_{\lambda, F}(f) = \{w \in \operatorname{int} F \,:\, \text{the probe } p_\lambda(w) \text{ displaces some } y \in U_\lambda(f)\}.$$

Because each such probe $p_\lambda(w)$ meets $U_\lambda(f)$ less than half way along its length, we must have $C(W_{\lambda,F}(f), \lambda) \subset \operatorname{int} \Delta$. But this implies that $\lambda \notin F$ for any $F \supset f$, i.e. $\lambda \notin \operatorname{Star}(f)$. Also $-\lambda \notin \operatorname{star} f$, since if it were the initial points w of the probes would lie in $\operatorname{star} f$ and not in the interior of a facet, as is required.

It remains to check that there is some $\lambda \in \Lambda(f)$ such that $-\lambda$ is in one of the facets F_i containing f. For this, it suffices that $-\lambda_i = -1$ for some $i \leq d$. But because $\lambda \notin F_i$ we know $-\lambda_i \in \{-1, 0\}$ for these i. And if $\lambda_i = 0$ for all i then λ would be parallel to $C(f, \{0\})$, or, if $d = n$, would be equal to $\{0\}$. Since $\lambda \in \partial\Delta$, the latter alternative is impossible. Therefore we may assume that $\dim f = n - d > 0$ and must check that there is an element in $\Lambda(f)$ that is not parallel to f.

To this end, we adapt the argument given above for facets. We shall suppose that the elements in $\Lambda(f)$ are all parallel to f and shall then show that there is a nondisplaceable point in $C(f, 0)$.

For fixed $t \in (0, 1]$ consider the polytope

$$f_t := \Delta \cap \{x_1 = \cdots = x_d = -t\},$$

so that $f = f_1$, and define the set I by

$$i \in I \Longleftrightarrow \Big(F_i \cap \{x_1 \in (-1, 0)\} \neq \emptyset, \text{ and } \ell_i \text{ is not constant on } f\Big).$$

Since the functions $\ell_i, i \in I$, are nonnegative on f_t and the boundary of f_t is the set where at least one ℓ_i vanishes, we may define a point $v_t \in f_t$ by applying the maximin construction of §2 to the restriction of the functions $\ell_i, i \in I$, to f_t. The argument in Lemma 2.7 shows that this point v_t is not displaceable by probes in f_t. (The only new element in the situation is that some of the ℓ_i may represent ghost facets, i.e. they may not vanish anywhere on f_t. But this does not affect any of these arguments; cf. Remark 2.8.)

The probes of interest to us have directions $\lambda \in \Lambda(f)$. Since these points λ lie in the plane $\{x_1 = 0\}$ there is $\varepsilon > 0$ such that each λ lies in a facet F of Δ that intersects f_t for all $t \in (0, \varepsilon)$. Therefore, as in Lemma 3.4, the directions $\lambda \in \Lambda(f)$ are integrally transverse to the facets of f_t for $t \leq \varepsilon$ when considered as probes in the plane $\{x_i = -t, 1 \leq i \leq d\}$ containing f_t. Hence the probes of Δ with directions $\lambda \in \Lambda(f)$ form a subset of the probes in f_t for $t \leq \varepsilon$. Therefore they cannot displace v_t. It remains to prove:

Claim: *$v_t \in C(f, 0) \cap f_t$ when $t \leq \varepsilon$.*

To see this, let $F_j, j \in J_f$, be the set of facets of Δ that intersect but do not contain f. Then $J_f \subset I$ and the facets of f are $f \cap F_j, j \in J_f$. Now observe that if $\langle x, \eta_j\rangle \leq 1$ then $\langle tx, \eta_j\rangle \leq t$, so that

$$\ell_j(tx) = 1 - \langle tx, \eta_j\rangle \geq 1 - t. \tag{3.1}$$

Applying this with $x = p_0 = (-1, \ldots, -1) \in \Delta$ we see that

$$\ell_i(tp_0) = 1 - t\langle\eta_i, p_0\rangle \geq 1 - t, \qquad \text{for all } i \in I.$$

Further, because we chose coordinates so that $f = \{x \in \Delta : x_i = -1, 1 \leq i \leq d\}$, equation (3.1) implies that when $t \in (0, 1]$ we have

$$C(f, 0) \cap f_t \quad = \quad \big\{y : \ell_j(y) \geq 1 - t, j \in J_f\big\} \cap \big\{y_i = -t, 1 \leq i \leq d\big\}.$$

Therefore the maximum value of the function $s_1(y) := \min_{i \in I} \ell_i(y)$ for $y \in f_t$ is at least $1 - t$, and because $J_f \subset I$ it is assumed in $C(f, 0) \cap f_t$. Thus P_1, and hence also $v_t \in P_1$, lies in $C(f, 0)$. This proves the claim, and completes the proof of the proposition. □

COROLLARY 3.9. *If Δ is a 2-dimensional monotone polytope then every point in $\operatorname{int}\Delta \setminus \{0\}$ may be displaced by a probe.*

PROOF. It suffices to check that the star Ewald condition holds, which is easy to do in each of the 5 cases (a square, or a standard simplex with i corners cut off, where $0 \leq i \leq 3$.) □

For the 3-dimensional version of this result see Proposition 4.7.

4. Low dimensional cases.

4.1. The 2-dimensional case. In this section we discuss the properties of arbitrary, not necessarily monotone, 2-dimensional polytopes. We begin by showing that there always is an inaccessible point near any *short odd edge*. Here we say that an edge ϵ is *odd* if its self-intersection number[9] is odd and negative, and that it is *short* if its affine length is at most half that of its neighbors.

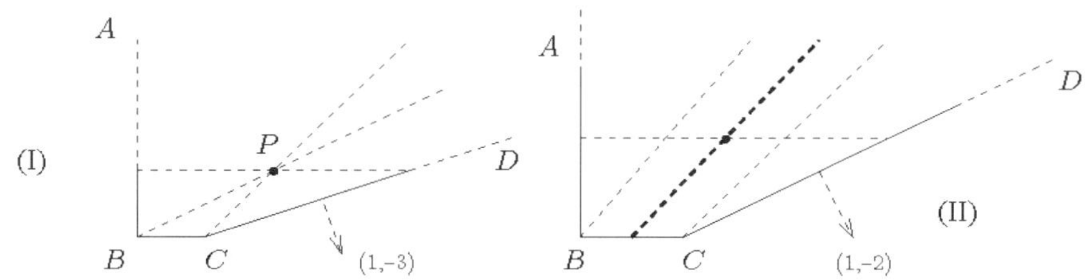

FIGURE 4.1. In (I), BC has self-intersection -3. P is the midpoint of the line parallel to BC and a distance $|BC|$ above it. It is not displaceable because the two probes from BC with good (i.e. integrally transverse) direction vectors have initial points at vertices. Figure (II) illustrates the case when BC has self-intersection -2; the heavy line consists of points midway between AB and DC. (This line is integrally transverse to BC because we are in the even case.) Points not on this line can be displaced by horizontal probes, while points on this line that are close to BC can be displaced by probing from BC parallel to it.

LEMMA 4.1. *Let A, B, C, D be four neighboring vertices on a smooth polygon such that the edge $\epsilon = BC$ of affine length d is short and odd. Then no probe displaces the midpoint P of the line parallel to ϵ and a distance d above it.*

PROOF. Suppose without loss of generality that $d = 1$. Choose coordinates so that B is at the origin and A, C are on the y, x-axes respectively as in Figure 4.1. Then the self-intersection condition implies that the normal to CD is $(1, -(1+2k))$ for some integer $k \geq 0$. The horizontal distance from P to BA is $k+1$. Since this is an integer, the only probes through P that start on BC have initial vertex at B or C. Therefore P cannot be displaced by such probes. But it also cannot be displaced by probes starting on AB since these must have direction $(1, a)$ for some $a \in \mathbb{Z}$. By symmetry, the same argument applies to probes from CD. Finally note that because BC is short, all probes starting on edges other than AB, BC or CD meet P at least half way along their length and so cannot displace P. □

[9]By this we mean the self-intersection number of the 2-sphere $\Phi^{-1}(\epsilon)$ in the corresponding toric manifold M^4. This is the Chern class of its normal bundle, and equals k, where we assume that ϵ has outward normal $(0, -1)$ and that its neighbor to the left has conormal $(-1, 0)$ and to the right has conormal $(1, k)$; cf. [**KKP**, §2.7].

Recall from the end of Section 1 that $\Delta(\kappa)$ is said to be *accessible* if all its points except for v_0 are displaceable by probes.

COROLLARY 4.2. *The only smooth polygons Δ such that $\Delta(\kappa)$ is accessible for all κ are triangles, and trapezoids with no odd edges.*

PROOF. As is illustrated in diagram (II) in Figure 4.1, the argument in Lemma 4.1 does not apply to even edges since then the line of midpoints is a good direction from BC. It follows easily that every trapezoid without an odd edge has only one nondisplaceable point. Every other smooth polygon with at least 4 sides can be obtained by blowup from the triangle or a trapezoid[10] and so has an edge ϵ of self-intersection -1, the result of the last blow up. Clearly, this edge can be made short. □

Denote by $\mathcal{ND}_p \subset \Delta$ the set of points $u \in \operatorname{int}\Delta$ that are not displaceable by probes, and by $\mathcal{ND}_{HF} \subset \Delta$ the set of points $u \in \operatorname{int}\Delta$ for which $HF_*(L_u, \chi) \neq 0$ for some χ.

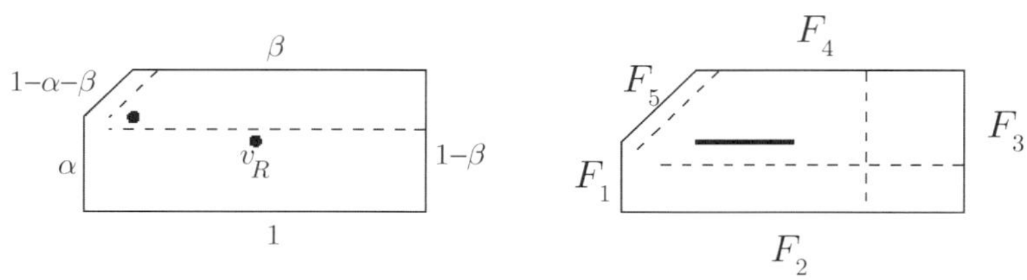

FIGURE 4.2. Some possibilities for $\mathcal{ND}_{HF}$ when Δ is a 2-point blow up of $\mathbb{C}P^2$. Here $\mathcal{ND}_p$ is depicted by the dark dots and heavy lines; the dotted lines show permissible directions of probes.

EXAMPLE 4.3. Let Δ be the moment polytope of a 2-point blow up of $\mathbb{C}P^2$ as in Figure 4.2. Then the three consecutive edges F_4, F_5, F_1 are odd. Denote their affine lengths by $L(F_i)$. We normalize the lengths of the edges of the triangle T formed by F_2, F_3, F_5 to be 1 and denote $\alpha := L(F_1), \beta := L(F_4)$, so that $L(F_5) = 1-\alpha-\beta$. Without loss of generality, we assume that $\alpha \leq \beta$. We denote by v_T the center of gravity of the triangle T and by v_R the center of gravity of the rectangle R with facets $F_1, \dots, F_4$.

The first question is: where is v_0? If $L(F_5) \leq \beta$ (as in both cases of Figure 4.2), then one can check that $v_0 = v_R$. In this case, one should think of Δ as the blow up of the rectangle R. On the other hand, if $L(F_5) = 1-\alpha-\beta > \beta = L(F_4)(\geq \alpha)$ then v_R can be displaced from F_5, and one should think of Δ as the blow up of the triangle T. If in addition $(\alpha \leq)\beta \leq \frac{1}{3}$, then v_T cannot be displaced from F_4 since it is at least as close to F_2 as to F_4, and it follows that $v_0 = v_T$. However, if $1-\alpha-\beta > \beta > \frac{1}{3}$, then v_T can be displaced from F_4. One can check in this case that v_0 is on the median of T through the point p where the prolongations of F_3 and F_5 meet, half way between the parallel edges F_2 and F_4.

Now consider the other points in $\mathcal{ND}_p$. We will say that an odd edge is *short enough* if it is shorter than its odd neighbors and has at most half the length of its

[10]This is well known; see Fulton [**F**, §2.5], or [**KKP**, Lemma 2.16] where the blowup process is called "corner chopping".

even neighbors. Because Δ has so few edges, one can check that the statement in Lemma 4.1 holds for all short enough edges in Δ.

Throughout the following discussion we assume that $\alpha \le \beta$. As α, β vary, precisely one of the following cases occurs.

(i) $L(F_5) < L(F_1)$. If F_5 is short, then as in the left hand diagram in Figure 4.2, $\mathcal{ND}_p$ consists of two points, namely v_0 (which coincides with v_R) and the point P corresponding to the short edge. An analogous statement continues to hold as long as $L(F_5) < L(F_1)(\le L(F_4))$, i.e. as long as F_5 is short enough: the proof of Lemma 4.1 shows that P cannot be displaced from the facets F_4, F_5 or F_1 and it cannot be displaced from F_2 or F_3 because they are too far away. Therefore $\mathcal{ND}_p$ consists of P and $v_0 = v_R$, as in the left hand diagram in Figure 4.2.

(ii) $L(F_5) = L(F_1) < L(F_4)$. Now there are no short edges and $\mathcal{ND}_p$ is an interval, with $v_0 = v_R$ as its "middle" end point; cf. the right hand diagram in Figure 4.2.

(iii) $L(F_5) = L(F_1) = L(F_4) = \frac{1}{3}$ (the monotone case). Again there are no short edges; $\mathcal{ND}_p$ is the single point $v_0 = v_R = v_T$.

(iv) $L(F_4) \ge L(F_5) > L(F_1)$. Note that $L(F_2) = L(F_4) + L(F_5) > 2L(F_1)$. Hence, F_1 is short enough and $\mathcal{ND}_p$ consists of $v_0 = v_R$ and the point P corresponding to F_1.

(v) $L(F_5) > L(F_4) \ge L(F_1)$. As we saw above, the position of v_0 varies depending on the relative sizes of $L(F_4) = \alpha$ and $\frac{1}{3}$. Further F_1 is always short enough, while F_4 may or may not be. Correspondingly, $\mathcal{ND}_p$ consists of two or three points.

This example was discussed in detail in [**FOOO1**, Examples 5.7, 10.17, 10.18] and in [**FOOO2**, §5], where the authors showed that $\mathcal{ND}_{HF} = \mathcal{ND}_p$ in all the above cases. On the other hand, in [**FOOO1**, Examples 8.2] the authors calculated Floer homology groups in the case of Hirzebruch surfaces and, in the case when the negative curve has self-intersection $-k \le -2$, appear to find only *one* point $u \in \mathcal{ND}_{HF}$ (with 4 corresponding deformation parameters y). In other words, the inaccessible point P described in Lemma 4.1 when k is odd is not in $\mathcal{ND}_{HF}$. This seems to be the simplest example where the two sets are different.[11]

It is shown in [**FOOO1**, §10] that if one moves the facets of Δ to be in general position (so that the Landau–Ginzburg potential function is nondegenerate), then $\mathcal{ND}_{HF}$ is finite. We now show that $\mathcal{ND}_p$ sometimes contains an open subset.

LEMMA 4.4. *There is a 2-dimensional smooth polytope with an open set of points that are not displaceable by probes.*

PROOF. The triangle ABC in Figure 4.3 has vertices $A = (0,5), B = (0,0)$ and $C = (3,0)$, and so is not smooth. The points inside the triangles ABG and CBG can be displaced by probes in the directions $\pm(-1,1)$, and all but a short segment of BG can be displaced by vertical probes from BC. It is easy to check that this is the most one can displace by probes starting on AB or BC. On the other hand, the best probes from AC are either parallel to AZ in the direction $(1,-2)$ or are parallel to CW in the direction $(-2,3)$. (In fact, the latter set of probes displaces no new points.) Therefore this triangle contains an open region that cannot be reached by probes.

To get a smooth example, blow up at the vertices A and C along the directions indicated in figures (I) and (II). Probes starting from these new edges will reach

[11] Of course, this point might be detected by more elaborate versions of Floer homology.

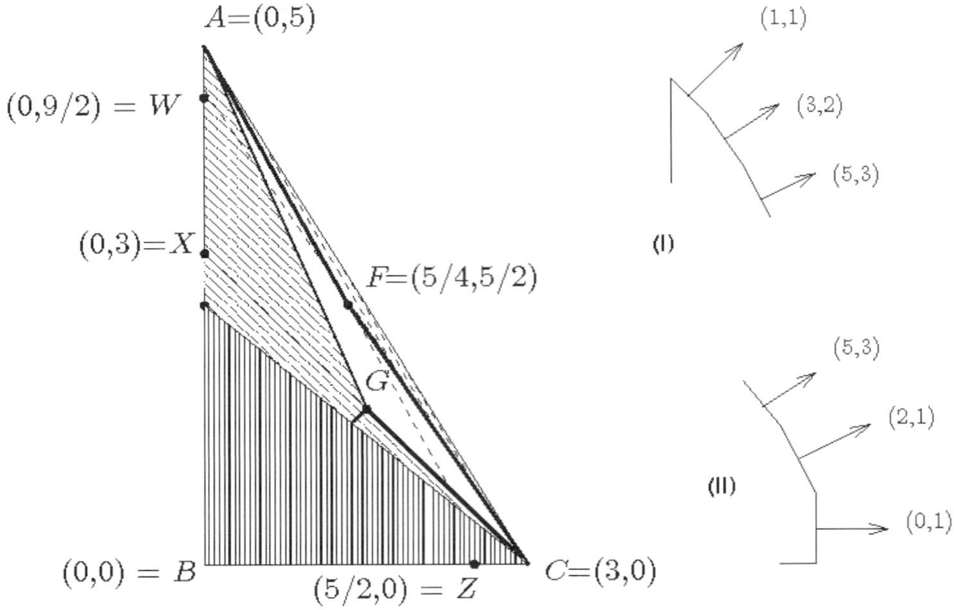

FIGURE 4.3. The shaded regions in the triangle ABC can be displaced by probes parallel to the shading; the heavy lines and open region cannot be so displaced. Here $G = (3/2, 3/2)$ is the midpoint of CX, while $F = (5/4, 5/2)$ is the midpoint of AZ. Figures (I) and (II) show the cuts needed to smooth the vertices A and C.

some more points, but these probes must be in one of a finite number of directions. (For example, from the edge near A with normal $(1,1)$ one reaches some new points by probes in the direction $(0,-1)$.) Hence, since the new edges can be arbitrarily short, the newly accessible regions can have arbitrarily small area. □

We leave it up to the reader to construct similar examples in higher dimensions. Note that the reason why one gets an open set of nondisplaceable points is that in the above example most probes exit through facets that are not integrally transverse to the direction of the probe. For example, in the triangle above the horizontal probes from AB exit through AC which is not integrally transverse to $(1,0)$. Figure 4.4 illustrates two more possibilities.

REMARK 4.5. (i) To see why the two sets $\mathcal{ND}_{HF}$ and $\mathcal{ND}_p$ do not always agree, notice that for a point u to be displaceable by a probe it must be "geometrically visible" from some nearby facet. On the other hand, by [**FOOO1**, §11] as u varies in Δ the properties of the Floer homology of L_u are governed by the sizes $\omega(\beta_i) = \ell_i(u)$ of the discs of Maslov index 2 that are transverse to $\Phi^{-1}(F_i)$ and have boundary on L_u. These holomorphic discs always exist, no matter where u is in Δ. Moreover, according to [**FOOO1**, §9] in order for $HF_*(L_u, \chi)$ to be nonzero for some perturbation χ one needs there to be more than one i for which $\ell_i(u)$ is a minimum. Therefore the set $\mathcal{ND}_{HF}$ always lies in the polytope P_1 defined in §2.2.

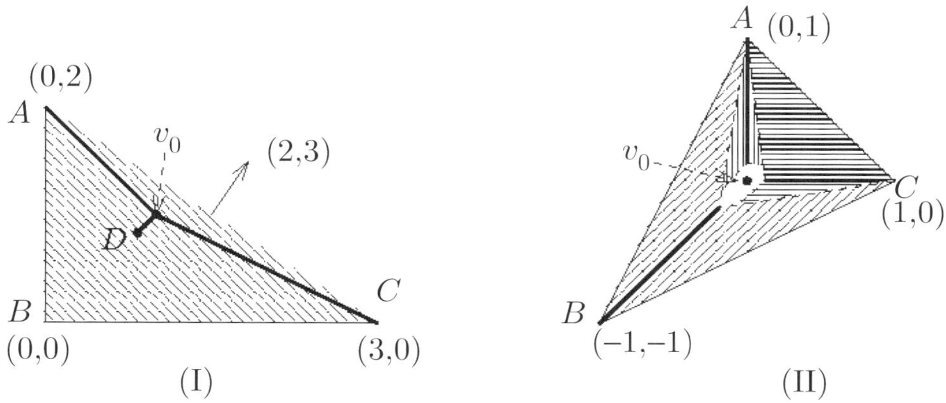

FIGURE 4.4. The shaded regions can be displaced by probes parallel to the shading. In (I), $\mathcal{ND}_p$ consists of the (open) heavy lines and the points $v_0 = (1,1)$ and $D = (\frac{6}{7}, \frac{6}{7})$. Points below D on the ray v_0D can be displaced horizontally from AB. In (II), $\mathcal{ND}_p$ is a hexagon containing $v_0 = (0,0)$ together with parts of the lines $x = y$, $x = 0$ and $y = 0$.

To illustrate these differences in the example of Figure 4.3, choose small $\varepsilon > 0$ and make the triangle Δ smooth by introducing the new edges

$$F_1 := \{x_1 + x_2 = 5 - \varepsilon\}, \qquad F_2 := \{3x_1 + 2x_2 = 10 - \varepsilon\},$$
$$F_3 := \{2x_1 + x_2 = 6 - \varepsilon\}, \qquad F_4 := \{x_1 = 3 - \varepsilon\}.$$

Then it is not hard to see that the calculation of $\inf\{\ell_i(u); u \in \operatorname{int} \Delta\}$ is dominated by the distances to the four facets $\{x_1 = 0\}$, $\{x_2 = 0\}$, F_3 and F_4; the other facets are simply too far away. In fact, the set P_1 defined in §2.2 is the vertical line segment between the points $\big(\frac{1}{2}(3-\varepsilon), \frac{1}{2}(3-\varepsilon)\big)$ and $\big(\frac{1}{2}(3-\varepsilon), \frac{1}{2}(3+\varepsilon)\big)$. In other words, as far as the calculation of v_0 is concerned, our polytope might as well be a trapezoid.

REMARK 4.6 (Reflexive polygons). Both triangles in Figure 4.4 are reflexive. (They are Examples 3 and 6d on the list in [**N**, §4].) In (I), the direction $(-1,1)$ is integrally transverse to all facets, so that probes in this direction or its negative displace all points except for those on certain lines. Because B is smooth, all points near B can be displaced. But there are lines of nondisplaceable points near the nonsmooth vertices A and C. The point $(1,1)$ has affine distance 1 from each facet, and so is the central point v_0. The line segment v_0D is contained in the line $x = y$. By way of contrast, the triangle (II) does not have one direction that is integrally transverse to all edges, although each pair of edges has an integrally transverse direction.

4.2. The 3-dimensional case. This section is devoted to proving the following result.

PROPOSITION 4.7. *Every 3-dimensional monotone polytope Δ satisfies the star Ewald condition. Hence all its points except for $\{0\}$ are displaceable by probes.*

This result is included in the computer check by Paffenholz that shows that all monotone polytopes of dimension ≤ 5 are star Ewald. However, we shall give a more conceptual proof to illustrate the kind of ideas that can be used to analyze this problem.

We begin with some lemmas. Throughout, we choose coordinates so that $v = (-1,-1,-1)$ is a vertex of Δ and so that the facets through v are $F_i := \{x_i = -1\}$. We shall say that a facet F of Δ is *small* if it is a triangle with one (and hence all) edges of length 1.

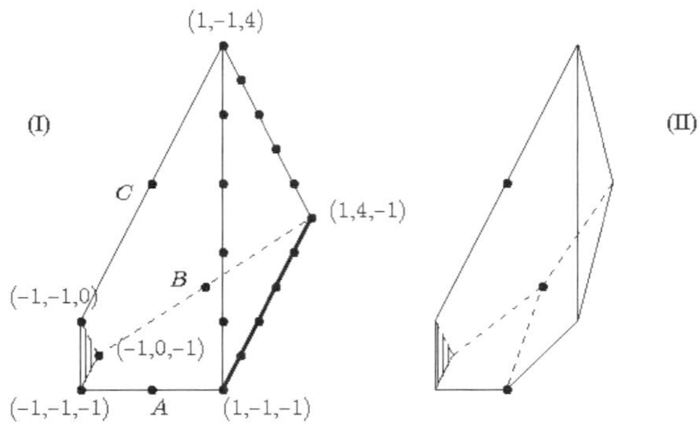

FIGURE 4.5. The truncated pyramid (I) on the left is a Δ_1-bundle over Δ_2 with one small triangular facet; cf. §5. The integral points on its edges are marked. Polytope (II) is its monotone blow up along the heavy edge.

LEMMA 4.8. *If Δ has a small facet F, it is one of the two polytopes illustrated in Fig. 4.5.*

PROOF. We may suppose that the vertices of $F = F_1$ are $v_1 = (-1,-1,-1)$, $v_2 := (-1,0,-1)$ and $v_3 := (-1,-1,0)$. The vertex-Fano condition at v_2 implies that the edge through v_2 transverse to F must have direction $e_2' = (1,2,0)$ while that through v_3 transverse to F_1 must have direction $e_3' = (1,0,2)$. Therefore $\Delta \cap \{x_1 = 0\}$ contains the points $A := (0,-1,-1), B := (0,2,-1)$, and $C := (0,-1,2)$.

Claim 1: *If none of A, B, C are vertices, the points* $(1,-1,-1), (1,4,-1)$ *and* $(1,-1,4)$ *are vertices of Δ and Δ is the polytope in part* (I) *of Figure 4.5.*

Let $y = (y_1, y_2, y_3)$ be a vertex on the edges through A, B, C (but not on F) with the smallest coordinate y_1. Without loss of generality we may suppose that y lies on the edge through A. By hypothesis, $y_1 \ge 1$. Let e_{y1}, e_{y2} be the primitive vectors along the edges from y that do not go through A. Then the x_1-coordinates of the vectors e_{yi} are nonnegative and (by the vertex-Fano condition) sum to $-y_1 + 1 \le 0$. It follows that $y_1 = 1$ and that the vectors e_{yi} lie in the plane $x_1 = 1$. It is now easy to see that the polytope must be as illustrated in (I).

Claim 2: *If at least one of A, B, C is a vertex, then Δ is the polytope in part* (II) *of Figure 4.5.*

Without loss of generality, suppose that A is a vertex. Then, the vertex-Fano condition implies that just one of the primitive edge vectors from A has positive x_1-coordinate. In Figure 4.5 (II) we illustrate the case when this edge vector e lies in F_2. Then the edge through A in F_3 ends at B, and $e = (1,0,1)$. It follows that C cannot be a vertex. Arguing similarly at B, we see that the slice $\Delta \cap \{x_1 \in [-1,1]\}$ must equal the polytope (II). Moreover, if any of the three edges of the slice $\Delta \cap \{x_1 \in (0,1)\}$ have a vertex on $\{x_1 = 1\}$, then the vertex-Fano condition at this vertex implies that $\Delta =$ (II). On the other hand, there must be some vertex of this kind. For if not, we can get a contradiction as in Claim 1 by looking at the vertex y on these edges with minimal x_1-coordinate. □

REMARK 4.9. Polytope (II) is the monotone blow up of (I) along an edge. Note that to make a monotone blow up along an edge ϵ of a monotone 3-dimensional polytope one must cut out a neighborhood of size 1. Thus, there is such a blow up of ϵ exactly if all edges meeting ϵ have length at least 2. Similarly, there is a monotone blow up of a vertex v of a monotone 3-dimensional polytope if all edges meeting v have length at least 3. Thus, up to permutation, there is only one monotone blow up of (I).

COROLLARY 4.10. *A 3-dimensional polytope can have at most one small facet. Moreover, if F is small, $F \cap \Delta_{\mathbb{Z}} \subset \mathcal{S}(\Delta)$.*

PROOF. The first statement is obvious; the second follows by inspection. □

Given a vertex v in some facet F of Δ we define the *special point $s_{v,F}$ of v in the facet F* to be $v + \sum e_j$ where e_j ranges over the primitive integral vectors along the edges from v that lie in F. With our choice of coordinates, we get for $v = (-1,-1,-1)$ the three points $s_1 := s_{v,F_1} = (-1,0,0)$, $s_2 := s_{v,F_2} = (0,-1,0)$ and $s_3 := s_{v,F_3} = (0,0,-1)$. Note that:

- v satisfies the star Ewald condition if and only if one of these three points lies in $\mathcal{S}(\Delta)$;
- $s_{v,F} \in \Delta$ unless F is a small facet.

Next, given $v \in F_i$ we define *the facet opposite to F_i at v* as follows: if ϵ_i is the edge from v transverse to F_i, and z_i is the other vertex of ϵ_i then F_i' is the facet through z_i not containing ϵ_i. Consider the special point $s_i' = s_{z_i,F_i'}$. Then $s_i' = -s_i$ because the vertex-Fano condition implies that $s_i + e_i = 0 = s_i' - e_i$. Therefore if there is i such that s_i, s_i' are both in Δ, the star Ewald condition is satisfied at v.

Proof of Proposition 4.7.

Step 1: *Every vertex of Δ satisfies the star Ewald condition.*

Without loss of generality, consider the vertex $v = (-1,-1,-1)$ as above. If neither of F_1, F_1' are small then $s_1 \in \mathcal{S}(\Delta)$ and v is star Ewald. But if one of these facets is small, then Lemma 4.8 implies that Δ is one of the polytopes in Figure 4.5 and one can check directly that in these cases every vertex lies on some non-small facet F whose opposite facet at v is also not small.

Step 2: *Every facet F of Δ satisfies the star Ewald condition.*

If F is small, then its three vertices lie in $\mathcal{S}(\Delta)$ by Corollary 4.10. Otherwise, let F_w' be the facet opposite to F at some vertex $w \in F$. If F_w' is small for all choices of w then F must be the large triangular facet in the polytope (I) of Figure 4.5, and so it contains points in $\mathcal{S}(\Delta)$ by Corollary 4.10. The remaining case is when there is w such that F_w' is not small. But then $s_{w,F} \in \mathcal{S}(\Delta)$.

Step 3: *Every edge ϵ of Δ satisfies the star Ewald condition.*

If Δ has no small facets, we may suppose that $\epsilon = F_2 \cap F_3$. The proof of Step 1 shows that $s_{v,F_2} = (0,-1,0) \in F_2$ lies in $\mathcal{S}(\Delta)$. Further, because $\mathrm{star}(\epsilon) = \epsilon$, the conditions $s_{v,F_2} \in \mathrm{Star}^*(\epsilon)$ and $-s_{v,F_2} \in \mathrm{Star}(\epsilon)$ are obviously satisfied.

On the other hand, if Δ has a small facet, then Δ is (I) or (II). In these cases, one can check that for each edge of Δ at least one of the two facets containing it, say F, is not small and has an opposite facet F' at some point $w \in F$ that is also not small. Therefore, again $s_{w,F} \in \mathcal{S}(\Delta)$ satisfies the star Ewald condition at ϵ.

This completes the proof of Proposition 4.7.

To extend this kind of argument to higher dimensions, one would have to understand facets that are small in the sense that they do not contain some (or all) of their special points $s_{v,F}$.

5. Bundles.

In [**OT**] Ostrover–Tyomkin construct an 8-dimensional monotone toric manifold M_{OT} whose quantum homology is not semisimple. As pointed out in [**FOOO1**] the properties of $QH_*(M)$ are closely related to the nondisplaceable points in Δ. Nevertheless, we show that in the case of M_{OT} the special fiber is a stem. The manifold M_{OT} is a toric bundle over $\mathbb{C}P^1 \times \mathbb{C}P^1$ with fiber the 3-point blow up of $\mathbb{C}P^2$. Hence this example is covered by Corollary 5.6 below.

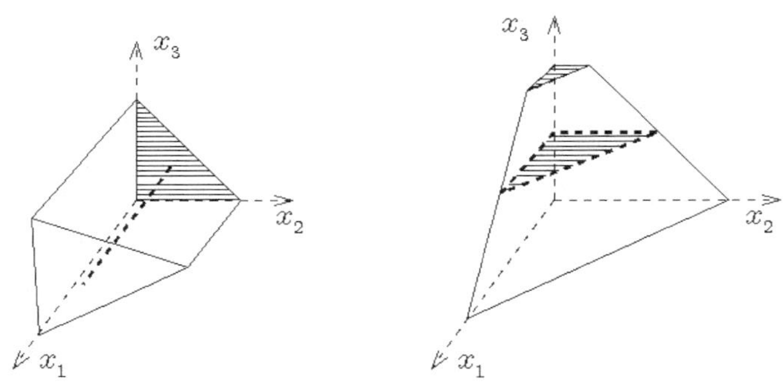

FIGURE 5.1. The polytope (a) is a Δ_2-bundle over Δ_1, while (b) is a Δ_1-bundle over Δ_2. The shaded facet in (a) is one of the base facets $\widehat{F}_i'$ and is affine equivalent to the fiber, while the top shaded facet in (b) represents a section of the bundle and is one of the two fiber facets $\widetilde{F}_j'$. The heavy dotted lines enclose the central slice $\widehat{\Delta}_0$ described in Lemma 5.2.

Recall that a smooth locally trivial fiber bundle $\widetilde{M} \to M \stackrel{\pi}{\to} \widehat{M}$ whose total space is a toric manifold (M,T) is said to be a toric bundle if the action of T permutes the fibers of π. It follows that there is a corresponding quotient homomorphism $T \to \widehat{T} := T/\widetilde{T}$ whose kernel $\widetilde{T}$ induces a toric action on the fiber $\widetilde{M}$ and whose image $\widehat{T}$ induces a toric action on the base $\widehat{M}$. Because the moment polytope Δ lives in the dual Lie algebra $\mathfrak{t}^*$ of T, there is no natural projection map from $\Delta \subset \mathfrak{t}^*$ to $\widehat{\Delta} \subset \widehat{\mathfrak{t}}^*$. Rather, the projection π induces an inclusion $\pi^* : \widehat{\mathfrak{t}}^* \to \mathfrak{t}^*$, and the natural projection is $\mathfrak{t}^* \to \widetilde{\mathfrak{t}}^*$. From now on, we will identify $\mathfrak{t}^*$ with

$\widehat{\mathfrak{t}}^* \times \widetilde{\mathfrak{t}}^* = \mathbb{R}^k \times \mathbb{R}^m$ in such a way that $\mathbb{R}^k = \pi^*(\widehat{\mathfrak{t}}^*)$, and will translate Δ so that its special point is at $\{0\}$. This implies that Δ is obtained by slicing a polytope $\widetilde{\Delta}'$ (which is affine isomorphic to $\mathbb{R}^k \times \widetilde{\Delta}$ and so has facets parallel to the first k-coordinate directions) by hyperplanes that are in bijective correspondence with the facets of the base $\widehat{\Delta}$.

Here is the formal definition of bundle. Note that here we consider the normals to the facets as belonging to the Lie algebra of the appropriate torus.

DEFINITION 5.1. Let $\widehat{\Delta} = \bigcap_{i=1}^{\widehat{N}}\{x \in \widehat{\mathfrak{t}}^* \mid \langle \widehat{\eta}_i, x\rangle \le \widehat{\kappa}_i\}$ and $\widetilde{\Delta} = \bigcap_{j=1}^{\widetilde{N}}\{y \in \widetilde{\mathfrak{t}}^* \mid \langle \widetilde{\eta}_j, y\rangle \le \widetilde{\kappa}_j\}$ be simple polytopes. We say that a simple polytope $\Delta \subset \mathfrak{t}^*$ is a **bundle** with **fiber** $\widetilde{\Delta}$ over the **base** $\widehat{\Delta}$ if there exists a short exact sequence

$$0 \to \widetilde{\mathfrak{t}} \stackrel{\iota}{\to} \mathfrak{t} \stackrel{\pi}{\to} \widehat{\mathfrak{t}} \to 0$$

so that

- Δ is combinatorially equivalent to the product $\widehat{\Delta} \times \widetilde{\Delta}$.
- If $\widetilde{\eta}_j{}'$ denotes the outward normal to the facet $\widetilde{F}_j{}'$ of Δ which corresponds to $\widehat{\Delta} \times \widetilde{F}_j \subset \widehat{\Delta} \times \widetilde{\Delta}$, then $\widetilde{\eta}_j{}' = \iota(\widetilde{\eta}_j)$ for all $1 \le j \le \widetilde{N}$.
- If $\widehat{\eta}_i{}'$ denotes the outward normal to the facet $\widehat{F}_i{}'$ of Δ which corresponds to $\widehat{F}_i \times \widetilde{\Delta} \subset \widehat{\Delta} \times \widetilde{\Delta}$, then $\pi(\widehat{\eta}_i{}') = \widehat{\eta}_i$ for all $1 \le i \le \widehat{N}$.

The facets $\widetilde{F}_1{}', \dots, \widetilde{F}_{\widetilde{N}}{}'$ correspond bijectively to the facets of the fiber and will be called **fiber facets**, while the facets $\widehat{F}_1{}' \dots, \widehat{F}_{\widehat{N}}{}'$ (which correspond bijectively to the facets of the base) will be called **base facets**.

The fiber facets $\widetilde{F}_j{}'$ are all parallel to the k-plane $\pi^*(\widehat{\mathfrak{t}}^*)$ in $\mathfrak{t}^*$ and (when $m > 1$) are themselves bundles over $\widehat{\Delta}$. For each vertex $\widetilde{v}_\beta = \cap_{j\in\beta}\widetilde{F}_j$ of $\widetilde{\Delta}$, the intersection

$$\bigcap_{j\in\beta} \widetilde{F}_j{}' =: \widehat{\Delta}_{\widetilde{v}_\beta}$$

of the corresponding fiber facets is an k-dimensional polytope, that is parallel to $\pi^*(\widehat{\mathfrak{t}}^*)$ and projects to a polytope in $\widehat{\mathfrak{t}}^*$ that is *analogous* to $\widehat{\Delta}$. In other words $\widehat{\Delta}_{\widetilde{v}_\beta}$ has the same normals $\widehat{\eta}_i$ as $\widehat{\Delta}$ but usually *different* structure constants $\widehat{\kappa}_i$. For example, in the polytope (a) in Figure 5.1 the $\widehat{\Delta}_{\widetilde{v}_\beta}$ are edges of various lengths that are parallel to the x_1-axis, while on the right they are the top and bottom triangles. In contrast, it is not hard to see that the faces

$$\bigcap_{i\in\alpha} \widehat{F}_i{}' =: \widetilde{\Delta}_{\widehat{v}_\alpha}$$

of Δ corresponding to the vertices $\widehat{v}_\alpha = \cap_{i\in\alpha}\widehat{F}_i$ of the base are all affine equivalent to the fiber $\widetilde{\Delta}$. Thus, if a given polytope Δ is a $\widetilde{\Delta}$-bundle over $\widehat{\Delta}$, the fiber polytope $\widetilde{\Delta}$ is completely determined by Δ while the base polytope is only determined modulo the structure constants (though these must remain in the same chamber so that the intersection pattern of the facets, i.e. the fan, does not change).

Observe also that the polytope Δ is the union of k-dimensional parallel slices

$$\widehat{\Delta}_y := \Delta \cap (\mathbb{R}^k \times \{y\}), \quad y \in \widetilde{\Delta}; \tag{5.1}$$

cf. Figures 5.1 and 5.2. The following lemma shows that if Δ is monotone then so is its fiber. Moreover, as we see in parts (ii) and (iii), Δ also determines a particular monotone base polytope.

LEMMA 5.2. *Suppose that the monotone polytope Δ is a $\widetilde{\Delta}$ bundle over $\widehat{\Delta}$ with special point $u_0 = 0$. Then:*
(i) *The fiber $\widetilde{\Delta}$ is monotone.*
(ii) *For each $y \in \widetilde{\Delta}$, the slice $\widehat{\Delta}_y$ in equation (5.1) does not depend on the choice of splitting $\mathfrak{t}^* = \widehat{\mathfrak{t}}^* \times \widetilde{\mathfrak{t}}^*$. It is integral whenever y is.*
(iii) *The slice*

$$\widehat{\Delta}_0 = \Delta \cap \left(\mathbb{R}^k \times \{0\}\right)$$

through the special point of $\widetilde{\Delta}$ is monotone.
(iv) *For any $w \in \mathcal{S}(\widetilde{\Delta})$ the intersection $\widehat{\Delta}_{w\mathbb{R}} := \Delta \cap \left(\mathbb{R}^k \times \{cw : c \in [-1,1]\}\right)$ is a (smooth, integral) Δ_1-bundle over $\widehat{\Delta}$ and is monotone.*

PROOF. Consider a vertex $v_{\alpha\beta}$ of Δ corresponding to the pair of vertices $\widehat{v}_\alpha, \widetilde{v}_\beta$, where $\widehat{v}_\alpha = \cap_{i\in\alpha}\widehat{F}_i \in \widehat{\Delta}$ for some k element subset $\alpha \subset \{1,\dots,\widehat{N}\}$ and similarly for $\widetilde{v}_\beta \in \widetilde{\Delta}$. The edge vectors at $v_{\alpha\beta}$ divide into two groups. There are k primitive edge vectors $e_i^{\alpha\beta}$ in the plane $\mathbb{R}^k \times \widetilde{v}_\beta$ that are parallel to the edge vectors of $\widehat{\Delta}$ through $\widehat{v}_\alpha$ (and hence are independent of the choice of $\widetilde{v}_\beta$). These lie in the face $\widehat{\Delta}_{\widetilde{v}_\beta}$. Similarly, there are m others $e_j^{\alpha\beta}$ that lie in the face $\widetilde{\Delta}_{\widehat{v}_\alpha}$ and project to the edge vectors $\widetilde{e}_j{}^\beta$ of $\widetilde{\Delta}$ through $\widetilde{v}_\beta$, but may also have nonzero components $\widehat{e}_j{}^{\alpha\beta}$ in the $\mathbb{R}^k$ direction that depend on α, β. Moreover, we may label these edges so that $e_i^{\alpha\beta}$ is transverse to the facet $\widehat{F}_i{}'$ for each $i \in \alpha$ and $e_j^{\alpha\beta}$ is transverse to the facet $\widetilde{F}_j{}'$ for each $j \in \beta$.

Now consider (i). Since $v_{\alpha\beta} + \sum_{i\in\alpha} e_i^{\alpha\beta} + \sum_{j\in\beta} e_j^{\alpha\beta} = 0 \in \mathbb{R}^n$, we find

$$w_{\alpha\beta} := v_{\alpha\beta} + \sum_{j\in\beta} e_j^{\alpha\beta} = (\widehat{w}_{\alpha\beta}, 0), \quad \text{where} \quad \widehat{w}_{\alpha\beta} + \sum_{i\in\alpha} e_i^{\alpha\beta} = 0.$$

In particular, the projection of $v_{\alpha\beta} + \sum_{j\in\beta} e_j^{\alpha\beta}$ onto $\widetilde{\mathfrak{t}}^*$ vanishes. Hence each vertex of $\widetilde{\Delta}$ satisfies the vertex-Fano condition with respect to $\{0\}$. Therefore $\{0\} \in \operatorname{int}\widetilde{\Delta}$, and by Remark 3.2 it must be the unique interior integral point in $\widetilde{\Delta}$. This proves (i). (Note that (i) is immediately clear if one thinks of the corresponding fibration of toric manifolds.)

The first statement in (ii) holds because the $\widehat{\Delta}_y$ are the intersections of Δ with the k-dimensional affine planes parallel to $\pi^*(\widehat{\mathfrak{t}}^*)$ and so do not depend on any choices.

Now suppose that y is integral. There is one vertex of $\widehat{\Delta}_y$ corresponding to each vertex $\widehat{v}_\alpha$ of $\widehat{\Delta}$, namely the intersection $\widehat{\Delta}_y \cap \widetilde{\Delta}_{\widehat{v}_\alpha}$, and we must show that this is integral. Since $\widetilde{\Delta}$ is monotone, y can be written as $\widetilde{v}_\beta + \sum k_j \widetilde{e}_j$, where $\widetilde{v}_\beta$ is a vertex, $k_j \ge 0$ are integers, and the $\widetilde{e}_j$ are the primitive integral edge vectors at $\widetilde{v}_\beta$. As explained above, there is a vertex $v_{\alpha\beta}$ of Δ corresponding to the pair α, β, and there are corresponding primitive edge vectors $e_j^{\alpha\beta}$ of Δ at $v_{\alpha\beta}$ that project onto $\widetilde{e}_j$ and lie in the face $\widetilde{\Delta}_{\widehat{v}_\alpha}$. It follows immediately that $v_{\alpha\beta} + \sum k_j e_j^{\alpha\beta}$ is an integral point that lies in $\widetilde{\Delta}_{\widehat{v}_\alpha}$ and projects to y. Hence it equals the intersection $\widehat{\Delta}_y \cap \widetilde{\Delta}_{\widehat{v}_\alpha}$. This proves (ii).

Claim (iii) will follow once we show that $w_{\alpha\beta} = (\widehat{w}_{\alpha\beta}, 0)$ is the vertex of $\widehat{\Delta}_0$ corresponding to $\widehat{v}_\alpha$. But, by construction, each edge $e_j^{\alpha\beta}$ lies in every facet $\widehat{F}_i', i \in$

α, as does $v_{\alpha\beta}$. Therefore

$$w_{\alpha\beta} := v_{\alpha\beta} + \sum_{j\in\beta} \widehat{e}_j{}^{\alpha\beta} \in \bigcap_{i\in\alpha} \widehat{F}_i{}'.$$

Thus it projects to $\widehat{v}_\alpha$ as claimed. This proves (iii).

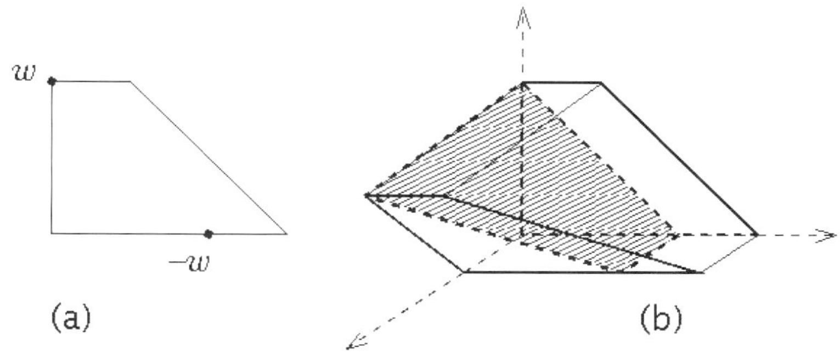

FIGURE 5.2. (a) shows the two points $\pm w \in \mathcal{S}(\widetilde{\Delta})$ where $\widetilde{\Delta}$ is the monotone one point blow up of Δ_2. (b) illustrates a $\widetilde{\Delta}$ bundle over Δ_1, with the corresponding Δ_1-bundle $\Delta_{w\mathbb{R}}$ shaded.

Now consider (iv). It is geometrically clear that $\widehat{\Delta}_{w\mathbb{R}}$ is a Δ_1-bundle over $\widehat{\Delta}$; cf. Figure 5.2. It is integral by (ii).

To see that it is monotone, we shall check the vertex-Fano condition. To this end, note that for every vertex $\widehat{v}_\alpha$ of $\widehat{\Delta}$ there are two vertices of $\widehat{\Delta}_{w\mathbb{R}}$. Call them $v_{\pm\alpha}$, where $v_{+\alpha} \in \widehat{\Delta}_w$. Denote by e the primitive vector along the edge from $v_{+\alpha}$ that projects to $-w \in \mathbb{R}^m$. Then $v_{+\alpha} + e =: \widehat{v}_{0\alpha} \in \widehat{\Delta}_0$. The other edge vectors $\widehat{e}_j{}^{+\alpha}$ of $\widehat{\Delta}_{w\mathbb{R}}$ through $v_{+\alpha}$ are parallel to the edge vectors $\widehat{e}_j{}^{\alpha 0}$ of $\widehat{\Delta}_0$ at the vertex $\widehat{v}_{\alpha 0}$. Hence

$$\widehat{v}_{\alpha 0} + \sum_j \widehat{e}_j{}^{\alpha 0} = \widehat{v}_{+\alpha} + e + \sum_{j\in\alpha} \widehat{e}_j{}^{+\alpha} = \{0\}$$

as required. A similar argument applies to $v_{-\alpha}$. □

It follows that if a monotone polytope Δ is a bundle we may identify its base with the slice $\widehat{\Delta}_0$. Since its fiber and its base are monotone, it makes sense to consider the star Ewald condition for these polytopes.

PROPOSITION 5.3. *Let Δ be a monotone polytope that is a $\widetilde{\Delta}$-bundle over $\widehat{\Delta}$. Suppose that $\widetilde{\Delta}$ and all monotone Δ_1-bundles over $\widehat{\Delta}$ satisfy the star Ewald condition. Then Δ satisfies the star Ewald condition.*

PROOF. Lemma 5.2 for Δ implies that $\widehat{\Delta}$ is monotone. Hence so is the product $\widehat{\Delta} \times \Delta_1$ where $\Delta_1 = [-1, 1]$ is monotone. Using the fact that $\widehat{\Delta} \times \Delta_1$ is star Ewald, one easily deduces that $\widehat{\Delta}$ is as well.

Now, consider the face

$$f = \left(\bigcap_{i\in\alpha} \widehat{F}_i'\right) \cap \left(\bigcap_{j\in\beta} \widetilde{F}_j'\right) =: F_{\alpha\beta}$$

of Δ. Then $\mathrm{Star} f \cap \widehat{\Delta}_0 \neq \emptyset$ provided that $\alpha \neq \emptyset$. Further, because $\{0\}$ is in the interior of all faces of $\widetilde{\Delta}$, the intersection $\mathrm{star} f \cap \widehat{\Delta}_0$ may be identified with $\mathrm{star}\widehat{F}_\alpha$ in $\widehat{\Delta}$, where, as usual, $\widehat{F}_\alpha := \cap_{i\in\alpha}\widehat{F}_i$. Thus, when $\alpha \neq \emptyset$ the star Ewald condition for the face $\widehat{F}_\alpha$ in $\widehat{\Delta}$ implies that for $F_{\alpha\beta}$ in Δ.

Now consider $F_\beta := F_{\emptyset\beta}$, and write $\widetilde{F}_\beta = \cap_{j\in\beta}\widetilde{F}_j$. Thus $F_\beta \subset \mathbb{R}^k \times \widetilde{F}_\beta$ is a union of the slices $\widehat{\Delta}_y, y \in \widetilde{F}_\beta$. By the star Ewald condition for $\widetilde{\Delta}$ there is $w \in \mathcal{S}(\widetilde{\Delta})$ such that $w \in \mathrm{Star}^*\widetilde{F}_\beta$ and $-w \notin \mathrm{Star}\widetilde{F}_\beta$. Then $\mathrm{Star}^* F_\beta$ contains the facet $\widehat{\Delta}_w$ of the polytope $\widehat{\Delta}_{w\mathbb{R}}$ considered in part (iv) of Lemma 5.2. (In fact, $\mathrm{Star}^* F_\beta$ can be identified with the union of all slices $\widehat{\Delta}_y$ with $y \in \mathrm{Star}^*\widetilde{F}_\beta$.) By hypothesis, the facet $\widehat{\Delta}_w$ has the star Ewald condition in $\widehat{\Delta}_{w\mathbb{R}}$. That is, there is an element $\lambda \in \mathcal{S}(\widehat{\Delta}_{w\mathbb{R}})$ in $\widehat{\Delta}_w$. By construction, λ has the form (x, w) for some $x \in \mathbb{R}^k$. Therefore $-\lambda = (-x, -w)$ projects to $-w \notin \mathrm{Star}\widetilde{F}_\beta$ and so $-\lambda \notin \mathrm{Star}F_\beta$. Since $\mathcal{S}(\widehat{\Delta}_{w\mathbb{R}}) \subset \mathcal{S}(\Delta)$, the result follows. □

The following result was proved in the course of the above argument.

LEMMA 5.4. *If $\widehat{\Delta}$ is star Ewald, then a Δ_1-bundle over $\widehat{\Delta}$ is also star Ewald provided that one (and hence both) of its fiber facets contains a symmetric point, i.e. intersects $\mathcal{S}(\Delta)$.*

COROLLARY 5.5. *A product $\widehat{\Delta} \times \widetilde{\Delta}$ of monotone polytopes is star Ewald if and only if its two factors are.*

PROOF. This is easy to check directly. However, the proof of Proposition 5.3 allows one to reduce the proof of the "if" statement to the case of $\Delta_1 \times \Delta_1$. □

COROLLARY 5.6. *Let $\widetilde{\Delta}$ be star Ewald. Then every monotone $\widetilde{\Delta}$-bundle over the simplex Δ_k is star Ewald.*

PROOF. By Proposition 5.3 it suffices to consider the case $\widetilde{\Delta} = \Delta_1$, and to show in this case that the two fiber facets $\widehat{\Delta}_\pm$ are star Ewald. We may choose coordinates $(x, y) \in \mathbb{R}^k \times \mathbb{R}$ so that Δ is given by the inequalities

$$-1 \le y \le 1, \;\; x_i \ge -1, i = 1, \dots, k, \;\; \sum_{i=1}^{k} x_i \le 1 + \alpha y,$$

for some integer $\alpha \ge 0$. Then the fiber facets $\widehat{\Delta}_+ = \{y = 1\}$ and $\widehat{\Delta}_- = \{y = -1\}$ are integral simplices with side lengths $k + 1 + \alpha$ and $k + 1 - \alpha$ respectively. Since the case $\alpha = 0$ is obvious we may suppose that $1 \le \alpha \le k$. Let $v = (v_1, \dots, v_{k+1})$ be a point with $v_{k+1} = 1$ and with precisely α of the coordinates $v_1, \dots, v_k$ equal to 1, while the others are 0. Then $v \in \widehat{\Delta}_+$ while $-v \in \widehat{\Delta}_-$. □

REMARK 5.7. (i) One should be able to prove the analog of Corollary 5.6 for all bases $\widehat{\Delta}$. However, this does not seem easy. The following discussion explains what the problem is, and proves a special case.

Let $\widehat{\eta}_i, i \in \{1, \dots, \widehat{N}\}$, be the (outward) normals of a smooth k-dimensional polytope $\widehat{\Delta}$, chosen so that the first k are the negatives of a standard basis. Then, with (x, y) as above, every Δ_1-bundle over $\widehat{\Delta}$ has the form

$$-1 \le y \le 1, \;\; x_j \ge -1, j = 1, \dots, k, \;\; \langle x, \eta_i \rangle \le \kappa_i - a_i y, \; i > k.$$

If $\widehat{\Delta}$ is monotone, we may take $\kappa_i = 1$ for all i so that Δ is determined by the $\widehat{N} - k$ integers $a_i, i > k$, where we set $a_i = 0, i \le k$. It is easy to check that Δ is monotone provided that it is combinatorially equivalent to a product. This will be the case exactly when the top and bottom facets $\{y = \pm 1\}$ of Δ are analogous. However, it is not clear in general what conditions this imposes on the constants a_i, except that they cannot be too large.[12] More precisely, if the edge $e_{i\ell}$ meets the facets $\widehat{F}_i$ and $\widehat{F}_\ell$ transversally, then then, because the length $L(e^{\pm}_{i\ell})$ of the corresponding edges in $\{x = \pm 1\}$ must be at least 1, we must have

$$L(e_{i\ell}) \ge |a_i - a_\ell| + 1, \quad \forall i, \ell.$$

Thus, if $\widehat{\Delta}$ has many short edges then it supports few monotone bundles. One easy case is when the $|a_i|$ are all ≤ 1. For then the points $(0, \dots, 0, \pm 1)$ lie in $\mathcal{S}(\Delta)$ so that Δ is star Ewald exactly if $\widehat{\Delta}$ is.

One can rephrase these conditions by making a different choice of coordinates. If there is a symmetric point w in $\{y = 1\}$ then we can use w instead of $(0, \dots, 0, 1)$ as the last basis vector, keeping the others unchanged. Then Δ is given by equations of the form

$$-1 \le y \le 1, \quad x_i \ge -1 + b_i y, i = 1, \dots, k, \quad \langle x, \eta_i \rangle \le 1 - b_i y, \ i > k, \tag{5.2}$$

where we must have $|b_i| \le 1$ for all $1 \le i \le \widehat{N}$ because now $(0, \dots, 0, \pm 1) \in \Delta$ by our choice of coordinates. It follows easily that, assuming $\widehat{\Delta}$ is star Ewald, then Δ is star Ewald if and only if it may be given by equations of the form (5.2).

(ii) The problem considered in (i) above is a special case of the following question. Consider a symplectic S^2 bundle $S^2 \to (X, \omega) \xrightarrow{\pi} \widehat{X}$ with symplectic base $(\widehat{X}, \widehat{\omega})$, where we assume that ω is nondegenerate on each fiber. If a subset $L \subset \widehat{X}$ is displaceable by a Hamiltonian isotopy, is it true that its inverse image $\pi^{-1}(L)$ is displaceable in (X, ω)? At first glance, one might think this is obviously true. However one cannot assume that there is a simple relation between ω and $\pi^*(\widehat{\omega})$, and so there is no obvious way to lift an isotopy of $\widehat{X}$ to one of X. In fact the awkwardness of the definition of toric bundle is one indication of the subtlety of this relation.

References

[Au] M. Audin, *The topology of torus actions on symplectic manifolds*, 2nd ed. (2004), Birkhäuser, Basel.

[BEP] P. Biran, M. Entov and L. Polterovich, *Calabi quasimorphisms for the symplectic ball.* Commun. Contemp. Math. **6** (2004), no. 5, 793–802.

[Ch] C.-H. Cho, *Holomorphic discs, Spin structures and Floer Cohomology of the Clifford torus*, Int. Math. Res. Not. **2004** (2004), 1803–1843.

[D] T. Delzant, *Hamiltoniens périodiques et image convexe de l'application moment.* Bull. Soc. Math. France. **116** (1988), 315–339.

[EP06] M. Entov and L. Polterovich, *Quasi-states and symplectic intersections* Comment. Math. Helv. **81** (2006), 75–99.

[EP09] M. Entov and L. Polterovich, *Rigid subsets of symplectic manifolds*, Compos. Math. **145** (2009), no. 3, 773–826.

[Ew] G. Ewald, *On the classification of toric Fano varieties*, Discrete Comput. Geom. **3** (1988), 49–54.

---

[12]The constants a_i determine and are determined by the Chern class of the corresponding $\mathbb{P}^1$ bundle over $\widehat{M}$.

[FOOO1] K. Fukaya, Y.-G. Oh, H. Ohta, K. Ono, *Lagrangian Floer theory on compact toric manifolds, I*, Duke Math. J. **151** (2010), no. 1, 23–175.

[FOOO2] K. Fukaya, Y.-G. Oh, H. Ohta, K. Ono, *Lagrangian Floer theory on compact toric manifolds, II: Bulk deformations*, arxiv-math:0810.5654, to appear in Selecta Math.

[F] W. Fulton, *Introduction to Toric Varieties*, Ann. of Math. Stud. **131**, Princeton University Press, Princeton, 1993.

[KKP] Y. Karshon, L. Kessler, and M. Pinsonnault, *A compact symplectic 4-manifold admits only finitely many inequivalent torus actions*, J. Symplectic Geom. **5** (2007), no. 2, 139–166.

[MT1] D. McDuff and S. Tolman, *Polytopes with mass linear functions, part I* Int. Math. Res. Not. **2010** (2010), no. 8, 1506–1574.

[MT2] D. McDuff and S. Tolman, *Polytopes with mass linear functions II: the 4-dimensional case*, in preparation.

[N] B. Nill, *Gorenstein toric Fano varieties*, Manuscripta Math. **116** (2005), 183–210.

[Ob] M. Øbro, Classification of smooth Fano polytopes, Ph. D. thesis, University of Aarhus 2007.

[OT] Y. Ostrover and I. Tyomkin, *On the quantum homology algebra of toric Fano manifolds*, Selecta Math. (N.S.) **15** (2009), no. 1, 121–149.

Department of Mathematics, Barnard College, Columbia University
E-mail address: dusa@math.columbia.edu

Proceedings of Symposia in Pure Mathematics
Volume **82**, 2011

Equivariant Bredon Cohomology and Čech hypercohomology

Haibo Yang

ABSTRACT. We give a new construction of $RO(G)$-graded equivariant cohomology theories from the sheaf-theoretic viewpoint. It induces a natural isomorphism between the Čech hypercohomology and the $RO(G)$-graded Bredon cohomology with some proper coefficients.

1. Introduction

For a finite group G, Illman ([**Ill78**]) showed that every smooth G-manifold admits a smooth equivariant triangulation onto a regular simplicial G-complex. With this result we extend to the equivariant context a well-known theorem [**BT82**, p. 42] about the existence of a good cover on a smooth manifold.

THEOREM 1.1 (Theorem 2.11). *Every smooth G-manifold has an equivariant good cover. Moreover, the equivariant good covers are cofinal in the set of all open covers of a G-manifold X.*

On the other hand, an ordinary $RO(G)$-graded cohomology theory is defined on any G-space X [**M$^+$96, LMM81**]. It is a cohomology theory on X with coefficients in a Mackey functor M and is one that is graded by the real orthogonal representation ring $RO(G)$ of G. Since ordinary equivariant cohomology was first defined by Bredon [**Bre72**], we call it $RO(G)$-graded Bredon cohomology theory. We may apply this theory to a G-manifold X with a coefficient system $\underline{M}$ associated to a discrete $\mathbb{Z}[G]$-module M. The Mackey functor $\underline{M}$ is defined by $\underline{M}(G/H) \overset{\text{def}}{=} M^H$, the H-fixed point set of M, and the value of the contravariant part on the projection $G/H \to G/K$, for $H \leq K \leq G$, is the inclusion of M^K into M^H, while that of the covariant part is $M^H \to M^K : x \mapsto \sum k_i x$, where $\{k_i\}$ are a set of representatives of left cosets K/H. For any finite representation V of G we define a cochain complex of presheaves $M(V)$ and show that, for any equivariant good cover $\mathscr{U}$ of X and for any $n \in \mathbb{Z}$, there is a natural isomorphism

$$\check{\mathbb{H}}^n_G(\mathscr{U}, M(V)) \cong H^{V+n-\dim(V)}_{\mathrm{Br}/G}(X, \underline{M}).$$

2010 *Mathematics Subject Classification.* Primary 55N91; Secondary 55N30 55N05 .

Key words and phrases. RO(G)-graded equivariant cohomology, Cech hyper-cohomology, sheaf cohomology.

The author was partially supported by a grant from Nanchang Hangkong University, China (NCHU Research Grant No. EA201007057; Renovation Project Grant No. 207004 22057).

Here for a complex of presheaves $\mathcal{F}^*$ and an open cover $\mathscr{U}$ of a topological space X, $\check{\mathbb{H}}^n_G(\mathscr{U}, \mathcal{F}^*)$ denotes the n-th Čech (equivariant) hypercohomology group with coefficients in $\mathcal{F}^*$. It is obtained by first forming a double complex $\mathscr{C}^{**}$ over $\mathcal{F}^*$, where $\mathscr{C}^{pq} = \mathscr{C}^p(\mathscr{U}, \mathcal{F}^q)$ and $\mathscr{C}^*(\mathscr{U}, \mathcal{F}^q)$ is the Čech resolution of $\mathcal{F}^q$ for each q. We then apply the global section functor $\Gamma(X, -)$ to $\mathscr{C}^{**}$, and use D^{**} to denote the resulted new double complex. The n-th Čech hypercohomology group $\check{\mathbb{H}}^n_G(\mathscr{U}, \mathcal{F}^*)$ is by definition the n-th cohomology group of the total complex of D^{**}. For details, see [**Har77**, Section 3.4] and [**Wei94**, Chapter 5].

Since equivariant good covers are cofinal, we have the following main theorem.

THEOREM 1.2 (Theorem 4.5). *There is a natural isomorphism*

$$\check{\mathbb{H}}^n_G(X, M(V)) \cong H^{V+n-\dim(V)}_{Br/G}(X, \underline{M}),$$

where $\check{\mathbb{H}}^n_G(X, \mathcal{F}^*)$ *is the direct limit of the system* $\check{\mathbb{H}}^n_G(\mathscr{U}, \mathcal{F}^*)$ *over equivariant good covers* $\mathscr{U}$ *on* X.

The last part of this paper mentions also some applications of our results.

2. Equivariant good cover of a G-manifold

Recall that an open cover $\mathscr{U} = \{U_\alpha\}$ of a smooth manifold M is called a *good cover* if all nonempty finite intersections $U_{\alpha_0 \dots \alpha_n} = U_{\alpha_0} \cap \cdots \cap U_{\alpha_n}$ are contractible. There is a classical theorem (c.f. [**BT82**, Theorem 5.1]) stating that every smooth manifold M has a good cover when considering the geodesic convex balls for a Riemannian metric on M ([**GHL04**, Corollary 2.89], [**dC92**, p. 70]). We extend this theorem to the equivariant case for a finite group G.

Let us start with briefly describing some notations and important properties about usual simplicial complexes needed for the latter part of this section. Since there are many excellent textbooks on this subject, we will not go into any details. The reader can refer to [**Pra06**], [**Bre72**] and [**Rot88**].

Let K be a simplicial complex. We write $\mathrm{Vert}(K)$ for the vertex set of K and $|K|$ for the associated *polyhedron* or the *underlying space* of K. K is *locally finite* if every point $x \in |K|$ has a neighborhood intersecting only finite many simplices of K ([**Mun66**, p. 69]). Here for our purpose it suffices to assume that all simplicial complexes are locally finite.

Pick any $x \in |K|$. The *carrier* $\mathrm{carr}(x)$ of x is defined to be the (unique) smallest simplex of K containing x. In some cases we write $\mathrm{carr}_K(x)$ for $\mathrm{carr}(x)$ in order to emphasize K. On the other hand, if v is a vertex of K then the *open star* of v is $\mathrm{st}_K(v) \overset{\text{def}}{=} \{x \in |K| \mid v \in \mathrm{carr}(x)\}$.

PROPOSITION 2.1. *Let K be a simplicial complex. Then*

(1) $x \in \mathrm{st}_K(v)$ *if and only if* $v \in \mathrm{carr}(x)$ *and for* $x, y \in |K|$, $y \in \mathrm{carr}(x)$ *implies* $\mathrm{carr}(y) \subset \mathrm{carr}(x)$.

(2) If $v_0, \dots, v_n$ *are vertices of a simplicial complex* K *then* $\bigcap_i \mathrm{st}_K(v_i) \neq \varnothing$ *if and only if* $\langle v_0, \dots, v_n \rangle$ *is a simplex of* K.

(3) For each vertex v *of* K, $\mathrm{st}_K(v) = \bigcup_{\substack{s \in K \\ v \in \mathrm{Vert}(s)}} s^\circ$. *Furthermore, the set* $\mathscr{U} = \{\mathrm{st}_K(v) \mid v \in \mathrm{Vert}(K)\}$ *is a good cover of* K.

Now let us consider an action of a group G on the simplicial complexes.

DEFINITION 2.2. Let G be a finite group.

(1) A *simplicial G-complex* consists of a simplicial complex K together with a G-action on K such that for every $g \in G$ the map $g : K \to K$ is a simplicial homeomorphism.
(2) A simplicial G-complex K is a *regular G-complex* if the following conditions are satisfied.
    (R1) If vertices v and gv belong to the same simplex then $v = gv$.
    (R2) If $s = \langle v_0, \dots, v_n \rangle$ is a simplex of K and $s' = \langle g_0 v_0, \dots, g_n v_n \rangle$, where $g_i \in G$, $i = 0, \dots, n$, also is a simplex of K then there exists $g \in G$ such that $gv_i = g_i v_i$, for $i = 0, \dots, n$.

REMARK 2.3. (a) If K is a simplicial G-complex, then the underlying space $|K|$ carries a natural G-action so that $|K|$ is a G-space.
(b) In fact the condition (R2) implies (R1) since if v and gv belong to some simplex, then $\langle v, v \rangle$ and $\langle v, gv \rangle$ are simplices of K, so for some g', $v = g'v = gv$.

PROPOSITION 2.4. *Let K be a simplicial G-complex. Then for any vertex v of K and any $g \in G$, we have*

$$\mathrm{st}_K(gv) = g(\mathrm{st}_K(v)).$$

PROOF. Simple modification to Proposition 2.1. □

The following proposition shows that any simplicial G-complex becomes regular after passing to the second barycentric subdivision. So restricting to regular G-complexes is not seriously harmful. The following proposition comes from [**Bre72**].

PROPOSITION 2.5 ([**Bre72**]). *If K is a simplicial G-complex, then the induced action on the barycentric subdivision K' satisfies (R1). If (R1) is satisfied for K, then (R2) is satisfied for K'.*

For a subgroup H of G, we define $K^H \stackrel{\mathrm{def}}{=} |K|^H$, the fixed point set of $|K|$ by H. The next proposition shows that when K is a regular G-complex, K^H a subcomplex of K. The proof is a straight application of Proposition 2.1.

PROPOSITION 2.6. *Let K be a regular G-complex.*

(1) For any subgroup $H \leq G$, K^H is a (nonequivariant) subcomplex of K.
(2) For $x \in K^H$, $\mathrm{carr}_{K^H}(x) = \mathrm{carr}_K(x)$. Moreover, if v is a vertex of K^H, then $\mathrm{st}_{K^H}(v) = \mathrm{st}_K(v) \cap K^H$.

Let $\mathscr{U} = \{U_\alpha\}_{\alpha \in I}$ be an open cover of a paracompact G-space X. Then for any $g \in G$, the set $g\mathscr{U} \stackrel{\mathrm{def}}{=} \{gU_\alpha \mid U_\alpha \in \mathscr{U}\}$ is still an open cover of X. If $g\mathscr{U} = \mathscr{U}$ for all g, we say $\mathscr{U}$ is *G-invariant* or just *invariant* for simplicity. In this case there is an induced action of G on the index set I defined by $g\alpha$ being the unique index with $U_{g\alpha} = gU_\alpha$.

If $\mathscr{U}$ and $\mathscr{V} = \{V_\beta\}_{\beta \in J}$ are open covers, then

$$\mathscr{U} \cap \mathscr{V} = \{U_\alpha \cap V_\beta \mid U_\alpha \in \mathscr{U}, V_\beta \in \mathscr{V}\}$$

is an open cover which refines both $\mathscr{U}$ and $\mathscr{V}$. Clearly (note that G is finite) $\bigcap_{g \in G} g\mathscr{U}$ is an invariant cover refining $\mathscr{U}$. Moreover this is locally finite if $\mathscr{U}$ is. Thus, for X paracompact, the locally finite invariant covers are cofinal in the set of all covers of X.

Now let $\mathscr{U} = \{U_\alpha\}_{\alpha \in I}$ be a locally finite invariant cover of X and let $f = \{f_\alpha\}_{\alpha \in I}$ be a partition of unity subordinate to $\mathscr{U}$ (in particular, $\operatorname{supp}(f_\alpha) \subset U_\alpha$). Then f is called a *G-partition of unity* if $f_{g\alpha}(gx) = f_\alpha(x)$ for all g, x and α. Let $f = \{f_\alpha\}$ be an arbitrary partition of unity subordinate to the invariant cover $\mathscr{U}$. Define $\tilde{f}_\alpha(x) = \frac{1}{|G|}\sum_g f_{g\alpha}(gx)$. Then $\tilde{f} = \{\tilde{f}_\alpha\}_{\alpha \in I}$ is a G-partition of unity.

Let $\mathscr{U} = \{U_\alpha\}_{\alpha \in I}$ be an open cover of a topological space X. If the index set $\{\alpha\}_{\alpha \in I}$ is ordered we associate a simplicial set $\mathcal{N}(\mathscr{U})$, called the *nerve* of the cover $\mathscr{U}$. If $\sigma^n = (\alpha_0, \dots, \alpha_n) \in \mathcal{N}(\mathscr{U})_n$, we denote by U_{σ^n} or $U_{\alpha_0 \dots \alpha_n}$ the nonempty finite intersection $U_{\alpha_0} \cap \dots \cap U_{\alpha_n}$.

There is a simplicial complex $\operatorname{Comp}(\mathcal{N}(\mathscr{U}))$ associated to the nerve $\mathcal{N}(\mathscr{U})$ whose vertices $\{v_\alpha\}$ are in one-to-one correspondence with the index set $\{\alpha \mid \alpha \in I\}$. A set $\{v_{\alpha_0}, \dots, v_{\alpha_n}\}$ is a simplex of $\operatorname{Comp}(\mathcal{N}(\mathscr{U}))$ if and only if $U_{\alpha_0} \cap \dots \cap U_{\alpha_n} \neq \varnothing$, that is, if and only if $(\alpha_0, \dots, \alpha_n) \in \mathcal{N}(\mathscr{U})_n$.

Similarly, we define a simplicial space $\mathcal{N}^{Top}(\mathscr{U})$ as follows. Let

$$\mathcal{N}^{Top}(\mathscr{U})_n = \coprod_{(\alpha_0, \dots, \alpha_n) \in \mathcal{N}(\mathscr{U})_n} U_{\alpha_0 \dots \alpha_n}$$

with disjoint union topology. For each nondecreasing function $f : [m] \to [n]$, the induced map $f^* : \mathcal{N}^{Top}(\mathscr{U})_n \to \mathcal{N}^{Top}(\mathscr{U})_m$ is defined by $f^*|_{U_{\alpha_0 \dots \alpha_n}} : U_{\alpha_0 \dots \alpha_n} \to U_{\alpha_{f(0)} \dots \alpha_{f(m)}}$, where the latter is either an inclusion or the identity map.

DEFINITION 2.7. An invariant open cover $\mathscr{U}$ of a G-space X is a *regular G-cover* if the complex associated to its nerve $\operatorname{Comp}(\mathcal{N}(\mathscr{U}))$ is a regular G-complex, that is, if it satisfies the following two conditions.

(RC1) For $U_\alpha \in \mathscr{U}$ and $g \in G$, if $U_\alpha \cap gU_\alpha \neq \varnothing$ then $U_\alpha = gU_\alpha$.

(RC2) If $U_0, \dots, U_n$ are members of $\mathscr{U}$ and $g_0, \dots, g_n$ are elements in G, and if the intersections $U_0 \cap \dots \cap U_n$ and $g_0U_0 \cap \dots \cap g_nU_n$ are nonempty, then there exists $g \in G$ such that $gU_i = g_iU_i$ for all $i \leq n$.

THEOREM 2.8. *Let X be a paracompact G-space, where G is finite. Then the locally finite, regular G-covers of X are cofinal in the set of open covers of X.*

PROOF. Pick an invariant cover $\mathscr{U}$ of X. Let $\operatorname{Comp}(\mathcal{N}(\mathscr{U}))$ be the simplicial complex associated to the nerve of $\mathscr{U}$. Then $\operatorname{Comp}(\mathcal{N}(\mathscr{U}))$ is a simplicial G-complex. Let $f = \{f_\alpha\}$ be a G-partition of unity subordinate to $\mathscr{U}$ and let $\bar{f} : X \to |\operatorname{Comp}(\mathcal{N}(\mathscr{U}))|$ be the associated map with $\bar{f}(x) = \sum_\alpha f_\alpha(x)v_\alpha$. Then $\bar{f}$ is a well-defined G-map since all but finite $f_\alpha = 0$ and

$$\begin{aligned}\bar{f}(gx) = \sum_\alpha f_\alpha(gx)v_\alpha &= \sum_\alpha f_{g^{-1}\alpha}(x)v_\alpha \\ &= g\sum_\alpha f_{g^{-1}\alpha}(x)g^{-1}v_\alpha = g\sum_\alpha f_\alpha(x)v_\alpha = g\bar{f}(x).\end{aligned}$$

For any map $f : X \to |K|$ to a polyhedron, let $f^{-1}(\mathrm{st}_K)$ denote the open cover of X by inverse images of open stars of vertices of K. Suppose that K is a G-complex and that f is equivariant. Then $f^{-1}(\mathrm{st}_K)$ is an invariant cover by Proposition 2.4. Moreover, if K is a regular G-complex then $f^{-1}(\mathrm{st}_K)$ is a regular G-cover. This is from the fact that if $U_0 \cap \dots \cap U_n \neq \varnothing \neq g_0U_0 \cap \dots \cap g_nU_n$, where $U_i = f^{-1}(\mathrm{st}_K(v_i))$, then by Proposition 2.1, $\langle v_0, \dots, v_n \rangle$ and $\langle g_0v_0, \dots, g_nv_n \rangle$ are simplices of K. Now the regularity of K implies that $f^{-1}(\mathrm{st}_K)$ is regular.

Back to the equivariant map $\bar{f}$, note that $\bar{f}^{-1}(\mathrm{st}_{\mathrm{Comp}(\mathcal{N}(\mathscr{U}))})$ is a refinement of $\mathscr{U}$. Actually, for any α, $\bar{f}^{-1}(\mathrm{st}_{\mathrm{Comp}(\mathcal{N}(\mathscr{U}))}(v_\alpha)) = f_\alpha^{-1}((0,1]) \subset U_\alpha$. Let L be the second barycentric subdivision of $\mathrm{Comp}(\mathcal{N}(\mathscr{U}))$ such that $|L| = |\mathrm{Comp}(\mathcal{N}(\mathscr{U}))|$ and L is a regular G-complex by Proposition 2.5. So $\mathscr{V} = \bar{f}^{-1}(\mathrm{st}_L)$ is a regular G-cover which refines $\mathscr{U}$. □

PROPOSITION 2.9. *Let X be a smooth G-manifold. Then*

(1) *There exists a regular simplicial G-complex K and a smooth equivariant triangulation $h : K \to X$.*
(2) *If $h : K \to X$ and $h_1 : L \to X$ are smooth equivariant triangulations of X there exist equivariant subdivisions K' and L' of K and L, respectively, such that K' and L' are G-isomorphic.*

PROOF. See [**Ill78**]. □

Let $\mathscr{U} = \{U_\alpha\}_{\alpha \in I}$ be an open cover of G-space X. For any subgroup H of G and $\alpha \in I$, let $U_\alpha^H = U_\alpha \cap X^H = \{x \in U_\alpha \mid hx = x \text{ for all } h \in H\}$. Denote by $\mathscr{U}^H$ the collection of $\{U_\alpha^H\}_{\alpha \in I}$. It is clear that $\mathscr{U}^H$ is an open cover of X^H.

DEFINITION 2.10. $\mathscr{U}$ is called an *equivariant good cover* of X if it is a regular G-cover (see Definition 2.7) and $\mathscr{U}^H$ is a good cover of X^H for all subgroups $H \leq G$.

THEOREM 2.11. *Every smooth G-manifold has an equivariant good cover. Moreover, the equivariant good covers are cofinal in the set of open covers of a G-manifold X.*

PROOF. By Theorem 2.9 it is no loss to assume X is a realization of a *regular* simplicial G-complex K. Consider the open cover $\mathscr{W} = \{\mathrm{st}_K(v) \mid v \in \mathrm{Vert}(K)\}$. By Proposition 2.4 $\mathscr{W}$ is G-invariant. Moreover, We claim that $\mathscr{W}$ is a regular G-cover. The proof is as follows. Let $U = \mathrm{st}_K(v) \in \mathscr{W}$ and $g \in G$ with $\varnothing \neq U \cap gU = \mathrm{st}_K(v) \cap \mathrm{st}_K(gv)$. It follows that $\langle v, gv\rangle$ is a simplex in K by Proposition 2.1. The regularity of K yields $v = gv$ and hence $U = gU$. If for $i = 0, \dots, n$, $U_i = \mathrm{st}_K(v_i)$ are members of $\mathscr{W}$ and g_i are members of G such that $U_0 \cap \cdots \cap U_n$ and $g_0U_0 \cap \cdots \cap g_nU_n = \mathrm{st}_K(g_0v_0) \cap \cdots \cap \mathrm{st}_K(g_nv_n)$ are nonempty, then again by Proposition 2.1 there are two simplices in K: $\langle v_0, \dots, v_n\rangle$ and $\langle g_0v_0, \dots, g_nv_n\rangle$. Since K is regular, there exists $g \in G$ such that $gv_i = g_iv_i$ for all i which is equivalent to $gU_i = g_iU_i$ for all i. So by Definition 2.7 $\mathscr{W}$ is a regular G-cover.

For any subgroup H of G, the Proposition 2.6 (1) shows that K^H is a simplicial subcomplex of K and X^H is homeomorphic to K^H. Pick an element $U = \mathrm{st}_K(v)$ of $\mathscr{W}$. Consider the intersection $U \cap K^H = \mathrm{st}_K(v) \cap K^H$. If $v \in \mathrm{Vert}(K^H)$ then by Proposition 2.6 (2), $U \cap K^H = \mathrm{st}_{K^H}(v)$. If $v \notin K^H$, we claim that $U \cap K^H = \varnothing$. To justify this, assume $U \cap K^H \neq \varnothing$. Pick $x \in U \cap K^H$. Then $x \in U = \mathrm{st}_K(v)$ implies $v \in \mathrm{carr}_K(x)$ by Proposition 2.1 and $x \in K^H$ yields $\mathrm{carr}_K(x) \subset K^H$ by Proposition 2.6 (2). Hence $v \in K^H$, contradicting the assumption $v \notin K^H$. So $\mathscr{W}^H = \{U \cap K^H \mid U \in \mathscr{W}\} = \{\mathrm{st}_{K^H}(v) \mid v \in \mathrm{Vert}(K^H)\}$, and hence $\mathscr{W}^H$ is a good cover of K^H.

Note that a barycentric subdivision a regular G-complex is still regular. Then for any given open cover $\mathscr{U}$ of X, there exists an integer m such that the m-th barycentric subdivision $K^{(m)}$ of the above K has the properties that $\mathscr{V} = \{\mathrm{st}_{K^{(m)}}(v) \mid v \in \mathrm{Vert}(K^{(m)})\}$ refines $\mathscr{U}$ and that $\mathscr{V}$ is still an equivariant good cover since $\mathscr{V}$ is again the set of open stars of the regular G-complex $K^{(m)}$, which shows the cofinality of equivariant good covers in the set of open covers of X. □

COROLLARY 2.12. *Let X be a G-manifold. Then there is an open cover consisting of G-invariant subspaces such that every finite intersection of the elements in this open cover is homeomorphic to the orbit of a contractible space, i.e., a space of the form $G/H \times D$, where H is a subgroup of G and D is contractible.*

PROOF. By the proof of Theorem 2.11, the G-manifold X has an equivariant good cover $\mathscr{U} = \{U_\alpha\}$ such that each U_α is the star of a vertex $v_\alpha \in \mathrm{Vert}(K)$. Here K is a regular G-complex. Now define a new open cover $\mathscr{V} = \{V_\alpha\}$ by letting $V_\alpha = \bigcup_{g\in G} g(\mathrm{st}_K(v_\alpha))$. Then V_α is G-invariant and every finite intersection $V_{\alpha_0} \cap \cdots \cap V_{\alpha_p}$ is homeomorphic to $G/H \times D$ where D is the contractible space $\mathrm{st}_{K/G}(\overline{v}_{\alpha_0}) \cap \cdots \cap \mathrm{st}_{K/G}(\overline{v}_{\alpha_p})$. □

The nerve of an equivariant good cover carries a great deal of information on the G-homotopy structure of X. Let us first review the ideas of "fat realization" of a simplicial space introduced by Segal.

DEFINITION 2.13. Let A be a simplicial space. The *fat geometric realization* of A is the topological space

$$\|A\| \stackrel{\mathrm{def}}{=} \left(\coprod_{n\geq 0} A_n \times \Delta^n \right) / \sim$$

where Δ^n is the standard n-simplex and the relation is $(\partial_i(x), t) \sim (x, \partial^i(t))$, for $\partial^i : \Delta^n \to \Delta^{n+1}$ the inclusion as the ith face and $\partial_i : A_{n+1} \to A_n$ the face map for A.

For the following we use both fat and usual realization rather than usual realization only since the fat realization filters unnecessary duplicate information induced from degenerate maps while the usual one does not.

If A is a simplicial G-space, then the fat realization $\|A\|$ naturally carries a G-action so that $\|A\|$ is a G-space.

A simplicial map f between simplicial spaces A and A' induces a map $\|f\| : \|A\| \to \|A'\|$. If f is a simplicial G-map between simplicial G-spaces then $\|f\|$ is a G-map between topological G-spaces.

PROPOSITION 2.14. *Let A and A' be simplicial spaces and let $f : A \to A'$ be a simplicial map.*

(1) If $f_n : A_n \to A'_n$ is a homotopy equivalence for all n then $\|f\| : \|A\| \to \|A'\|$ is a homotopy equivalence.
(2) $\|A \times A'\|$ is homotopy equivalent to $\|A\| \times \|A'\|$.
(3) The ith degeneracy map $\eta_i : [n] \to [n-1]$ induces a map $s_i : A_{n-1} \to A_n$ and s_i maps A_{n-1} into A_n as a retraction. If the inclusion $s_i(A_{n-1}) \hookrightarrow A_n$ is a closed cofibration for all i and n, then $\|A\| \to |A|$ is a homotopy equivalence.

PROOF. See [**Seg74**]. □

Let $\mathscr{U} = \{U_\alpha\}_{\alpha\in I}$ be an open cover of a topological space X. Recall that if $\sigma^n = (\alpha_0, \dots, \alpha_n) \in \mathcal{N}(\mathscr{U})_n$, we denote by U_{σ^n} the nonempty finite intersection $U_{\alpha_0} \cap \dots \cap U_{\alpha_n}$. Let $X_\mathscr{U}$ be the fat realization $\|\mathcal{N}^{Top}(\mathscr{U})\|$, i.e.

$$X_\mathscr{U} = \left(\coprod_{\substack{n\geqslant 0 \\ \sigma^n \in \mathcal{N}(\mathscr{U})_n}} U_{\sigma^n} \times \Delta^n_{\sigma^n} \right) / \sim$$

where $\Delta^n_{\sigma^n}$ is the standard n-simplex with vertices $v_{\alpha_0}, \dots, v_{\alpha_n}$ and the equivalence relation is $(\partial_i(x), t) \sim (x, \partial^i(t))$, where $\partial^i : \Delta^{n-1} \to \Delta^n$ is the ith face map and ∂_i is the inclusion $U_{\alpha_0\dots\alpha_n} \to U_{\alpha_0\dots\widehat{\alpha_i}\dots\alpha_n}$.

Let $\pi : \coprod_{\sigma^n} (U_{\sigma^n} \times n_{\sigma^n}) \to X_\mathscr{U}$ be the quotient map.

PROPOSITION 2.15. *If $\mathscr{U} = \{U_\alpha\}_{\alpha\in I}$ is a locally finite open cover of a paracompact space X, then the fat realization $X_\mathscr{U} = \|\mathcal{N}^{Top}(\mathscr{U})\|$ is homotopy equivalent to X.*

PROOF. For each $\sigma^n = (\alpha_0 \dots \alpha_n) \in \mathcal{N}(\mathscr{U})$ let p_{σ^n} be the composite of maps $U_{\sigma^n} \times \Delta^n_{\sigma^n} \xrightarrow{p_1} U_{\sigma^n} \hookrightarrow X$, where p_1 is the first coordinate projection. The set of maps p_{σ^n} induces a map $p : \coprod_{\substack{n\geqslant 0 \\ \sigma^n \in \mathcal{N}(\mathscr{U})_n}} (U_{\sigma^n} \times \Delta^n_{\sigma^n}) \to X$. Easy to verify p preserves the equivalence relation, so there is a unique map $q : X_\mathscr{U} \to X$ such that $q\pi = p$.

$$\begin{array}{ccc} \coprod_{\sigma^n}(U_{\sigma^n} \times \Delta^n_{\sigma^n}) & \xrightarrow{\ p\ } & X \\ \Big\downarrow{\scriptstyle\pi} & \nearrow{\scriptstyle q} & \\ X_\mathscr{U} & & \end{array}$$

For every point $x \in X$, let $\{\alpha_0, \dots, \alpha_n\}$ be the set of *all* the indices such that $x \in U_{\alpha_i}$. That is, $x \notin U_\alpha$ for all $\alpha \neq \alpha_0, \dots, \alpha_n$. This set is finite since the cover is locally finite. Then the preimage $q^{-1}(x)$ is just the n-simplex $\{x\} \times \Delta^n_{\alpha_0\dots\alpha_n}$. Hence every point in $q^{-1}(x)$ can be represented as $\{x\} \times \sum_{i=1}^n t_i v_{\alpha_i}$, where $t_i \geq 0$, $\sum t_i = 1$.

Since X is paracompact, there exists a partition of unity $\{f_\alpha\}$ subordinate to the cover $\{U_\alpha\}$. In particular, $\mathrm{supp}(f_\alpha) \subset U_\alpha$ for each α. Pick $x \in X$ and let $\{\alpha_0, \dots, \alpha_n\}$ be the set of *all* the indices such that $x \in U_{\alpha_i}$. Then the set of α's such that $f_\alpha(x) > 0$ is a subset of $\{\alpha_0, \dots, \alpha_n\}$ and hence is finite. Now consider a map $s : X \to \coprod_{\sigma^n} U_{\sigma^n} \times \Delta^n_{\sigma^n}$, $x \mapsto \{x\} \times \sum f_{\alpha_i}(x) v_{\alpha_i} \in U_{\alpha_0\dots\alpha_n} \times \Delta^n_{\alpha_0\dots\alpha_n} \subset \coprod_{\sigma^n} U_{\sigma^n} \times \Delta^n_{\sigma^n}$, and let $r : X \to X_\mathscr{U}$ be the composite πs. Clearly, $qr = \mathrm{id}_X$. We need to verify that $rq \simeq \mathrm{id}_{X_\mathscr{U}}$. Suppose that a point x belongs to sets $U_{\alpha_0}, \dots, U_{\alpha_n}$ and does not belong to any other U_α. Then the points $y = \{x\} \times \sum t_i v_{\alpha_i}$ and $r(q(y)) = \{x\} \times \sum f_{\alpha_i}(x) v_{\alpha_i}$ belong to the simplex with vertices $v_{\alpha_0}, \dots, v_{\alpha_n}$. The required homotopy uniformly moves $r(q(y))$ to y along the segment joining these points. □

PROPOSITION 2.16. *If $\mathscr{U} = \{U_\alpha\}_{\alpha\in I}$ is a good cover of a topological space X, then the fat realization $X_\mathscr{U} = \|\mathcal{N}^{Top}(\mathscr{U})\|$ is homotopy equivalent to the usual geometric realization $|\mathcal{N}(\mathscr{U})|$ of the nerve $\mathcal{N}(\mathscr{U})$.*

PROOF. Proposition 2.14 (1) implies that if $\mathscr{U}$ is a good cover then $X_{\mathscr{U}}$ is homotopy equivalent to the fat realization $\|\mathcal{N}(\mathscr{U})\|$ of the nerve $\mathcal{N}(\mathscr{U})$. Here we identify simplicial sets with discrete simplicial spaces. On the other hand, as simplicial sets $\|\mathcal{N}(\mathscr{U})\|$ is homotopy equivalent to $|\mathcal{N}(\mathscr{U})|$ by Proposition 2.14 (3). □

COROLLARY 2.17. *If $\mathscr{U} = \{U_\alpha\}_{\alpha \in I}$ is a locally finite good cover of a paracompact space X, then the normal realization $|\mathcal{N}(\mathscr{U})|$ of the nerve $\mathcal{N}(\mathscr{U})$ is homotopy equivalent to X.*

Now let us turn to the equivariant case.

LEMMA 2.18. *If A is a simplicial G-space, then the realizations $|A|$ and $\|A\|$ inherit G-actions such that*

$$|A^H| = |A|^H \quad \text{and} \quad \|A^H\| = \|A\|^H$$

for all subgroups H of G.

THEOREM 2.19. *If $\mathscr{U} = \{U_\alpha\}_{\alpha \in I}$ is a locally finite equivariant good cover of a G-CW complex X, then the normal realization $|\mathcal{N}(\mathscr{U})|$ of the nerve $\mathcal{N}(\mathscr{U})$ is G-homotopy equivalent to X.*

PROOF. The realization $|\mathcal{N}(\mathscr{U})|$ is a G-space since $\mathscr{U}$ is G-invariant. With the natural CW complex structure on a realization, $|\mathcal{N}(\mathscr{U})|$ becomes a G-CW complex. So it is sufficient to show that $|\mathcal{N}(\mathscr{U})|$ is weakly G-homotopy equivalent to X. We prove this by showing that for any subgroup H of G, $|\mathcal{N}(\mathscr{U})|^H$ is homotopy equivalent to X^H.

By definition, if $\mathscr{U}$ is an equivariant good cover of X then $\mathscr{U}^H$ is a good cover of X^H, so by Corollary 2.17 $|\mathcal{N}(\mathscr{U}^H)|$ is homotopy equivalent to X^H. But $\mathcal{N}(\mathscr{U}^H) = \mathcal{N}(\mathscr{U})^H$. Hence, together with Lemma 2.18, we have

$$|\mathcal{N}(\mathscr{U})|^H = |\mathcal{N}(\mathscr{U})^H| = |\mathcal{N}(\mathscr{U}^H)| \simeq X^H. \qquad \square$$

REMARK 2.20. By the conclusions of the above theorem and Theorem 2.11, for every smooth G-manifold there exists an equivariant good cover such that the normal realization of the nerve of this cover is G-homotopy equivalent to the G-manifold.

3. Presheaves on G-Manifolds

Given a finite group G, let G-*Man* denote the category of smooth manifolds with smooth G-action and equivariant smooth morphisms. Here we always assume a manifold is paracompact and Hausdorff. Given $U \in G$-*Man*, let $\widehat{U}$ denote the full subcategory of G-*Man*$\downarrow U$ consisting of equivariant finite covering maps $p_X : X \to U$. The morphisms in $\widehat{U}$ from $X \xrightarrow{p_X} U$ to $X' \xrightarrow{p_{X'}} U$ are G-maps $\phi : X \to X'$ such that the following diagram commutes:

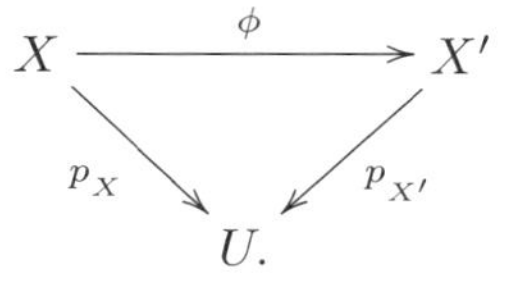

In particular, when $U = pt$, a one-point space, $\widehat{U}$ is the category G-Fin of finite G-sets.

PROPOSITION 3.1. *Let $f : V \to U$ be a morphism in G-Man. Given $p_X : X \to U$ in $\widehat{U}$, let $f^*X = V \times_U X$ be the pull-back of X along f. Then the morphism $p_{f^*X} : f^*X \to V$ is in $\widehat{V}$.*

$$\begin{array}{ccc} f^*X & \xrightarrow{q_X} & X \\ {\scriptstyle p_{f^*X}}\downarrow & & \downarrow{\scriptstyle p_X} \\ V & \xrightarrow[f]{} & U \end{array}$$

PROOF. The fiber $(p_{f^*X})^{-1}(b)$ on b is homeomorphic to $(p_X)^{-1}(f(b))$. $\square$

DEFINITION 3.2. A *Mackey presheaf* M on G-Man is a contravariant functor $M : G\text{-}Man \to Ab$ which, for each $U \in G$-Man, has additional covariant structure on the subcategory $\widehat{U}$. That is, for any morphism $p_X : X \to U$ in $\widehat{U}$, M induces two morphisms of abelian groups, $(p_X)_* : M(X) \to M(U)$ and $(p_X)^* : M(U) \to M(X)$, satisfying the following condition. If

$$\begin{array}{ccc} Y & \xrightarrow{q} & X \\ {\scriptstyle p_Y}\downarrow & & \downarrow{\scriptstyle p_X} \\ V & \xrightarrow[f]{} & U \end{array}$$

is a pull-back diagram with $p_X \in \widehat{U}$, then

$$\begin{array}{ccc} M(Y) & \xleftarrow{q^*} & M(X) \\ {\scriptstyle (p_Y)_*}\downarrow & & \downarrow{\scriptstyle (p_X)_*} \\ M(V) & \xleftarrow[f^*]{} & M(U) \end{array}$$

commutes.

A topological G-module M represents an abelian Mackey presheaf $\underline{M}$ on G-Man by sending a G-manifold X to $\underline{M}(X) \stackrel{\text{def}}{=} \mathrm{Hom}_{G\text{-}Top}(X, M)$, which we view as a discrete abelian group. If ϕ:

$$\begin{array}{ccc} X' & \xrightarrow{\phi} & X \\ & {\scriptstyle p_{X'}}\searrow \quad \swarrow{\scriptstyle p_X} & \\ & U & \end{array}$$

is a morphism in $\widehat{U}$, the covariant part of $\underline{M}$ on ϕ is $\phi_* : \mathrm{Hom}_{G\text{-}Top}(X', M) \to \mathrm{Hom}_{G\text{-}Top}(X, M)$ with

$$\phi_*(f)(a) = \sum_{a' \in \phi^{-1}(a)} f(a'),$$

where $f \in \mathrm{Hom}_{G\text{-}Top}(X', M)$ and $a \in X$.

Recall that an abelian presheaf on *G-Man* is by definition a contravariant functor $G\text{-}Man \to Ab$. Let $\mathcal{F}$ be an abelian presheaf on *G-Man* and let M be a Mackey presheaf. Given $U \in G\text{-}Man$, we denote by $\mathcal{F} \otimes_{\widehat{U}} M$ the coend

$$\left(\bigoplus_{\{X \xrightarrow{p_X} U\} \in \widehat{U}} \mathcal{F}(X) \otimes M(X) \right) / K_{\mathcal{F},M}(U)$$

in the category Ab, where $K_{\mathcal{F},M}(U)$ is the subgroup generated by elements of the form

$$(\phi^*_{\mathcal{F}} a) \otimes m' - a \otimes (\phi_M)_*(m')$$

where $\phi : X' \to X$ is a morphism in $\widehat{U}$, $a \in \mathcal{F}(X)$ and $m' \in M(X')$.

Given an abelian presheaf $\mathcal{F}$ on *G-Man*, for any nonnegative integer n, the presheaf $C^{-n}(\mathcal{F})$ is defined by $C^{-n}(\mathcal{F})(U) = \mathcal{F}(\Delta^n \times U)$, where Δ^n is the standard topological n-simplex with the trivial G-action. The natural cosimplicial structure (see [**Wei94**, Chapter 8]) of $\{\Delta^n \mid n \geq 0\}$ induces a simplicial abelian presheaf $C^\bullet(\mathcal{F})$ on *G-Man*. Denote the associated complex of presheaves by $C^*(\mathcal{F})$. For the convenience, let $C^i(\mathcal{F}) = 0$ for $i > 0$.

PROPOSITION 3.3. *Let $\mathcal{F}$ be an abelian presheaf and let M be a Mackey presheaf on G-Man. Then the assignment $U \mapsto \mathcal{F} \otimes_{\widehat{U}} M$ is a contravariant functor $G\text{-}Man \to Ab$.*

REMARK 3.4. In most category theory literature, the integral sign "$\int$" is used to denote a coend (e.g. [**Mac98**]). we adopt this by denoting the resulting abelian presheaf in the proposition by $\mathcal{F} \int M$, i.e., $\mathcal{F} \int M(U) \stackrel{\text{def}}{=} \mathcal{F} \otimes_{\widehat{U}} M$.

PROOF OF PROPOSITION 3.3. Let $f : V \to U$ be a morphism in *G-Man*. Given $p_X : X \to U$ in $\widehat{U}$, then by Proposition 3.1 $p_{f^*X} : f^*X \to V$ is an element in $\widehat{V}$, and the pull-back square

$$\begin{array}{ccc} f^*X & \xrightarrow{q_X} & X \\ {\scriptstyle p_{f^*X}} \downarrow & & \downarrow {\scriptstyle p_X} \\ V & \xrightarrow[f]{} & U \end{array} \tag{3.1}$$

implies there is a functor $f^* : \widehat{U} \to \widehat{V}$. This functor in turn, induces a morphism

$$\widehat{f^*} : \bigoplus_{\{X \xrightarrow{p_X} U\} \in \widehat{U}} \mathcal{F}(X) \otimes M(X) \to \bigoplus_{\{Y \xrightarrow{p_Y} V\} \in \widehat{V}} \mathcal{F}(Y) \otimes M(Y)$$

sending $a \otimes m \in \mathcal{F}(X) \otimes M(X)$ to $q^*_{X,\mathcal{F}}(a) \otimes q^*_{X,M}(m) \in \mathcal{F}(f^*X) \otimes M(f^*X)$, where the morphism $q^*_{X,\mathcal{F}}$ is just $\mathcal{F}(q_X)$ obtained by applying $\mathcal{F}$ to the pull-back diagram (3.1):

$$\begin{array}{ccc} \mathcal{F}(f^*X) & \xleftarrow{q^*_{X,\mathcal{F}}} & \mathcal{F}(X) \\ {\scriptstyle \mathcal{F}(p_{f^*X})} \uparrow & & \uparrow {\scriptstyle \mathcal{F}(p_X)} \\ \mathcal{F}(V) & \xleftarrow[\mathcal{F}(f)]{} & \mathcal{F}(U). \end{array}$$

Similarly, $q^*_{X,M} = M(q_X)$.

We claim that the homomorphism $\widehat{f^*}$ sends $K_{\mathcal{F},M}(U)$ to $K_{\mathcal{F},M}(V)$. The proof is as follows. Let ϕ

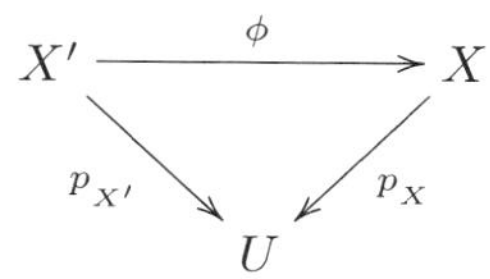

be a morphism in $\widehat{U}$. Consider the following diagram:

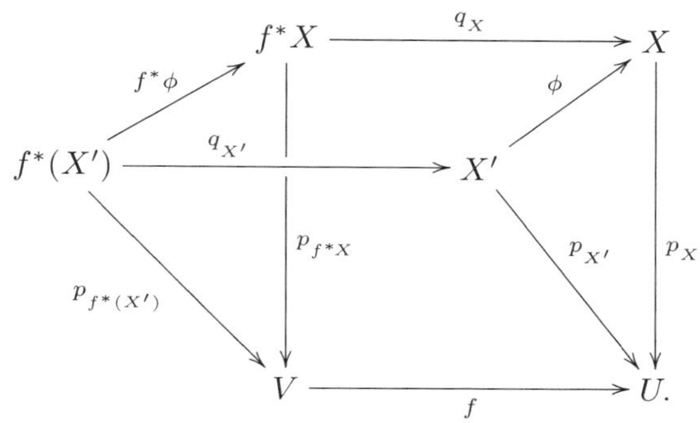

Since all of the squares and triangles are commutative except the top square, the top one is also commutative, i.e. $q_X \circ f^*\phi = \phi \circ q_{X'}$. After applying $\mathcal{F}$ and M to the top square, we have the following commutative diagrams.

$$\begin{array}{ccc} \mathcal{F}(f^*X') & \xleftarrow{q^*_{X',\mathcal{F}}} & \mathcal{F}(X') \\ \uparrow{\scriptstyle (f^*\phi)^*_{\mathcal{F}}} & & \uparrow{\scriptstyle \phi^*_{\mathcal{F}}} \\ \mathcal{F}(f^*X) & \xleftarrow[q^*_{X,\mathcal{F}}]{} & \mathcal{F}(X), \end{array}$$

and

$$\begin{array}{ccc} M(f^*X') & \xleftarrow{q^*_{X',M}} & M(X') \\ \downarrow{\scriptstyle ((f^*\phi)_M)_*} & & \downarrow{\scriptstyle (\phi_M)_*} \\ M(f^*X) & \xleftarrow[q^*_{X,M}]{} & M(X), \end{array}$$

where $\phi^*_{\mathcal{F}} = \mathcal{F}(\phi)$ and $(\phi_M)_* = M(\phi)$, etc.

Now given $a \in \mathcal{F}(X)$ and $m' \in M(X')$, we have

$$\begin{aligned} &\widehat{f^*}(\phi^*_{\mathcal{F}}(a), m') \\ =&q^*_{X',\mathcal{F}} \circ \phi^*_{\mathcal{F}}(a) \otimes q^*_{X',M}(m') \\ =&(f^*\phi)^*_{\mathcal{F}} \circ q^*_{X,\mathcal{F}}(a) \otimes q^*_{X',M}(m') \end{aligned}$$

and

$$\begin{aligned} &\widehat{f^*}(a, (\phi_M)_*(m')) \\ =&q^*_{X,\mathcal{F}}(a) \otimes q^*_{X,M} \circ (\phi_M)_*(m') \\ =&q^*_{X,\mathcal{F}}(a) \otimes ((f^*\phi)_M)_* \circ q^*_{X',M}(m'), \end{aligned}$$

where the latter equality follows from the fact that M is a Mackey presheaf and the top square in diagram (3.1) is commutative. Now the claim follows.

Using the claim we obtain a homomorphism

$$\mathbf{f}^* : \mathcal{F} \otimes_{\widehat{U}} M \to \mathcal{F} \otimes_{\widehat{V}} M.$$

It is easy to check that $\mathbf{f}^*$ makes the assignment $U \mapsto \mathcal{F} \otimes_{\widehat{U}} M$ a contravariant functor. □

Given a G-manifold X, let $\mathcal{Z}X$ be the abelian presheaf on G-*Man* defined by $\mathcal{Z}X(U) \stackrel{\text{def}}{=} \mathbb{Z}\mathrm{Hom}_{G\text{-}Man}(U, X)$.

Let $\mathcal{F}$ be an abelian presheaf on G-*Man* and $\underline{M}$ the Mackey presheaf associated to a G-module M. The *singular cochain complex* $C^*(\mathcal{F}, \underline{M})$ of $\mathcal{F}$ with coefficients in $\underline{M}$ is defined by

$$C^*(\mathcal{F}, \underline{M}) = C^*(\mathcal{F}) \int \underline{M}.$$

In particular, for a finite-dimensional representation space V of G, let $\mathcal{F} = \mathcal{Z}S^V$. Denote by $M(V)$ the shifted complex $C^*(\mathcal{Z}S^V, \underline{M})[-\dim(V)]$. Here we recall that, for an integer $q \in \mathbb{Z}$, the shifted complex $(C^*[q], \delta')$ of a cochain complex (C^*, δ) is again a cochain complex defined by $C^*[q]^n \stackrel{\text{def}}{=} C^{n+q}$ and the differential $\delta'^n = (-1)^q \delta^{n+q} : C^*[q]^n \to C^*[q]^{n+1}$ for each $n \in \mathbb{Z}$.

LEMMA 3.5. *As abelian presheaves on G-Man, $C^{-n}(\mathcal{Z}X)$ is naturally isomorphic to $C^{-n}(\underline{Hom}_{G\text{-}Top}(-, X))$ for any $n \geq 0$.*

The notation $\underline{Hom}$ denotes the Hom enriched over topological spaces.

PROOF. Pick a G-manifold U. We have

$$\begin{aligned} &C^{-n}(\mathcal{Z}X)(U) \\ &= \mathbb{Z}\mathrm{Hom}_{G\text{-}Top}(\Delta^n \times U, X) \\ &\cong \mathbb{Z}\mathrm{Hom}_{G\text{-}Top}(\Delta^n, \underline{Hom}_{Top}(U, X)) \\ &\cong \mathbb{Z}\mathrm{Hom}_{Top}(\Delta^n, (\underline{Hom}_{Top}(U, X))^G) \\ &\cong \mathbb{Z}\mathrm{Hom}_{Top}(\Delta^n, \underline{Hom}_{G\text{-}Top}(U, X)) \\ &= C^{-n}(\underline{Hom}_{G\text{-}Top}(U, X)). \end{aligned}$$

The naturality is clear. □

LEMMA 3.6. *Let $*$ be the one point set G/G. Then for any G-manifold X, we have*

$$\begin{aligned} C^{-n}(\mathcal{Z}X, \underline{M})(*) &= C^{-n}(\mathcal{Z}X) \otimes_{\hat{*}} \underline{M} \\ &\cong C^{-n}(\underline{Hom}_{G\text{-}Top}(-, X)) \otimes_{G\text{-}Fin} \underline{M} \end{aligned}$$

for every $n \geq 0$. Hence

$$H^{-n}(C^*(\mathcal{Z}X, \underline{M})(*)) \cong H_n^{Br,G}(X, \underline{M}) \tag{3.2}$$

where the right hand side is the ordinary n-th Bredon homology groups.

PROOF. Since the category $\hat{*}$ is exactly the category G-*Fin*, there is an isomorphism $C^*(\mathcal{Z}X) \otimes_{\hat{*}} \underline{M} \cong C^*(\mathcal{Z}X) \otimes_{G\text{-}Fin} \underline{M}$. By Lemma 3.5, the right hand side is isomorphic to $C^*(\underline{Hom}_{G\text{-}Top}(-, X)) \otimes_{G\text{-}Fin} \underline{M}$. The cohomology groups of this complex are exactly by definition the Bredon homology groups. □

In many cases we need to consider actions of various groups at the same time. This leads important functors of restriction and induction.

Let H be a subgroup of G. Restricting the group action from G to H induces the functor

$$\mathrm{Res}_H^G : G\text{-}Top \to H\text{-}Top$$

which is called *restriction.* There is also a functor

$$\mathrm{Ind}_H^G : H\text{-}Top \to G\text{-}Top$$

defined as follows. Pick any H-space A. The cartesian product $G \times A$ carries an H-action

$$(h, (g, a)) \mapsto (gh^{-1}, ha),$$

and define $\mathrm{Ind}_H^G(A)$ to be the H-orbit space $G \times_H A \overset{\text{def}}{=} G \times A/H$. The G-action $(g', (g, a)) \mapsto (g'g, a)$ on $G \times A$ induces a G-action on $\mathrm{Ind}_H^G(A) = G \times_H A$. For an H-map $f : A \to B$, there is an induced map

$$\mathrm{Ind}_H^G(f) = G \times_H f : G \times_H A \to G \times_H B, \ (g, a) \mapsto (g, f(a)).$$

The functors Res and Ind are adjoint pairs as shown in the next proposition.

PROPOSITION 3.7. *Pick a G-space Y and an H-space A. Then there is a natural bijection*

$$\mathrm{Hom}_{G\text{-}Top}(G \times_H A, Y) \cong \mathrm{Hom}_{H\text{-}Top}(A, \mathrm{Res}_H^G(Y)).$$

PROOF. See [**tD87**, p. 32, Proposition 4.3]. □

Apply this to the category of presheaves. For a presheaf $\mathcal{F}$ on G-*Man* and a subgroup K of G, let $\mathrm{Res}_K^G \mathcal{F}$ be the presheaf on the K-*Man* defined on $U \in K$-*Man* by

$$(\mathrm{Res}_K^G \mathcal{F})(U) = \mathcal{F}(G \times_K U) = \mathcal{F}(\mathrm{Ind}_K^G U). \tag{3.3}$$

LEMMA 3.8. *(1) $\mathrm{Res}_K^G(\underline{M})$ is a Mackey presheaf on K-Man.*
(2) Let $\mathcal{F}$ be an abelian presheaf on G-Man. Then for any K-manifold U there is a natural isomorphism

$$(\mathrm{Res}_K^G \mathcal{F}) \otimes_{\widehat{U}} (\mathrm{Res}_K^G \underline{M}) \cong \mathcal{F} \otimes_{\widehat{G \times_K U}} \underline{M}. \tag{3.4}$$

PROOF. It is easy to check that there is a natural G-homeomorphism $G \times_K X \cong G/K \times X$ for any G-space X. For a K-map $f : X \to Y$ of K-manifold X and Y, the induced map $\mathrm{Ind}_K^G f : G \times_K X \to G \times_K Y$ is just the map $1_{G/K} \times f$.

(1) For any K-map $f : X \to Y$, the contravariant part $f^* : \mathrm{Res}_K^G(\underline{M})(Y) \to \mathrm{Res}_K^G(\underline{M})(X)$ is just the induced map $(1 \times f)^* : \mathrm{Hom}_{G\text{-}Top}(G/K \times Y, M) \to \mathrm{Hom}_{G\text{-}Top}(G/K \times X, M)$. The covariant part ϕ_* for a map

$$\begin{array}{ccc} X & \xrightarrow{\ \phi\ } & Y \\ & {\scriptstyle p_X}\searrow \quad \swarrow{\scriptstyle p_Y} & \\ & U & \end{array}$$

in K-$Man{\downarrow}U$ is just the covariant part $(1\times\phi)_*$ of the Mackey presheaf $\underline{M}$ for the map $1\times\phi$:

$$\begin{array}{ccc} G/K\times X & \xrightarrow{1\times\phi} & G/K\times Y \\ & {\scriptstyle 1\times p_X}\searrow \quad \swarrow{\scriptstyle 1\times p_Y} & \\ & G/K\times U. & \end{array}$$

Furthermore, a pullback diagram

$$\begin{array}{ccc} Z & \xrightarrow{q} & Y \\ {\scriptstyle p}\downarrow & & \downarrow{\scriptstyle f} \\ Y' & \xrightarrow[m]{} & U \end{array}$$

in K-Man induces a pullback diagram in G-Man

$$\begin{array}{ccc} G/K\times Z & \xrightarrow{1\times q} & G/K\times Y \\ {\scriptstyle 1\times p}\downarrow & & \downarrow{\scriptstyle 1\times f} \\ G/K\times Y' & \xrightarrow[1\times m]{} & G/K\times U. \end{array}$$

Hence the pullback condition for $\mathrm{Res}^G_K(\underline{M})$ comes from that of $\underline{M}$.

(2) For any G-manifold X with (surjective) structure map $p: X\to G/K\times U$, let $Y=Y_1=p^{-1}(eK\times U)$. If $\{g_1=e,g_2,\ldots,g_l\}$ is a set of representatives of left cosets G/K, then $X=\amalg Y_i$ where $Y_i=p^{-1}(g_iK\times U)$ and $X=G/K\times Y$. This implies each $p: X\to G\times_K U\in\widehat{G\times_K U}$ is one to one correspondence to $\tilde{p}: Y\to U\in\widehat{U}$. Then easy to show, by definition, $(\mathrm{Res}^G_K\mathcal{F})\otimes_{\widehat{U}}(\mathrm{Res}^G_K\underline{M})\cong\mathcal{F}\otimes_{\widehat{G\times_K U}}\underline{M}$. □

COROLLARY 3.9. *Let K be a subgroup of G. Then for any G-manifold X, there is an isomorphism*

$$C^{-n}(\mathcal{Z}X,\underline{M})(G/K)\cong C^{-n}(\underline{Hom}_{K\text{-}Top}(-,X))\otimes_{K\text{-}Fin}\underline{M}$$

for every $n\geq 0$. Hence

$$H^{-n}(C^*(\mathcal{Z}X,\underline{M})(G/K))\cong H_n^{Br,K}(X,\underline{M}). \tag{3.5}$$

PROOF. For every $U\in K$-Man,

$$\begin{aligned} &\mathrm{Res}^G_K(C^{-n}(\mathcal{Z}X))(U)\\ &=C^{-n}(\mathcal{Z}X)(\mathrm{Ind}^G_K U) && \text{(by (3.3))}\\ &=\mathbb{Z}\mathrm{Hom}_{G\text{-}Top}(\Delta^n\times\mathrm{Ind}^G_K U,X)\\ &\cong\mathbb{Z}\mathrm{Hom}_{G\text{-}Top}(\Delta^n,\underline{Hom}_{Top}(\mathrm{Ind}^G_K U,X))\\ &\cong\mathbb{Z}\mathrm{Hom}_{Top}(\Delta^n,(\underline{Hom}_{Top}(\mathrm{Ind}^G_K U,X))^G)\\ &\cong\mathbb{Z}\mathrm{Hom}_{Top}(\Delta^n,\underline{Hom}_{G\text{-}Top}(\mathrm{Ind}^G_K U,X))\\ &\cong\mathbb{Z}\mathrm{Hom}_{Top}(\Delta^n,\underline{Hom}_{K\text{-}Top}(U,X)) && \text{(Proposition 3.7)}\\ &\cong C^{-n}(\underline{Hom}_{K\text{-}Top}(U,X)). \end{aligned}$$

So

$$\begin{aligned} & C^{-n}(\mathcal{Z}X, \underline{M})(G/K) \\ & = C^{-n}(\mathcal{Z}X) \otimes_{\widehat{G/K}} \underline{M} \\ & \cong \operatorname{Res}_K^G(C^{-n}(\mathcal{Z}X)) \otimes_{\widehat{K/K}} \operatorname{Res}_K^G \underline{M} \qquad \text{(by (3.4))} \\ & = C^{-n}(\underline{Hom}_{K\text{-}Top}(-, X)) \otimes_{K\text{-}Fin} \operatorname{Res}_K^G \underline{M}. \end{aligned}$$

Then the isomorphism (3.5) is again from the definition of the Bredon homology groups. □

Pick a G-manifold X and let $\mathscr{U}$ be an open G-cover of X. For any complex of presheaves $\mathcal{F}^*$ on G-*Man*, denote by $\check{\mathbb{H}}^n_G(\mathscr{U}, \mathcal{F}^*)$ the n-th Čech equivariant hypercohomology of $\mathscr{U}$ with coefficients in $\mathcal{F}^*$ (see Section 1). Let $\check{\mathbb{H}}^n_G(X, \mathcal{F}^*) = \varinjlim_{\mathscr{U}} \check{\mathbb{H}}^n_G(\mathscr{U}, \mathcal{F}^*)$.

Recall the definition of the complex $\underline{M}(V)$. Given a Mackey presheaf $\underline{M}$ associated to a discrete G-module M, let $C^*(S^V, \underline{M}) \stackrel{\text{def}}{=} C^*(\mathcal{Z}S^V) \int \underline{M}$ be the singular cochain complex of presheaves on G-*Man* and denote by $M(V)$ the shifted complex $C^*(S^V, \underline{M})[-\dim(V)]$.

DEFINITION 3.10. A presheaf $\mathcal{F}$ is *homotopy invariant* if for every space X the induced map $p^* : \mathcal{F}(X) \to \mathcal{F}(X \times I)$ of projection $p : X \times I \to X$ is an isomorphism.

REMARK 3.11. As $p : X \times I \to X$ has a section, p^* is always split injective. Thus homotopy invariance of $\mathcal{F}$ is equivalent to p^* being onto.

LEMMA 3.12. *Let $i_t : X \to X \times I, x \mapsto (x, t)$ be the inclusion map. A presheaf $\mathcal{F}$ is homotopy invariant if and only if $i_0^* = i_1^* : \mathcal{F}(X \times I) \to \mathcal{F}(X)$ for all X.*

PROOF. One way is obvious. Now suppose $i_0^* = i_1^*$ for all X. Applying $\mathcal{F}$ to the multiplication map $m : I \times I \to I, (s, t) \mapsto st$, yields the following diagram

$$\begin{array}{ccccc} & & \mathcal{F}(X \times I) & \xrightarrow{i_0^*} & \mathcal{F}(X) \\ & \swarrow^{1} & \downarrow{\scriptstyle (1_X \times m)^*} & & \downarrow{\scriptstyle p^*} \\ \mathcal{F}(X \times I) & \xleftarrow[(i_1 \times 1_I)^*]{} & \mathcal{F}(X \times I \times I) & \xrightarrow[(i_0 \times 1_I)^*]{} & \mathcal{F}(X \times I) \end{array}$$

Hence $p^* i_0^* = (i_0 \times 1_I)^*(1_X \times m)^* = (i_1 \times 1_I)^*(1_X \times m)^* = \text{id}$. Since $i_0^* p^* = \text{id}$, p^* is an isomorphism. □

LEMMA 3.13. *Let $\mathcal{F}$ be a presheaf. Then the maps $i_0^\#, i_1^\# : C^*\mathcal{F}(X \times I) \to C^*\mathcal{F}(X)$ are chain homotopic for all X.*

PROOF. For all $i = 0, \ldots, n$, define $\theta_i : \Delta^{n+1} \to \Delta^n \times I$ to be the map that sends the vertex v_j to $v_j \times \{0\}$ for $j \leq i$ and to $v_{j-1} \times \{1\}$ otherwise. The maps θ_i induce maps

$$h_i = (1_X \times \theta_i)^* : C^{-n}\mathcal{F}(X \times I) \to C^{-n-1}\mathcal{F}(X)$$

The h_i form a simplicial homotopy from $i_1^{\#} = \partial_0 h_0$ to $i_0^{\#} = \partial_{n+1} h_n$, so the alternating sum $s = \sum(-1)^i h_i$ is a chain homotopy from $i_1^{\#}$ to $i_0^{\#}$.

$$\begin{array}{ccccccccc}
\cdots \longrightarrow & C^{-n-1}\mathcal{F}(X \times I) & \xrightarrow{d} & C^{-n}\mathcal{F}(X \times I) & \xrightarrow{d} & C^{-n+1}\mathcal{F}(X \times I) & \longrightarrow \cdots \\
& \Big\downarrow i_1^{\#} - i_0^{\#} & \swarrow s & \Big\downarrow i_1^{\#} - i_0^{\#} & \swarrow s & \Big\downarrow i_1^{\#} - i_0^{\#} & \\
\cdots \longrightarrow & C^{-n-1}\mathcal{F}(X) & \xrightarrow[d]{} & C^{-n}\mathcal{F}(X) & \xrightarrow[d]{} & C^{-n+1}\mathcal{F}(X) & \longrightarrow \cdots
\end{array}$$

□

Corollary 3.14. *For any presheaf $\mathcal{F}$ the complex $C^*\mathcal{F}$ has homotopy invariant cohomology presheaves. That is, for every p, $\mathcal{H}^p(C^*\mathcal{F})$ is homotopy invariant. In particular, $M(V)$ has homotopy invariant cohomology presheaves.*

We apply this corollary to some suitable open covers of a G-manifold in the following sections.

4. Equivariant Bredon cohomology and Čech hypercohomology

Given a group G, let $\mathfrak{h}_G^*$ be a generalized reduced $RO(G)$-graded equivariant cohomology theory which is defined by a G-spectrum $\{E_V | V \in RO(G)\}$. That is, for any G-space X,

$$\mathfrak{h}_G^V(X) \stackrel{\text{def}}{=} \varinjlim_{W \supset V} [S^{W-V} \wedge X_+, E_W]_G^0.$$

As a special case, fix a finite dimensional representation V of G and a Mackey functor $\underline{M}$ associated to a discrete $\mathbb{Z}[G]$-module M. Define the functors h^p ($p \in \mathbb{Z}$) on G-CW complexes X graded by $\mathbb{Z}$ by $h^p(X) \stackrel{\text{def}}{=} H_{\text{Br}}^{p+V-\dim(V)}(X, \underline{M})$, which is just isomorphic to the homotopy classes of maps $[S^{\dim(V)} \wedge X_+, K(\underline{M}, p+V)]_G^0$. Here a $K(\underline{M}, V)$ space is by definition a classifying space for the functor $H_{\text{Br}}^V(-, \underline{M})$.

The functors h^* satisfy the following cohomology axioms:

(i) *Homotopy invariance.* If $f, g : X \to Y$ are G-homotopic, then $f^* = g^* : h^*(Y) \to h^*(X)$.

(ii) *Exact sequence* for G-CW pairs (X, A). This is from the standard G-cofibration sequence associated to (X, A).

(iii) *Suspension.* Clear from the homotopy representation.

Let $A_\bullet$ be a simplicial G-space. We denote by A_p^d the degenerate part of A_p, i.e. the union of the images of all maps $A_r \to A_p$ with $r < p$, and by A_p^{nd} the non-degenerate part of A_p. The geometric realization $|A_\bullet|$ has a natural skeleta filtration:

$$|A_\bullet| \supset \cdots \supset |A_\bullet|^{(p)} \supset |A_\bullet|^{(p-1)} \supset \cdots \supset |A_\bullet|^{(0)} \supset \{*\}$$

and it gives rise to an associated spectral sequence which is first formulated in [**Seg68**]. Here we apply it to the equivariant case.

Lemma 4.1. *The filtration of $|A_\bullet|$ induces a natural spectral sequence converging to $h^*(|A_\bullet|)$ with $E_1^{pq} = h^q(A_p/A_p^d) = h^q(A_p^{nd})$. Moreover, under the natural map $E_1^{pq} \to h^q(A_p)$ the differential $d_1^{pq} : E_1^{pq} \to E_1^{p+1,q}$ is compatible with that of the cochain complex $h^q(A_\bullet)$.*

PROOF. The filtration on $|A_\bullet|$ yields a spectral sequence converging to $h^*(|A_\bullet|)$ with $E_1^{pq} = h^{p+q}(|A_\bullet|^{(p)}/|A_\bullet|^{(p-1)})$. There is a homeomorphism

$$A_p \times \Delta^p/((A_p^d \times \Delta^p) \cup (A_p \times \partial\Delta^p)) \to |A_\bullet|^{(p)}/|A_\bullet|^{(p-1)}$$

where $\partial\Delta^p$ is the $(p-1)$-skeleton of the simplex Δ^p. Thus the space $|A_\bullet|^{(p)}/|A_\bullet|^{(p-1)}$ can be identified with the p-fold suspension of A_p/A_p^d, and accordingly $E_1^{pq} \cong h^q(A_p/A_p^d)$.

Next, the compatibility of the differentials, i.e. the commutativity of the diagram

$$\begin{array}{ccc} E_1^{pq} & \longrightarrow & h^q(A_p) \\ {\scriptstyle d_1}\big\downarrow & & \big\downarrow{\scriptstyle d} \\ E_1^{p+1,q} & \longrightarrow & h^q(A_{p+1}) \end{array}$$

follows from the commutativity of the following diagram

$$\begin{array}{ccc} h^n(|A_\bullet|^{(p)}/|A_\bullet|^{(p-1)}) & \xrightarrow{\quad d_1 \quad} & h^n(|A_\bullet|^{(p+1)}/|A_\bullet|^{(p)}) \\ \big\downarrow & & \big\downarrow \\ h^n(A_p \times \Delta^p/A_p \times \partial\Delta^p) & \xleftarrow[S^p\wedge -]{\cong} & h^{n-p}(A_p) \\ {\scriptstyle \theta\times 1}\big\downarrow & & \big\downarrow{\scriptstyle \theta} \quad (d) \\ \prod_p h^n(A_{p+1} \times \Delta^p/A_{p+1} \times \partial\Delta^p) & \xleftarrow[S^p\wedge -]{\cong} & \prod_p h^{n-p}(A_{p+1}) \\ {\scriptstyle \cong}\big\uparrow & & \big\downarrow{\scriptstyle \Sigma} \\ h^n(A_{p+1} \times \partial\Delta^{p+1}/A_{p+1} \times \partial^2\Delta^{p+1}) & & h^{n-p}(A_{p+1}) \\ & {\scriptstyle d_1}\searrow \quad \searrow{\scriptstyle S^{p+1}\wedge -,\ \cong} & \\ & & h^{n+1}(A_{p+1} \times \Delta^{p+1}/A_{p+1} \times \partial\Delta^{p+1}) \end{array}$$

where $\partial^2\Delta^p$ means the $(p-2)$-skeleton of Δ^p. The maps θ are induced by the $p+2$ face maps $[p] \to [p+1]$, and Σ denotes the alternative sum, so that the composite $\Sigma \circ \theta$ is the differential d. □

REMARK 4.2. By the the proof of the above Lemma, the E_2-term of the spectral sequence is $E_2^{p,q} \cong H^p(h^q(A_\bullet^{nd}))$.

Pick an equivariant good cover $\mathscr{U}$ of a smooth G-manifold X. Then for every $p \geq 0$ the nonempty finite intersection $U_{\sigma^p} = U_{\alpha_0} \cap \cdots \cap U_{\alpha_p}$ has the form $(G/J_{\sigma^p}) \times D$ where D is a contractible space and J_{σ^p} is a subgroup of G. Applying Theorem 2.19 and Lemma 4.1 to h^* and $\mathcal{N}(\mathscr{U})$ yields

LEMMA 4.3. *Given X and $\mathscr{U}$ as above, there is a spectral sequence converging to $H_{Br}^{V+*-\dim(V)}(X, \underline{M})$ whose E_1-term is*

$$E_1^{pq} = \prod_{\sigma^p \in \mathcal{N}(\mathscr{U})_p^{nd}} H_{Br,J_{\sigma^p}}^{q+V-\dim(V)}(*, \underline{M})$$

where $*$ denotes the trivial coset J/J for subgroups J of G. Moreover, the differential of E_1 is compatible with that of the complex $\prod_{\sigma^* \in \mathcal{N}(\mathscr{U})_*} H_{Br}^{q+V-\dim(V)}(U_{\sigma^*}, \underline{M})$.

LEMMA 4.4. *Given X and $\mathscr{U}$ as above, there is a spectral sequence converging to $\check{\mathbb{H}}^*_G(\mathscr{U}, M(V))$ whose E_1-term is*

$$E_1^{pq} = \prod_{\sigma^p \in \mathcal{N}(\mathscr{U})_p} \mathcal{H}^q(M(V))(U_{\sigma^p}) \cong \prod_{\sigma^p \in \mathcal{N}(\mathscr{U})_p} H_{Br, J_{\sigma^p}}^{q+V-\dim(V)}(*, \underline{M}).$$

PROOF. The standard filtration on the double complex $\check{C}^*(\mathscr{U}, M(V))$ yields

$$\begin{aligned} E_1^{pq} &= H^q(\check{C}^p(\mathscr{U}, M(V))) = \check{C}^p(\mathscr{U}, \mathcal{H}^q(M(V))) \\ &= \prod_{\sigma^p \in \mathcal{N}(\mathscr{U})_p} \mathcal{H}^q(M(V))(U_{\sigma^p}). \end{aligned}$$

For any homogeneous space G/J we proved in Corollary 3.9 that

$$\begin{aligned} (\mathcal{H}^q(M(V)))(G/J) &= H^q(M(V)(G/J)) \\ &\cong H_{\mathrm{Br}, J}^{q+V-\dim(V)}(*, \underline{M}). \end{aligned}$$

Now since U_{σ^p} is homeomorphic to $(G/J_{\sigma^p}) \times D$, the homotopy invariant property of $\mathcal{H}^q(M(V))$ implies $\mathcal{H}^q(M(V))(U_{\sigma^p}) \cong H_{\mathrm{Br}, J_{\sigma^p}}^{q+V-\dim(V)}(*, \underline{M})$. So

$$E_1^{pq} = \prod_{\sigma^p \in \mathcal{N}(\mathscr{U})_p} \mathcal{H}^q(M(V))(U_{\sigma^p}) \cong \prod_{\sigma^p \in \mathcal{N}(\mathscr{U})_p} H_{\mathrm{Br}, J_{\sigma^p}}^{q+V-\dim(V)}(*, \underline{M}). \qquad \square$$

THEOREM 4.5. *There is a natural isomorphism*

$$\check{\mathbb{H}}^n_G(X, M(V)) \cong H_{Br/G}^{V+n-\dim(V)}(X, \underline{M}).$$

PROOF. If denote by E_1 and E_1' the E_1 terms in Lemma 4.3 and Lemma 4.4, respectively, the natural map $f_1 : E_1 \to E_1'$ induces a morphism f of spectral sequences (by the compatibility stated in Lemma 4.3 and $f^{pq} : E_1^{pq} \cong E'^{pq}_1$. Hence f induces an isomorphism on E_∞ pages. $\square$

5. Examples and Applications

One of the applications to algebraic geometry of our results is to develop a version of bigraded cohomology and Deligne cohomology for real varieties. The reader can find the work of dos Santos and Lima-Filho on this topic in [**dS03, dSLF07, dSLF08**].

DEFINITION 5.1. A real algebraic variety X is a complex algebraic variety endowed with an anti-holomorphic involution $\sigma : X \to X$. A morphism of real varieties $(X, \sigma) \to (X', \sigma')$ is a morphism of complex varieties $f : X \to X'$ such that f is compatible with the involution, i.e. $f \circ \sigma = \sigma' \circ f$.

Let $\mathfrak{G} \stackrel{\text{def}}{=} Gal(\mathbb{C}/\mathbb{R})$, the Galois group of $\mathbb{C}$ over $\mathbb{R}$. It is isomorphic to the group $\mathbb{Z}/2$. If (X, σ) is a real variety, the anti-holomorphic involution σ induces a $\mathfrak{G}$-action on X. The fixed point set $X^{\mathfrak{G}}$ of this action is called the set of *real points* of X and denoted by $X(\mathbb{R})$. On the other hand, we use $X(\mathbb{C})$ to denote the set of *complex-valued points* of X.

In this section we mainly consider the case $G = \mathfrak{S} \cong \mathbb{Z}/2$. The real orthogonal representation ring of $\mathfrak{S}$ is $RO(\mathfrak{S}) = \mathbb{Z}\cdot\mathbf{1}\oplus\mathbb{Z}\cdot\xi$, where $\mathbf{1}$ is the trivial representation and ξ is the sign representation. Furthermore, we use the bigraded cohomology notation $H^{r,s}_{\mathrm{Br}}(X, \underline{M})$ for the $\mathfrak{S}$-equivariant Bredon cohomology with coefficients $\underline{M}$ in dimension $(r-s)\cdot\mathbf{1} + s\cdot\xi$, i.e.

$$H^{r,s}_{\mathrm{Br}}(X,\underline{M}) \overset{\text{def}}{=} H^{(r-s)\cdot\mathbf{1}+s\cdot\xi}_{\mathfrak{S}}(X,\underline{M}).$$

Recall that in Section 3 we defined the complex of presheaves $C^*(\mathcal{F})$ for any presheaf $\mathcal{F}$ on G-*Man* whose $(-n)$-th term is

$$C^{-n}(\mathcal{F}) : U \mapsto \mathcal{F}(\Delta^n \times U),\ n \geq 0.$$

Also, given a G-manifold X, the abelian presheaf $\mathcal{Z}X$ on G-*Man* was defined by $\mathcal{Z}X(U) \overset{\text{def}}{=} \mathbb{Z}\mathrm{Hom}_{G\text{-}Man}(U,X)$. We then defined the *singular cochain complex* $C^*(\mathcal{F},\underline{M})$ of $\mathcal{F}$ with coefficients in $\underline{M}$ by

$$C^*(\mathcal{F},\underline{M}) = C^*(\mathcal{F})\int \underline{M}.$$

In particular, for a finite-dimensional representation space V of $G = \mathfrak{S}$, let $\mathcal{F} = \mathcal{Z}S^V$. We denoted by $\mathbb{Z}(V)$ the shifted complex $C^*(\mathcal{Z}S^V,\underline{\mathbb{Z}})[-\dim(V)]$.

In [**dSLF08**], a complex of presheaves called *Bredon complex* is defined as follows. First denote

$$(\mathbb{C}^\times)^{p-1}_i \overset{\text{def}}{=} \mathbb{C}^\times \times\cdots\times 1\times\cdots\times\mathbb{C}^\times \subset \mathbb{C}^{\times p},$$

where 1 appears in the i-th coordinate.

DEFINITION 5.2 ([**dSLF08**]). Given a $\mathfrak{S}$-manifold X, let

$$J_{X,p} : \bigoplus_{i=1}^{p} C^*(\mathcal{Z}((\mathbb{C}^\times)^{p-1}_i \times X)) \longrightarrow C^*(\mathcal{Z}(\mathbb{C}^{\times p}\times X))$$

be the map induced by the inclusions and denote

$$C^*(\mathcal{Z}_0(S^{p,p}\wedge X_+)) \overset{\text{def}}{=} \mathrm{cone}(J_{X,p}).$$

We denote $\mathrm{cone}(J_{X,p})$ by $C^*(\mathcal{Z}_0(S^{p,p}))$ when $X=\varnothing$. The *p-th Bredon complex with coefficients in $\mathbb{Z}$* is the complex of presheaves

$$\mathbb{Z}(p)_{\mathcal{B}r} := C^*(\mathcal{Z}_0(S^{p,p}))\int\underline{\mathbb{Z}}\,[-p].$$

PROPOSITION 5.3. *Pick an integer $p\geq 0$. Let V be the representation space $p\cdot\xi$ of $\mathfrak{S}$. Then there is a natural quasi-isomorphism $f : \mathbb{Z}(p)_{\mathcal{B}r} \to \mathbb{Z}(V)$.*

We proceed the proof by two lemmas. First we define a complex of presheaves similar to the Bredon complex.

DEFINITION 5.4. Let $S^\xi \subset \mathbb{C}$ be the unit circle. Denote

$$(S^\xi)^{p-1}_i \overset{\text{def}}{=} S^\xi\times\cdots\times 1\times\cdots\times S^\xi \subset (S^\xi)^p,$$

where 1 appears in the i-th coordinate. Let

$$K_p : \bigoplus_{i=1}^{p} C^*(\mathcal{Z}((S^\xi)^{p-1}_i)) \longrightarrow C^*(\mathcal{Z}((S^\xi)^p))$$

be the map induced by the inclusions. Define $L(p)$ to be the complex of presheaves

$$L(p) \stackrel{\text{def}}{=} \text{cone}(K_p) \int \underline{\mathbb{Z}}\, [-p].$$

LEMMA 5.5. *The map of complexes*

$$\varphi : \mathbb{Z}(p)_{\mathcal{B}r} \to L(p)$$

induced by the retraction $r : \mathbb{C}^\times \to S^\xi$ *is a quasi-isomorphism of complexes of presheaves.*

PROOF. Given $U \in G\text{-}Man$, for each j, $0 \le j \le p$, we have

$$\begin{aligned}
\mathbb{Z}(p)^j_{\mathcal{B}r}(U) &= C^j(\mathcal{Z}_0(S^{p,p})) \int \underline{\mathbb{Z}}\, [-p]\, (U) \\
&= \bigoplus_{\{T \xrightarrow{\pi} U\} \in \widehat{U}} \Big[\Big(\oplus_{i=1}^p C^{j+1-p}(\mathcal{Z}((\mathbb{C}^\times)^{p-1}_i))(T) \oplus C^{j-p}(\mathcal{Z}(\mathbb{C}^{\times p}))(T)\Big) \otimes \mathbb{Z}(T)\Big]/K \\
&= \bigoplus_{\{T \xrightarrow{\pi} U\} \in \widehat{U}} \Big[\Big(\oplus_{i=1}^p \mathbb{Z}\text{Hom}_{G\text{-}Man}(\Delta^{p-j-1} \times T, (\mathbb{C}^\times)^{p-1}_i) \\
&\qquad \oplus \mathbb{Z}\text{Hom}_{G\text{-}Man}(\Delta^{p-j} \times T, \mathbb{C}^{\times p})\Big) \otimes \text{Hom}_{G\text{-}Top}(T, \mathbb{Z})\Big]/K.
\end{aligned}$$

So elements in $\mathbb{Z}(p)^j_{\mathcal{B}r}(U)$ are represented by sums of pairs of the form $\alpha \otimes m = (a, f) \otimes m$ where a, f and m are equivariant maps satisfying

1. $a\colon \Delta^{p-j-1} \times T \to (\mathbb{C}^\times)^{p-1}_i \subset \mathbb{C}^{\times p}$ is smooth and $\pi\colon T \to U$ is a map in $\widehat{U}$;
2. $f\colon\ \Delta^{p-j} \times T \to (\mathbb{C}^\times)^p$ is a smooth map;
3. $m\colon\ T \to \mathbb{Z} \in \underline{\mathbb{Z}}(T)$ is locally constant (since $\mathbb{Z}$ has discrete topology).

With the same argument each element in $L(p)^j(U)$ is represented by sums of pairs of the form $\alpha' \otimes m' = (a', f') \otimes m'$ where equivariant maps a', f' and m' satisfy

1. $a'\colon \Delta^{p-j-1} \times T \to (S^\xi)^{p-1}_i \subset \mathbb{C}^{\times p}$ is smooth and $\pi\colon T \to U$ is a map in $\widehat{U}$;
2. $f'\colon\ \Delta^{p-j} \times T \to (S^\xi)^p$ is a smooth map;
3. $m'\colon\ T \to \mathbb{Z} \in \underline{\mathbb{Z}}(T)$ is locally constant.

The map $\varphi : \mathbb{Z}(p)_{\mathcal{B}r} \to L(p)$ induced by the retraction $r : \mathbb{C}^\times \to S^\xi$ is defined as follows. If $j < 0$ or $j > p$, let $\varphi = 0 : \mathbb{Z}(p)^j_{\mathcal{B}r} \to L(p)^j$. If $0 \le j \le p$, let $\varphi : \mathbb{Z}(p)^j_{\mathcal{B}r}(U) \to L(p)^j(U)$ be the map sending a representative element $(a, f) \otimes m$ to $(r_1 \circ a, r_2 \circ f) \otimes m$, where $r_1 : (\mathbb{C}^\times)^{p-1}_i \to (S^\xi)^{p-1}_i$ and $r_2 : (\mathbb{C}^\times)^p \to (S^\xi)^p$ are maps both induced by r. It is easy to check φ is a map of complexes.

Since both $\mathbb{Z}(p)_{\mathcal{B}r}$ and $L(p)$ have homotopy invariant cohomology presheaves by Corollary 3.14 and G-manifolds are locally contractible, in order to show φ induces an isomorphism of cohomology presheaves, it suffices to check $\varphi : \mathbb{Z}(p)^*_{\mathcal{B}r}(\text{pt}) \to L(p)^*(\text{pt})$ induces an isomorphism of cohomology groups. But in this case the map of complexes $\psi : L(p)^*(\text{pt}) \to \mathbb{Z}(p)^*_{\mathcal{B}r}(\text{pt})$ induced by the inclusion $\iota : S^\xi \to \mathbb{C}^\times$ serves as inverse of φ in the cohomology level. $\square$

LEMMA 5.6. *There is a quasi-isomorphism*

$$\varphi : L(p) \to \mathbb{Z}(V).$$

PROOF. Similar to the proof of Lemma 5.5. $\square$

COROLLARY 5.7. *Let X be a $\mathfrak{G}$-manifold. Then for all $n, p \geq 0$ there is a natural isomorphism*

$$\check{\mathbb{H}}^n_G(X, \mathbb{Z}(p)_{\mathcal{B}r}) \cong H^{n,p}_{Br}(X, \underline{\mathbb{Z}}).$$

PROOF. This comes from Theorem 4.5 and Proposition 5.3. Let $V = p \cdot \xi$ be a representation space of $\mathfrak{G}$. Then we have

$$\begin{aligned} &\check{\mathbb{H}}^n_G(X, \mathbb{Z}(p)_{\mathcal{B}r}) \\ &\cong \check{\mathbb{H}}^n_G(X, \mathbb{Z}(V)) && \text{(Proposition 5.3)} \\ &\cong H^{V+n-\dim(V)}_{\mathrm{Br}}(X, \underline{\mathbb{Z}}) && \text{(Theorem 4.5)} \\ &= H^{(n-p)\cdot\mathbf{1}+p\cdot\xi}_{\mathrm{Br}}(X, \underline{\mathbb{Z}}) \\ &= H^{n,p}_{\mathrm{Br}}(X, \underline{\mathbb{Z}}). \end{aligned}$$

□

All the results from this paper naturally apply to the Deligne cohomology for a real variety X. For the general theory of Deligne cohomology on a smooth manifold, see [**Bry93**, Section 1.5]. Here, for a real holomorphic proper manifold X, let $p \geq 0$ and we define the *Deligne cohomology* of X as the Čech hypercohomology groups

$$H^i_{\mathcal{D}/\mathbb{R}}(X, \mathbb{Z}(p)) \stackrel{\text{def}}{=} \check{\mathbb{H}}^i_G(X, \mathbb{Z}(p)_{\mathcal{D}/\mathbb{R}}),$$

where $\mathbb{Z}(p)_{\mathcal{D}/\mathbb{R}}$ is some equivariant Deligne complex. If $p < 0$, then define Deligne cohomology such that it coincides with equivariant Bredon cohomology. The author will address in the near future the generalization of this paper's results to Deligne cohomology.

References

[Bre72] G. Bredon, *Introduction to Compact Transformation Groups*, Academic Press, New York, NY, 1972.

[Bry93] J. L. Brylinski, *Loop Spaces, Characteristic Classes and Geometric Quantization*, Progress in Mathematics, vol. 107, Birkhäuser, Boston, MA, 1993.

[BT82] R. Bott and L. W. Tu, *Differential Forms in Algebraic Topology*, Graduate Texts in Mathematics, no. 82, Springer-Verlag, New York, NY, 1982.

[dC92] M. do Carmo, *Riemannian geometry*, Mathematics: Theory & Applications, Birkhäuser, Boston, MA, 1992.

[dS03] P. F. dos Santos, *Algebraic cycles on real varieties and $\mathbb{Z}/2$-equivariant homotopy theory*, Proc. London Math. Soc. **86** (2003), 513–544.

[dSLF07] P. F. dos Santos and P. Lima-Filho, *Bigraded equivariant cohomology of real quadrics*, Preprint, 2007.

[dSLF08] ———, *Integral Deligne cohomology for real varieties*, Preprint, 2008.

[GHL04] S. Gallot, D. Hulin, and J. Lafontaine, *Riemannian Geometry*, 3 ed., Universitext, Springer, New York, NY, 2004.

[Har77] R. Hartshorne, *Algebraic Geometry*, Graduate Texts in Mathematics, no. 52, Springer-Verlag, New York, NY, 1977.

[Ill78] S. Illman, *Smooth equivariant triangulations of G-manifolds for G a finite group*, Math. Ann. **233** (1978), 199–220.

[LMM81] L. G. Lewis, Jr., J. P. May, and J. E. McClure, *Ordinary $RO(G)$-graded cohomology*, Bull. Amer. Math. Soc. (N.S.) **4** (1981), 208–212.

[M+96] J. P. May et al., *Equivariant Homotopy and Cohomology Theory*, CBMS, no. 91, American Mathematical Society, 1996, With contributions by M. Cole, G. Comezana, S. Costenoble, A. D. Elmendorf, J. P. C. Greenlees, L. G. Lewis, Jr., R. J. Piacenza, G. Triantafillou, and S. Waner.

[Mac98] S. Mac Lane, *Categories for the Working Mathematician*, 2 ed., Graduate Texts in Mathematics, no. 5, Springer-Verlag, New York, NY, 1998.

[Mun66] James R. Munkres, *Elementary differential topology: Lectures given at Massachusetts Institute of Technology, Fall 1961*, revised ed., Annals of Mathematics Studies, vol. 54, Princeton University Press, Princeton, NJ, 1966.

[Pra06] V. V. Prasolov, *Elements of Combinatorial and Differential Topology*, Graduate Studies in Mathematics, vol. 74, American Mathematical Society, Providence, RI, 2006.

[Rot88] J. Rotman, *An Introduction to Algebraic Topology*, Graduate Texts in Mathematics, no. 119, Springer-Verlag, New York, NY, 1988.

[Seg68] G. Segal, *Classifying spaces and spectral sequences*, Publ. Math. I.H.E.S. **34** (1968), 105–112.

[Seg74] ———, *Categories and cohomology theories*, Topology **13** (1974), 293–312.

[tD87] T. tom Dieck, *Transformation groups*, de Gruyter Studies in Mathematics, no. 8, Walter de Gruyter, New York, 1987.

[Wei94] C. Weibel, *An Introduction to Homological Algebra*, Cambridge Studies in Advanced Mathematics, no. 38, Cambridge University Press, 1994.

Department of Mathematics, Texas A&M University, College Station, TX 77843-3368, USA.

Current address: School of Mathematics and Information Science, Nanchang Hangkong University, Jiangxi 330063, P. R. China.

E-mail address: hyang@neo.tamu.edu, hyang6k@gmail.com

Proceedings of Symposia in Pure Mathematics
Volume **82**, 2011

Sphere recognition lies in NP

Saul Schleimer

ABSTRACT. We prove that the three-sphere recognition problem lies in **NP**. Rubinstein [Haifa, 1992] and Thompson [Math. Res. Let., 1994] showed that the problem is decidable. Our result relies on Casson's version [MSRI, 1997] of their algorithm and recent results of Agol, Hass, and Thurston [STOC, 2002].

1. Introduction

The three-sphere recognition problem asks: given a triangulation T, is the underlying space $|T|$ homeomorphic to the three-sphere? To solve this problem, Rubinstein [**R92**] introduced *almost normal two-spheres* (see Section 4). Thompson [**T94**] greatly simplified Rubinstein's proof using Gabai's technique of *thin position* [**G87**].

THEOREM 1.1 (Rubinstein [**R92**], Thompson [**T94**]). *The three-sphere recognition problem lies in* **EXPTIME**.

Casson [**C97**] then introduced the novel idea of *crushing* triangulations along normal two-spheres (see Section 13). This reduced the space complexity.

THEOREM 14.1 (Casson [**C97**]). *The three-sphere recognition problem lies in* **PSPACE**.

Following Casson's algorithm and work of Agol, Hass, and Thurston [**AHT02**] (see Theorem 4.6) we show the following.

THEOREM 15.1. *The three-sphere recognition problem lies in* **NP**.

That is, any triangulation T of the three-sphere admits a polynomial-sized *certificate*: a proof that T is indeed a triangulation of the three-sphere (see Section 3). Theorem 15.1 has an immediate corollary.

COROLLARY 1.2. *The three-ball recognition problem lies in* **NP**.

PROOF. Theorem 3.4 gives a polynomial-time algorithm to verify that $|T|$ is a three-manifold. Suppose that T is a triangulation of the three-ball. First verify that that $S = \partial|T|$ is a two-sphere by checking connectedness and Euler characteristic. Next, build $D(T)$: the triangulation obtained by *doubling* across S. Next,

---

2010 *Mathematics Subject Classification.* Primary 57M40.

Theorem 15.1 gives a certificate that $|D(T)|$ is a three-sphere. Finally, Alexander's Theorem [**H01**, Theorem 1.1] implies that the two-sphere S bounds a three-ball in $|D(T)|$. □

A *surface vector* $v(S)$ is the vector of normal coordinates of S with respect to the triangulation T (Section 4.1). Here is a result closely related to Corollary 1.2.

COROLLARY 1.3. *The following problem lies in* **NP**: *given a triangulation T of a three-manifold and a surface vector $v(S)$, decide if S bounds a three-ball in $|T|$.*

PROOF. Using Lemma 4.5 to compute the Euler characteristic and using Theorem 4.6 to check connectedness, verify that S is a two-sphere. Crush the triangulation T along the surface S to obtain a triangulation T' (Section 13). Theorem 5.9 of Jaco and Rubinstein's paper [**JR03**] tells us that the connect sum of the components of $|T'|$ is homeomorphic to $|T|$, up to keeping track of lens space summands.

Following Casson, Barchechat's thesis [**B03**, page 50] gives a polynomial-time algorithm that reassembles these lens spaces and the components of $|T'|$, recovering $|T|$. Thus we only need to check that these lens spaces and the components of $|T'|$, arising as submanifolds of the ball bounded by S, are three-spheres. The former are dealt with as in [**B03**]. The latter are certified using Theorem 15.1. □

We next state a technical result, involved in the proof of Theorem 15.1, that may be of independent interest.

THEOREM 12.1. *There is a polynomial-time algorithm that, given a triangulation T of an oriented three-manifold and $v(S)$ where S is a transversely oriented almost normal surface, produces as output $v(\mathrm{norm}(S))$, the vector for the* normalization *of S.*

Corollary 1.3, Theorem 12.1, and the bounds given by [**S01**, Chapter 6] result in the following.

COROLLARY 1.4. *The following problem lies in* **NP**: *given a triangulation T of a closed orientable irreducible atoroidal three-manifold, decide if $|T|$ is a surface bundle over the circle.* □

Corollary 1.4, unfortunately, is very far from proving that bundle recognition lies in **NP**; certifying irreducibility or atoroidality are interesting and difficult open questions. Note that certifying *zero-efficiency* would in turn certify irreducibility. Corollary 1.4 is similar to a earlier result of Ivanov [**Iv01**]. He shows that recognition of the three-sphere, amongst the class of zero-efficient triangulations (there called *irreducible Q–triangulations*), lies in **NP**.

There are other problems in three-manifold topology lying in **NP**. Hass, Lagarias, and Pippenger [**HLP99**] have shown that the unknotting problem, first solved by Haken, lies in **NP**. Agol [**A**] has given a proof, using sutured manifold hierarchies, that the recognition of Haken manifolds lies in **NP** (see also [**JO84**]). Agol's algorithm requires, as the base case, some version of Corollary 1.3. Agol deduces that the unknotting problem lies in co-**NP**. Agol, Hass and Thurston [**AHT02**] have shown that the 3-manifold knot genus problem is **NP**-complete. For a discussion of algorithmic three-manifold topology we refer the reader to [**HLP99**] or to Matveev's book [**M03**].

Historical note. Three-sphere recognition (Theorem 1.1) is a fundamental result in low-dimensional topology. Accordingly, many expositions exist. A solution to the problem was first presented by Rubinstein [**R92**] in a series of lectures in Haifa, Israel, in 1992. Thompson [**T94**] gave a different proof, relying on Gabai's technique of *thin position* [**G87**] soon after. Rubinstein's papers [**R95, R97**] lay out his original ideas on the problem. Shortly after Thompson's paper appeared, Matveev [**M95**] gave an exposition of her algorithm using handle decompositions instead of triangulations; his paper gives a particularly elegant version of the *lightbulb* trick, replacing Thompson's "fluorescent light bulb trick" [**T94**, Section 3.1]. Next, Casson [**C97**], in lectures at MSRI, California, introduced the idea of crushing triangulations along normal two-spheres. An exposition of the Rubinstein-Thompson algorithm was given by Ivanov [**Iv01**], stated in the language of *Q–triangulations*. Barchechat's thesis [**B03**, Chapter 6.1] gives an exposition of Casson's algorithm; it has been implemented by Burton in his computer program Regina [**Bu99**].

Regarding the present work: the material in Sections 7 to 10 had its genesis as Chapter 4 of my thesis [**S01**], supervised by Andrew Casson; this material has not been otherwise published. Ian Agol, when we were both at the University of Illinois, Chicago, suggested that those techniques might bear on the computational complexity of three-sphere recognition. This paper, first posted to the arXiv in 2004, is the result. Another novelty, also introduced here, is to use the Agol-Hass-Thurston machinery [**AHT02**] to produce a normalization algorithm that runs in polynomial time (Theorem 12.1).

In 2008, Sergei Ivanov [**Iv08**] published a different proof of Theorem 15.1. As in this paper, his argument closely follows Casson's algorithm. Our polynomial-time normalization is, in his paper, replaced by a discussion of *vertex fundamental surfaces* (following Hass-Lagarias-Pippinger [**HLP99**]) together with the intriguing idea of crushing along almost normal two-spheres. It is an interesting question whether his method is more efficient than ours. His paper also shows that the problem of recognizing manifolds with compressible boundary lies in **NP**.

Acknowledgments. I thank both Andrew Casson and Ian Agol for many enlightening mathematical conversations. I thank the mathematics department at UIC for its support during the writing of this paper. I thank the referees for their comments and corrections.

2. Sketch of the proof of the main theorem

We closely follow Casson's algorithm [**C97**] for recognizing the three-sphere. Fix T, a triangulation of S^3. Produce a certificate $\{(T_i, v(S_i))\}_{i=0}^n$ as follows: The triangulation T_0 is equal to T. For every i, Lemma 4.13 provides S_i, a normal two-sphere in T_i that is not vertex-linking, if such exists. If T is zero-efficient then Lemma 4.13 provides S_i, an almost normal two-sphere in T_i. Definitions are given in Section 4.

If S_i is normal apply Theorem 13.1: T_{i+1} is obtained from T_i by crushing T_i along S_i. Briefly, we cut $|T_i|$ along S_i, cone the resulting two-sphere boundary components to points, and collapse non-tetrahedral cells of the resulting cell structure to obtain the triangulation T_{i+1}. This is discussed in Section 13, below.

If S_i is almost normal then obtain T_{i+1} from T_i by deleting the component of $|T_i|$ that contains S_i. Finally, the last triangulation T_n is empty, as is S_n.

That completes the construction of the certificate. We now turn to the procedure for checking a given certificate; we cite a series of polynomial-time algorithms that verify each part of the certificate. Begin by checking that T is a triangulation of a three-manifold that is a homology three-sphere, using Theorems 3.4 and 3.5. Next, check if $T = T_0$ using Lemma 3.1. For general i, verify that S_i is a two-sphere by checking its Euler characteristic (Lemma 4.5) and checking that it is connected (Theorem 4.6). Next, if S_i is normal verify that the triangulation T_{i+1} is identical to the triangulation obtained by crushing T_i along S_i. To do this in polynomial time requires Theorem 13.1 and Lemma 3.1. If S_i is almost normal then check that the component T' of T_i containing S_i satisfies $|T'| \cong S^3$ using Theorems 12.1 and 10.3.

Finally, by Theorem 13.2, for every i we have that $\#|T_i| \cong \#|T_{i+1}|$ where the connect sum on the left hand side ranges over the components of $|T_i|$ while the right hand side ranges over the components of $|T_{i+1}|$. By definition the empty connect sum is S^3, and this finishes the verification of the certificate.

3. Definitions

Complexity theory. Please consult [**GJ79, P94**] for more thorough treatments.

A *problem* P is a function from a set of finite binary strings, the *instances*, to another set of finite binary strings, the *answers*. If T is an instance we use size(T) to denote the length of T. A problem P is a *decision problem* if the range of P is the set $\{0, 1\}$. A *solution* for P is a Turing machine $\mathcal{M}$ that, given an instance T on its tape, computes and then halts with only the answer $P(T)$ on its tape. We will engage in the usual abuse of calling such a Turing machine an *algorithm* (or *procedure*) that *solves* the problem P.

An algorithm $\mathcal{M}$ runs in *polynomial time* if there is a polynomial q so that, for any instance T, the machine $\mathcal{M}$ halts in time at most $q(\text{size}(T))$. Computing q precisely, or even its degree, is a delicate question and sensitive to the exact model of computation. Thus one simply says that a decision problem *lies in* **P** if it has some polynomial-time solution.

A decision problem P lies in **NP** if there is a polynomial q with the following property: For all instances T with $P(T) = 1$ there is a proof of length at most $q(\text{size}(T))$ that $P(T) = 1$. Such a polynomial-length proof is a *certificate* for T. More concretely: Suppose that there is a polynomial q' and a Turing machine $\mathcal{M}'$ so that, for every instance T with $P(T) = 1$, there is a string C where $\mathcal{M}'$ run on (T, C) outputs the desired proof that $P(T) = 1$ in time less than $q'(\text{size}(T))$. Then, again, the problem P is in **NP** and we again call C a certificate for T.

A decision problem lies in **PSPACE** if there is a polynomial q so that on every instance T and for every step of the computation the distance between the first and last non-blank squares of the tape of $\mathcal{M}$ is at most $q(\text{size}(T))$.

A decision problem lies in **EXPTIME** if there is a polynomial q so that that the Turing machine $\mathcal{M}$ halts in time at most $\exp(q(\text{size}(T)))$ on every instance T. Note that $\mathbf{P} \subset \mathbf{NP} \subset \mathbf{PSPACE} \subset \mathbf{EXPTIME}$. At least one of these inclusions is strict as $\mathbf{P} \neq \mathbf{EXPTIME}$.

Triangulations. A *model tetrahedron* τ is a copy of the regular Euclidean tetrahedron of side length one with vertices labeled by 0, 1, 2, and 3. See Figure 1 for a picture. Label the six edges by their vertices $(0, 1)$, $(0, 2)$, and so on. Label the four faces by the number of the vertex they do *not* contain. The standard orientation on $\mathbb{R}^3$ induces an orientation on the model tetrahedron which in turn induces orientations on the faces.

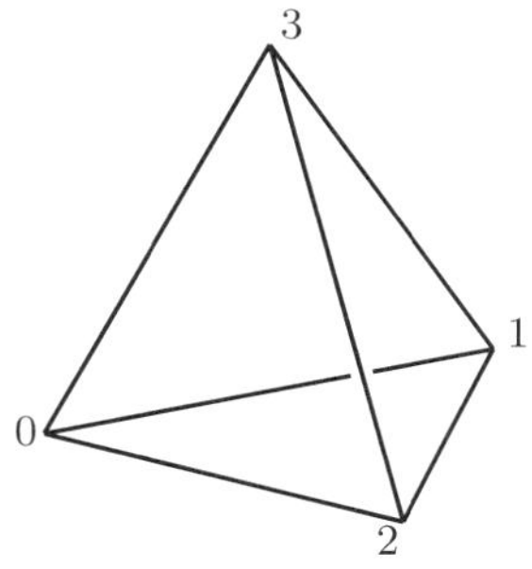

FIGURE 1. A regular Euclidean tetrahedron with all side-lengths equal to one.

A *labeled triangulation* T, of size n, is a collection of n model tetrahedra $\{\tau_i\}_{i=1}^n$, each with a unique name, and a collection of *face pairings.*

Here a face pairing is a triple (i, j, σ) specifying a pair of tetrahedra τ_i and τ_j as well as an isometry σ from a face of τ_i to a face of τ_j. We will omit the labelings when they are clear from the context.

A triangulation is not required to be a simplicial complex. However every face must appear in exactly two face pairings or in none. We do allow face pairings of the form (i, i, σ) but, as a matter of convenience, we do not allow a face to be glued to itself.

LEMMA 3.1. *There is a polynomial-time algorithm that, given triangulations T and T', decides whether or not T is identical to T'.*

PROOF. Recall that T and T' are labeled: all of the tetrahedra come equipped with names. To check for isomorphism check that every name appearing in T also appears in T' and that all of the face pairings in T and T' agree. □

REMARK 3.2. Note that, for unlabeled triangulations, there is still a quadratic algorithm that determines isomorphism of triangulations. This is because an isomorphism is determined by the image of a single tetrahedron.

Let $|T|$ be the *underlying topological space*; the space obtained from the disjoint union of the model tetrahedra by taking the quotient by the face pairings. Notice that $|T|$ is not, in general, a manifold.

At this point we should fix an encoding scheme which translates triangulations into binary strings. However we will not bother to do more than remark that there are schemes which require about $n \log(n)$ bits to specify a triangulation with n tetrahedra. (This blow-up in length is due to the necessity of giving the tetrahedra unique names.) Thus we will abuse notation and write $\text{size}(T) = n$ even though the representation of T as a binary string is somewhat longer.

Topology. Recall that the *three-sphere* is the three-manifold

$$S^3 = \{x \in \mathbb{R}^4 \mid \|x\| = 1\}.$$

The connect sum $M\#N$ of two connected oriented three-manifolds M and N is obtained by removing an open three-ball from the interior of each of M and N and gluing the resulting two-sphere boundary components with an orientation reversing homeomorphism. The connect sum naturally extends to a collection of connected, oriented three-manifolds; if M is the disjoint union of connected three-manifolds then $\#M$ denotes their connect sum.

Note that Alexander's Theorem [**H01**, Theorem 1.1] implies that $M\#S^3$ is homeomorphic to M, for any three-manifold M. We adopt the convention that the empty connect sum yields the three-sphere.

DEFINITION 3.3. Suppose that T is a triangulation and suppose that $p \in |T|$. Fix ϵ, sufficiently small, and take ϵ–neighborhoods about the preimages of p in the model tetrahedra. Each is a cone on a subsurface of the sphere. These fit together to form a cone on a two-complex S_p, the *link* of p. If $p \in T^0$ then S_p is called a *vertex link.*

We now relate several algorithms which take triangulations and check topological properties. See [**HLP99, JT95, M03**] for in-depth discussions.

THEOREM 3.4. *There is a polynomial-time algorithm that, given a triangulation T, decides whether or not $|T|$ is a three-manifold.*

PROOF. The underlying space $|T|$ is a quotient of a disjoint union of finitely many model tetrahedra. Furthermore, the face pairings are isometries. It follows that $|T|$ is second-countable and Hausdorff.

It remains only to verify that every point $p \in |T|$ has a neighborhood homeomorphic to a three-ball. Equivalently, every point p has link S_p being a sphere or a disk. This is automatic for points lying in the interior of tetrahedra. Since faces cannot be glued to themselves, any point in the interior of a face also has the desired link.

Now suppose that p lies in the interior of an edge. The link S_p is a union of spherical lunes. Thus S_p is D^2, S^2, or $\mathbb{RP}^2$. The latter may happen only at the midpoint of an edge.

Finally, suppose that p is a vertex. Now S_p is a union of spherical triangles. Again, p has the desired three-ball neighborhood if and only if S_p is a sphere or a disk.

Thus the algorithm need only check how tetrahedra are glued around an edge and the topology of each vertex link. In terms of size(T) there are at most linearly many edges and vertices. Checking each edge and each vertex link takes at most polynomial time. This is because there are at most $6 \cdot \text{size}(T)$ tetrahedra around any edge. Also, each vertex link is a union of at most $4 \cdot \text{size}(T)$ spherical triangles. □

Recall that a three-manifold M is a *homology three-sphere* if it has the same homology groups as S^3.

THEOREM 3.5. *There is a polynomial-time algorithm that, given a triangulation T of a three-manifold, decides whether or not $|T|$ is a homology three-sphere.*

PROOF. The homology groups $H_*(|T|, \mathbb{Z})$ may be read off from the Smith normal forms of the chain boundary maps: we refer the reader to [**DC91**, Section 2] for an accessible overview of algorithmic computation of homology. Finally, the Smith normal form of an integer matrix may be computed in polynomial time [**Il89**]. □

We record a few consequences of the homology three-sphere assumption.

LEMMA 3.6. *If M^3 is a homology three-sphere then M is connected, closed, and orientable. Every closed, embedded surface in M is orientable and separating. Every connect summand of M is also a homology three-sphere.* □

It follows that if N is a connect summand of a homology three-sphere and N is a lens space then $N \cong S^3$.

4. Normal and almost normal surfaces

In order to study triangulations we first discuss Haken's theory of *normal surfaces*. See [**HLP99**] for a detailed discussion, including references to the foundational work of Haken and Schubert. Other references on normal surfaces include [**JR03, B03**].

DEFINITION 4.1. An isotopy $H\colon |T| \times I \to |T|$ is a *normal isotopy* if, for all $s \in I$ and for every simplex σ in T, $H_s(\sigma) = \sigma$.

We make the same definition for faces, model tetrahedra, and subcomplexes of the triangulation. Two subsets of such are *normally isotopic* if there is a normal isotopy taking one to the other.

For example, suppose that f is a face of a model tetrahedron τ. There are three normal isotopy classes of properly embedded arcs with end points in distinct edges of f. Any such arc in f is called a *normal arc*. A simple closed curve $\alpha \subset \partial\tau$ is a *normal curve* if α is transverse to the one-skeleton of τ and α is a union of normal arcs. The *length* of a normal curve α is the number of normal arcs it contains. A normal curve α is called *short* if it has length three or four.

LEMMA 4.2. *Suppose that $\alpha \subset \partial\tau$ is a connected normal curve. The following are equivalent:*

- *α is short.*
- *α meets every edge of τ^1 at most once.*
- *α misses some edge of τ^1.*

PROOF. To see this, let $\{v_{ij} \mid 0 \leq i < j \leq 3\}$ be the number of intersections of α with each of the six edges of τ. There are twelve inequalities $v_{01} \leq v_{12} + v_{02}$, and so on. Additionally there are six congruences $v_{01} + v_{12} + v_{02} \equiv 0$, and so on, all modulo two. An easy argument now gives the desired results. □

In a model tetrahedron there are seven normal isotopy classes (or *types*) of *normal disk*, corresponding to the seven distinct short normal curves in $\partial\tau$. See Figure 2. These are the four *normal triangles* and three *normal quads*. The triangles are of type 0, 1, 2, or 3 depending on which vertex they cut off of the model tetrahedron, τ. The quads are of type 1, 2, or 3 depending on which vertex is with 0 when τ is cut by the quad.

DEFINITION 4.3. A surface S, properly embedded in $|T|$, is *normal* if for every model tetrahedron $\tau \in T$ the preimage of S in τ is a collection of normal disks.

There is also the almost normal octagon and almost normal annulus, defined by Rubinstein [**R97**]. See Figure 3 for examples. An octagon is a disk in the model tetrahedron bounded by a normal curve of length eight. An annulus is obtained by taking two disjoint normal disks and tubing them together along an arc parallel to

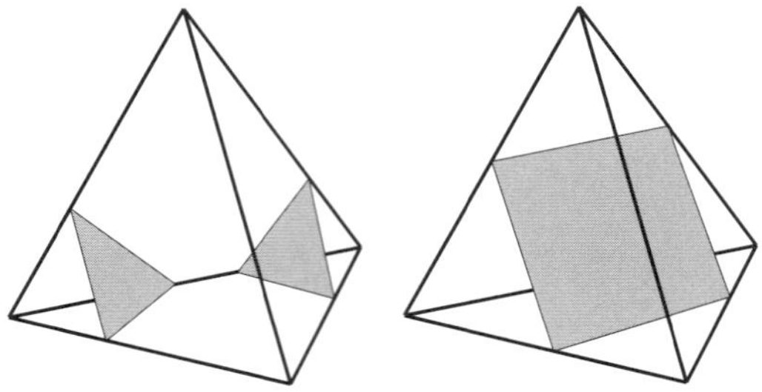

FIGURE 2. Two of the four triangles and one of the three quads.

an edge of the model tetrahedron. A surface S properly embedded in $|T|$ is *almost normal* if the preimage of S in τ is a collection of normal disks for every tetrahedron $\tau \in T$, except one. In the exceptional tetrahedron there is a collection of normal disks and exactly one almost normal piece.

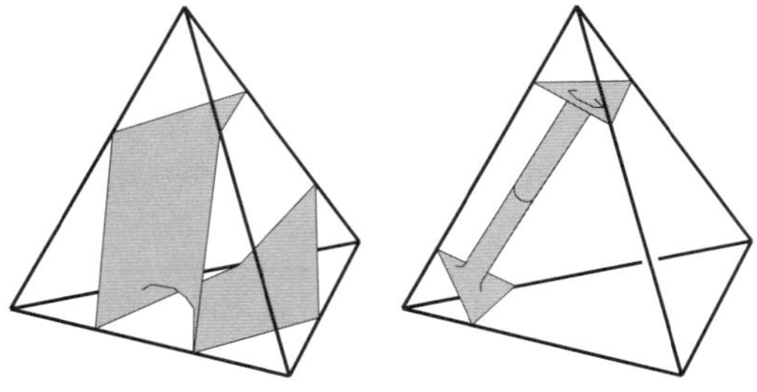

FIGURE 3. One of the three octagons and one of the 25 annuli.

REMARK 4.4. Following Jaco and Rubinstein [**JR03**, page 74] we do not allow, as an almost normal surface, parallel normal surfaces connected by an almost normal annulus contained in the product region between them. We also remark that in many cases of interest the almost normal annulus can be removed from the theory. For example, see the proof of Proposition 5.12 in [**JR03**].

4.1. Weight and Euler characteristic. For any surface $S \subset |T|$, transverse to the skeleta, define its *weight* to be the number of intersections between S and the one-skeleton T^1: $\text{weight}(S) = |S \cap T^1|$. We record a normal surface S as a *surface vector* $v(S) \in \mathbb{Z}^{7 \cdot \text{size}(T)}$. The first $4 \cdot \text{size}(T)$ coordinates describe the number of normal triangles of each type while the last $3 \cdot \text{size}(T)$ coordinates describe the number of normal quads of each type. At least two-thirds of these last $3 \cdot \text{size}(T)$ coordinates are zero as an embedded surface has only one type of normal quad in each tetrahedron.

For an almost normal surface S we again record the vector $v(S)$ of normal disks, as well as the type of the almost normal piece and the name of the tetrahedron containing it. (There is a small issue when the almost normal piece is an annulus obtained by tubing a pair of normal disks of the same type. Then $v(S)$ has length $7 \cdot \text{size}(T) + 1$ as one parallel collection of normal disks may be interrupted by the almost normal piece.) Note that two normal (or almost normal) surfaces have the same vector if and only if they are normally isotopic.

We now have a few results concerning normal and almost normal surfaces. We assume throughout that the triangulation T has underlying space a three-manifold. We first reproduce Algorithm 9.1 from [**JT95**]. See the end of Section 5 in [**AHT02**] for a similar treatment.

LEMMA 4.5. *There is a polynomial-time algorithm that, given a triangulation T and a normal or almost normal surface vector $v(S)$, computes the weight of S and the Euler characteristic of S.*

PROOF. To find the weight of S on a single edge e of T^1 count the number of normal disks meeting e (with multiplicity depending on how many times the containing tetrahedron meets e) and divide by the valency of e in T^2, the two-skeleton.

For the Euler characteristic simply use the formula $\chi(S) = F - E + V$ and the cell structure on S coming from its being a normal surface. (If S contains an almost normal annulus then we must add a single edge running between the two boundary components of the annulus.) The number of faces is the sum of the coordinates of $v(S)$. The number of edges is $3/2$ times the sum of the triangle coordinates plus twice the sum of the quad coordinates. The number of vertices of S can be computed from $v(S)$ and the degrees of the edges in T^1. Small corrections are necessary when S is almost normal. $\square$

THEOREM 4.6 (Agol-Hass-Thurston [**AHT02**]). *There is a polynomial-time algorithm that, given a triangulation T and a normal or almost normal surface vector $v(S)$, produces integers n_i and surface vectors $v(F_i)$ so that*

- $v(S) = \sum n_i \cdot v(F_i)$,
- *if* $i \neq j$ *then* $F_i \cap F_j = \emptyset$, *and*
- *if* $i \neq j$ *then* $v(F_i) \neq v(F_j)$.

PROOF. This is one application of the "extended counting algorithm" given in [**AHT02**]. See the proof of Corollary 17 of that paper. $\square$

4.2. Haken sums. Suppose S, F, G are three non-empty normal surfaces with $v(S) = v(F) + v(G)$. Then we say that F and G are *compatible*: in every tetrahedron where both F or G have quads, these quads are of the same type. After a normal isotopy of F and G we find that S is the *Haken sum* of F and G; there is a cut-and-paste of F and G constructing S. It follows that $\chi(S) = \chi(F) + \chi(G)$.

Likewise, suppose S and F are almost normal with identical almost normal piece, G is normal, the quads of G are disjoint from the almost normal piece of F, and the normal coordinates add: $v(S) = v(F) + v(G)$. Again we say that S is a Haken sum. If F contains an octagon then we may normally isotope G so that no normal triangle of G meets the octagon. If F contains an annulus we may normally isotope G so that no triangle of G meets the annulus in a meridian of the tube. In either case we may perform cut-and-paste and find $\chi(S) = \chi(F) + \chi(G)$.

When S is a Haken sum as above we write $S = F + G$. If S is not a Haken sum then S is *fundamental.*

LEMMA 4.7. *If $S = F + G$, where G is a vertex link, then S is not connected.* $\square$

LEMMA 4.8. *If $S \subset |T|$ is a fundamental normal or almost normal surface then the largest entry of $v(S)$ is at most* $\exp(\text{size}(T))$.

PROOF. There is a constant c (not depending on T or S) such that the largest entry of $v(S)$ is less than $2^{c \cdot \text{size}(T)}$. This lemma is proved for normal surfaces in [**HLP99**, Lemma 6.1]. In the almost normal case, when the almost normal piece is an octagon, we obtain similar bounds using the system of linear equations provided by [**T94**, Section 5]. The case of the annulus is similar. □

LEMMA 4.9. *Suppose T is a triangulation of a homology three-sphere. Suppose T contains a non-vertex-linking normal two-sphere. Then T contains a fundamental non-vertex-linking normal two-sphere.*

PROOF. This is similar to work of Haken and Schubert; our statement follows directly from Proposition 5.7 of [**JR03**]. The essential points are that Euler characteristic is additive under Haken sum, that T does not contain any normal $\mathbb{RP}^2$ or $\mathbb{D}^2$ (by Lemma 3.6), and that no summand is vertex-linking (by Lemma 4.7). □

DEFINITION 4.10. Fix a triangulation T so that $|T|$ is a three-manifold. If every normal two-sphere is vertex linking then, following Jaco and Rubinstein [**JR03**], we say that T is *zero-efficient.*

LEMMA 4.11. *Suppose T is a zero-efficient triangulation of a homology three-sphere. Suppose T contains an almost normal two-sphere. Then T contains a fundamental almost normal two-sphere.*

PROOF. This is identical to the proof of Lemma 4.9, except that S cannot have a normal two-sphere summand as T is zero-efficient. □

Of a much different level of difficulty is the following.

THEOREM 4.12 (Rubinstein [**R92**], Thompson [**T94**]). *If $|T| \cong S^3$ then T contains an almost normal two-sphere.* □

We end this section with a useful lemma.

LEMMA 4.13. *There is an exponential-time algorithm that, given a triangulation T of a closed three-manifold*

- *produces the surface vector of a fundamental non-vertex-linking normal two-sphere or, if none exists,*
- *produces the surface vector of a fundamental almost normal two-sphere or, if neither exists,*
- *reports that $|T|$ is not homeomorphic to the three-sphere.*

PROOF. We only sketch a proof – the interested reader should consult [**HLP99**], [**JR03**, page 66] or [**B03**, page 83]. If T admits a non-vertex-linking normal two-sphere or an almost normal two-sphere then, by Lemmas 4.9 and 4.11 there is a fundamental such surface. This surface can now be found by enumerating all fundamental surfaces (a finite list, by work of Haken) and checking Euler characteristics (Lemma 4.5). On the other hand, if no almost normal two-sphere exists then Theorem 4.12 implies that $|T|$ is not the three-sphere.

As presented the running time of the algorithm is unclear. The time depends on the number of fundamental surfaces. However, in both cases (normal or almost normal) fundamental solutions may be replaced by *vertex fundamental* surfaces [**JT95, C97**]. This gives an algorithm with running-time at most a polynomial multiplied by $3^{\mathrm{size}(T)}$. □

5. Blocked submanifolds

Normal (and almost normal) surfaces cut a triangulated manifold into pieces. These submanifolds have natural polyhedral structures which we now investigate.

Let τ be a model tetrahedron, and suppose that $S \subset \tau$ is a embedded collection of normal disks and at most one almost normal piece. Let B be the closure of any component of $\tau - S$. We call B a *block*. See Figure 4.

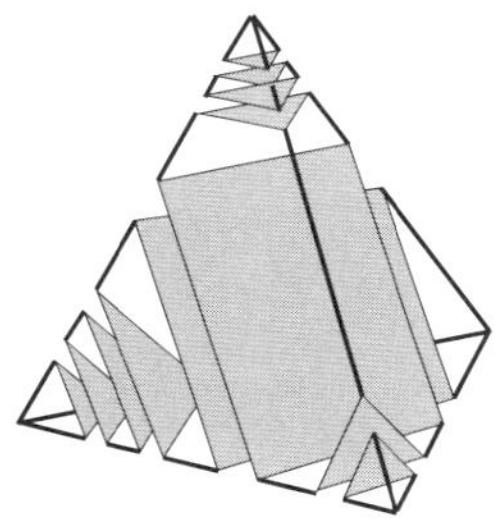

FIGURE 4. The tetrahedron τ cut along S. Note that in this example there are two blocks of the form "normal disk cross interval".

A block cobounded by two normally isotopic normal disks is called a *product block*. All other blocks are called *core blocks*. Note that there are only seven classes (or *types*) of product block, corresponding to the seven types of normal disks. Likewise there is a bounded number of core blocks. Five such are shown in Figure 4, but many more are possible. Most of these meet an almost normal annulus.

Suppose that B is a block. The components of ∂B meeting S are the *horizontal* boundary components of B, denoted $\partial_h B$. All other faces of B (the faces of B which lie in the two-skeleton) are $\partial_v B$, the *vertical* boundary.

Suppose now that T is a triangulation of a three-manifold and $S \subset |T|$ is a normal or almost normal surface. For simplicity, suppose that S is transversely oriented and separating. Let N_S be the closure of the component of $|T| - S$ pointed at by the transverse orientation.

We call N_S a *blocked submanifold* of $|T|$. Let $\widehat{N}_P$ be the union of all product blocks in N_S and let $\widehat{N}_C$ be the union of all core blocks in N_S.

REMARK 5.1. In any blocked submanifold the number of core blocks is at most linear in size(T). In fact there are at most six in each tetrahedron plus possibly two more coming from the almost normal annulus.

Note that $\widehat{N}_P$ and $\widehat{N}_C$ need not be submanifolds of $|T|$. To produce submanifolds let N_P be a closed regular neighborhood of $\widehat{N}_P$, taken inside of N_S. Also, take N_C to be the closure of $N_S - N_P$. Note the asymmetry between the definitions of

N_P and N_C: we have $\widehat{N}_P \subset N_P$ while $N_C \subset \widehat{N}_C$. As above define $\partial_h N_P = N_P \cap S$ and $\partial_v N_P = \overline{\partial N_P - \partial_h N_P}$. The horizontal and vertical boundaries $\partial_h N_C$ and $\partial_v N_C$ are defined similarly.

We now give an efficient encoding of a component $N \subset N_P$, similar to the surface vectors $v(S)$. As usual, two blocks B, B' of N are of the same *type* if there is a normal isotopy of the model tetrahedron sending B to B'. A *stack* k is all blocks of N of a fixed type. The size of k is the number of blocks it contains. Orienting k gives *positions* to the blocks in k. The *block vector* $v(N) = (v_k(N))$ is a vector in $\mathbb{Z}^{7 \cdot \text{size}(T)}$ recording the size of each stack.

We also record the vertical boundary of N; note that $\partial_v N$, as it is the boundary of a regular neighborhood of $\widehat{N}$, is subdivided into a linear number of vertical rectangles (Remark 5.1). Along with the annuli $\partial_v N$ we record the position of every block (or *position*) in every stack giving rise to a rectangle in $\partial_v N$. Taken all together, the block vector, the vertical boundary, the positions in the stacks, and the matching equations suffice to recover N up to normal isotopy.

THEOREM 5.2. *There is a polynomial-time algorithm that, given a triangulation T and a surface vector $v(S)$ for a transversely oriented S, produces the block vector $v(N)$, the vertical boundary $\partial_v N$, and the stack positions for every connected component $N \subset N_P \subset N_S$. In addition, for every N, every stack k, and every position p in k (not necessarily coming from N) the algorithm computes how p partitions the number $v_k(N)$.*

PROOF. When S is not separating replace S by the horizontal boundary of a regular neighborhood $S \times I \subset |T|$, with transverse orientation pointing outward. Take N_S equal to the closure of $|T| - (S \times [0,1])$. In what follows we assume that S is separating.

By Remark 5.1 we may explicitly build the core N_C for N_S. The vertical boundary of N_C is the desired collection of annuli.

Every normal disk meets an edge $(0i)$ of the containing model tetrahedron, minimizing i. The transverse orientation on S is equivalent to a bit vector $(\epsilon_k) \in \{0,1\}^{7 \cdot \text{size}(T)}$, as follows. The disk (of type k in S) closest to the vertex 0 has transverse orientation pointing away from 0 if and only if $\epsilon_k = 0$. The transverse orientation of the mth disk of type k now depends only on the parity of m. Verifying that (ϵ_k) gives a consistent transverse orientation on the whole of S is a polynomial number of parity calculations.

Recall that the transverse orientation points into N_S. Let $v(N_P)$ be the block vector. If $v_k(S)$ is odd then $v_k(N_P) = (v_k(S) - 1)/2$. If $v_k(S)$ is even then $v_k(N_P) = (v_k(S) - 2\epsilon_k)/2$. We may now label every rectangle in $\partial_v N_P$ by its position in the corresponding stack. As in Theorem 4.6 use the extended counting algorithm [**AHT02**] to split $v(N_P)$ as a sum of block vectors together with vertical boundaries. For any component N of N_P the counting algorithm can also detect how any position p in any stack k partitions the number $v_k(N)$. We do this by modifying the *weight vector* to record partitions instead of just weights. See the second paragraph of the proof of Corollary 17 in [**AHT02**]. □

REMARK 5.3. If S is connected then the number of connected components of N_P is at most a linear in size(T). This is because $\partial_v N_P = \partial_v N_C$ and the latter has at most linearly many components. (See Remark 5.1.) This is in pleasant contrast to Theorem 4.6.

6. Normalizing slowly

In this section we discuss a restricted version of Haken's *normalization* procedure for producing normal surfaces. This material appeared first in an unpublished preprint of mine and later in my thesis [**S01**]. I thank Danny Calegari for reading an early version of this work. I also thank Bus Jaco for several enlightening conversations regarding *barrier surfaces*. For example Lemma 8.1 can also be deduced, via a careful induction argument, from Lemma 3.1 of [**JR03**]. In addition to Jaco and Rubinstein [**JR03**], other authors have independently produced versions of these ideas such as King [**K01**] and Barchechat [**B03**].

DEFINITION 6.1. Take S a closed orientable surface. Let $C_0 = S \times [0,1]$. Choose a disjoint collection of simple closed curves in some component of $S \times \{0\}$ and attach two-handles in the usual fashion along these curves. Cap off some (but not necessarily all) of any resulting two-sphere boundary components with three-handles. The final result, C, is a *compression body*. Set $\partial_+ C = S \times \{1\}$ and set $\partial_- C = \partial C - \partial_+ C$.

Our definition differs from others (for example [**CG87**]) in that two-sphere components in $\partial_- C$ are allowed. The reasons for this are explained in Remark 9.3.

Fix T a triangulation of a closed, orientable, connected three-manifold. Let $S \subset |T|$ be a transversely oriented, almost normal surface.

DEFINITION 6.2. A compression body $C_S \subset |T|$ is *associated* to S if $\partial_+ C_S = S$, $\partial_- C_S$ is normal, the transverse orientation points into C_S, and any normal surface S' disjoint from S may be normally isotoped to be disjoint from C_S.

As a bit of notation take $\mathrm{norm}(S) = \partial_- C_S$ and call this the *normalization* of S. This is well-defined by the following.

THEOREM 10.1. *Given a transversely oriented almost normal surface S there exists an associated compression body C_S and it is unique up to normal isotopy. Furthermore there is a algorithm that, given the triangulation T and the surface vector $v(S)$, computes the surface vector $v(\mathrm{norm}(S))$.*

REMARK 6.3. As in Theorem 5.2, when S is not separating we add a parallel copy S' and transversely orient away from the parallel region between S and S'. Henceforth, we will assume that S is in fact separating.

The proof of Theorem 10.1 spans Sections 7 to 10. We here give the necessary definitions. In Section 7 we discuss the tightening procedure. In Section 8 we show that the tightening procedure gives an embedded isotopy. We discuss the capping off procedure in Section 9. The proof is finished in Section 10.

6.1. Non-normal surfaces. Let S be a surface properly embedded in a triangulated three-manifold $|T|$ and suppose that S is transverse to the skeleta of T. Denote the i-skeleton of T by T^i.

We characterize some of the ways S can fail to be normal. A *facial curve* of S is a simple closed curve of intersection between S and the interior of some triangular face $f \in T^2$. A *bent arc* of S is a properly embedded arc of intersection between S and the interior of some triangular face $f \in T^2$ with both endpoints of the arc contained in a single edge of f. Both of these are drawn in Figure 5.

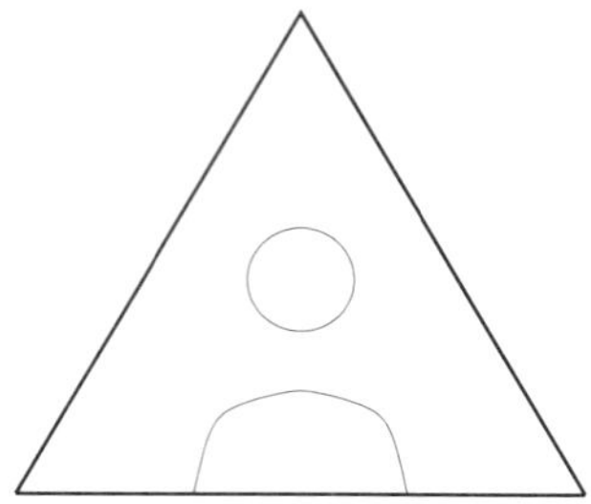

FIGURE 5. A facial curve and a bent arc.

6.2. Surgery and tightening disks.

DEFINITION 6.4. An embedded disk $D \subset |T|$ is a *surgery disk* for S if

- $D \cap S = \partial D$,
- $D \subset T^2$ or $D \cap T^2 = \emptyset$, and
- $D \cap T^1 = \emptyset$.

There is a *surgery* of S along D: Remove a small neighborhood of ∂D from S and cap off the boundaries thus created with disjoint, parallel copies of D. Note that we do not require ∂D to be essential in S. A facial curve of $S \cap T^2$ is *innermost* if it is the boundary of a surgery disk embedded in a triangle of T^2.

A *bigon* is an disk D with given subarcs $\alpha, \beta \subset \partial D$ so that $\alpha \cup \beta = \partial D$ and $\alpha \cap \beta = \partial\alpha = \partial\beta$.

DEFINITION 6.5. An embedded bigon $D \subset |T|$ is a *tightening disk* for S if

- $D \cap S = \alpha$,
- $D \subset T^2$ or $D \cap T^2 = \beta$,
- $D \cap T^1 = \beta$, and
- $D \cap T^0 = \emptyset$.

There is a *tightening isotopy* of S across D: Push α along D, via ambient isotopy of S supported in a small neighborhood of D, until α moves past β. This procedure reduces weight(S) by exactly two. A bent arc of S is *outermost* if it lies on the boundary of a tightening disk embedded in a triangle of T^2.

Suppose S contains an almost normal octagon, $A \subset \tau$. Then there are two tightening disks on opposite sides of A both giving tightening isotopies of S to possibly non-normal surfaces of lesser weight. To see these disks, consult the left-hand side of Figure 3. Notice there are two edges of τ, say e and e', so that $|A \cap e| = |A \cap e'| = 2$. The first tightening disk has boundary running along e between the points of intersection with A; then the boundary runs along an arc in the interior of A. The second disk similarly meets e'. We arrange matters so that the tightening disks lying in the interior of τ meet each other in a single point.

The above disks are the *exceptional tightening disks* associated to A. If S contains an almost normal annulus then the tube is parallel to at least one edge of the containing tetrahedron. See the right-hand side of Figure 3. For every such edge there is an exceptional tightening disk. Also, the disk which surgers the almost normal annulus will be called the *exceptional surgery disk*.

7. Tightening

This section discusses the *tightening* procedure which will yield an embedded isotopy. This is proved in Lemma 8.1 below. Fix T is a triangulation of a three-manifold. Suppose that $S \subset |T|$ is a transversely orientable separating almost normal surface. We wish to isotope S off of itself while steadily reducing the weight of S.

Suppose that D is an exceptional tightening disk for S. Choose the transverse orientation for S which points into the component of $|T| - S$ which meets D. The F*-tightening procedure* constructs a map $\mathcal{F}\colon S \times [0, n] \to |T|$ as follows:

(1) Let $F_0 = S$. Take $\mathcal{F}_0\colon S \times \{0\} \to |T|$ to be projection to the first factor. Let $D_0 = D$.
(2) Do a small normal isotopy of F_0 in the transverse direction while tightening F_0 along D_0. This extends $\mathcal{F}_0$ to a map $\mathcal{F}_1\colon S \times [0,1] \to |T|$, with $F_t = \mathcal{F}_1(S \times \{t\})$. Note that the surface F_1 inherits a transverse orientation from F_0. Arrange matters so that $F_{\frac{1}{2}}$ is the only level which is not transverse to T^2. Furthermore $F_{\frac{1}{2}}$ only has a single tangency with T^1 and this tangency occurs in the middle of $\partial D_0 \cap T^1$.
(3) Suppose, at step $k \geq 1$, that F_k has an outermost bent arc α with the transverse orientation of F_k pointing into the tightening disk D_k. So D_k is cut out of T^2 by α. Extend $\mathcal{F}_k$ to $\mathcal{F}_{k+1}\colon S \times [0, k+1] \to |T|$ by doing a small normal isotopy of F_k in the transverse direction while tightening F_k across D_k, the kth tightening disk. So $\mathcal{F}_k = \mathcal{F}_{k+1}|S \times [0,k]$ and $F_t = \mathcal{F}_{k+1}(S \times \{t\})$. Note that the surface F_{k+1} inherits a transverse orientation from F_k. Arrange matters so that $F_{k+\frac{1}{2}}$ is the $k+1^{\text{th}}$ level which is not transverse to T^2. Furthermore $F_{k+\frac{1}{2}}$ only has a single tangency with T^1 and this tangency occurs in the middle of $\partial D_k \cap T^1$.
(4) Suppose, at step $k \geq 1$, that there is no outermost bent arc $\alpha \subset F_k$. Set $n = k$ and halt.

REMARK 7.1. As $\text{weight}(F_{k+1}) = \text{weight}(F_k) - 2$ this tightening procedure terminates after at most weight(S) step. Note also that $\mathcal{F}_n$ is far from unique – at any stage in the procedure there may be many tightening disks to choose from.

We will show in Lemma 8.1 that the map $\mathcal{F}_n\colon S \times [0, n] \to M$ is an embedding. Note that, by construction, $S = F_0 = \mathcal{F}_n(S \times \{0\})$ and in general $F_t = \mathcal{F}_n(S \times \{t\})$. To simplify notation set $\mathcal{F} = \mathcal{F}_n$.

8. Tracking the isotopy

Let $S \subset |T|$, $\mathcal{F}$, $\mathcal{F}_k$, and F_t be as defined in Section 7. Suppose that f is any face of any model tetrahedron $\tau \in T$. In this section we analyze how the preimage (in τ) of the image of $\mathcal{F}_k$ (in $|T|$) intersect f. We will abuse notation by writing $f \cap \mathcal{F}_k$ for this intersection.

LEMMA 8.1. *For every k, the map $\mathcal{F}_k$ is an embedding. Furthermore, for $k > 0$ and for every face $f \in T^2$, the connected components of $f \cap \mathcal{F}_k$ are given, up to symmetry, by Figures 6 and 7.*

Before proceeding to the proof note that the normal arcs, bent arcs, and facial curves bounding the components shown Figures 6 and 7 inherit a transverse

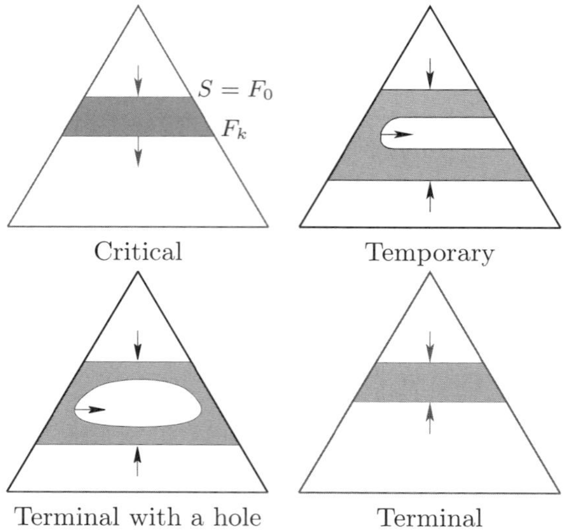

FIGURE 6. The rectangles

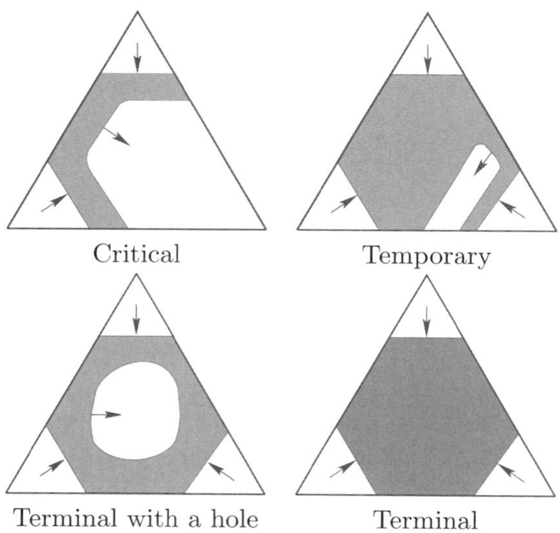

FIGURE 7. The hexagons

orientation from S or F_k. In all cases the transverse orientation on S points into the intersection $f \cap \mathcal{F}_k$ while the transverse orientation on F_k points away. The components of intersection containing a normal arc of F_k are called *critical*. Those with a single bent arc of F_k are called *temporary*. Any component containing a single facial curve of F_k is called *terminal with a hole*. Finally, components of $f \cap \mathcal{F}_k$ which are completely disjoint from F_k are simply called *terminal*. Again, refer to Figures 6 and 7.

The tightening procedure combines the critical components in various ways. However, a temporary component always results in a terminal (possibly with a hole) and these are stable. Note also that there is a second critical rectangle with the opposite transverse orientation. The non-critical components may be foliated by the levels of $\mathcal{F}_k$ in multiple ways, depending on the order of the tightening isotopies.

PROOF OF LEMMA 8.1. We induct on k; both claims are trivial for $k = 0$. Now to deal with $k = 1$. The exceptional tightening disk D_0 has interior disjoint from $S = F_0$. It follows that $\mathcal{F}_1$ is an embedding. To verify the second claim for $k = 1$ note that the image of $\mathcal{F}_1|S \times [0, \epsilon]$ intersects all faces $f \in T^2$ only in critical rectangles. Up to $t = \frac{1}{2}$ the image of $\mathcal{F}_1|S \times [0, t]$ intersected with f is combinatorially constant. Crossing $t = \frac{1}{2}$ adds a regular neighborhood of D_0 to the image. This only intersects f in a regular neighborhood of $\partial D_0 \cap T^1$. So the pieces of $f \cap \mathcal{F}_1$ are unions of critical rectangles connected by small neighborhoods of sub-arcs of T^1. Also these sub-arcs only meet the F_t side of the critical rectangles. As each critical rectangle meets two edges of the face f it follows that at most three critical rectangles are joined together to form a component of $f \cap \mathcal{F}_1$. We list all possible cases – consulting Figures 6 and 7 will be helpful:

(1) Two critical rectangles in f combine to produce a temporary rectangle, a terminal rectangle with a hole, or a critical hexagon.
(2) Three critical rectangles in f combine to produce a temporary hexagon or a terminal hexagon with a hole.

Now to deal with the general case: Suppose that both hypotheses hold at stage k. Suppose that $\alpha \subset F_k$ is the bent arc on the boundary of $D_k \subset f \in T^2$, the next tightening disk in the sequence. Suppose that interior(D_k) meets image$(\mathcal{F}_k)$. By the second induction hypothesis there is a component, C, of $f \cap \mathcal{F}_k$ which meets interior(D_k) and appears among those listed in Figures 6 and 7. Observe that each component of $f \cap \mathcal{F}_k$, and hence C, meets at least two edges of f. The bent arc α meets only one edge of f. It follows that the interior of C must meet α. Thus $\mathcal{F}_k$ was not an embedding, a contradiction.

It follows that $D_k \cap \text{image}(\mathcal{F}_k) = \alpha$. Since the $k + 1^{\text{th}}$ stage of the isotopy is supported in a small neighborhood of $F_k \cup D_k$ it follows that $\mathcal{F}_{k+1}$ is an embedding.

Now, the transverse orientation on F_k gives rise to a transverse orientation on F_{k+1}. To verify the second hypothesis we again list the possible cases:

(1) Two critical rectangles in f combine to produce a temporary rectangle, a terminal rectangle with a hole, or a critical hexagon.
(2) Three critical rectangles f combine to produce a temporary hexagon or a terminal hexagon with a hole.
(3) A critical rectangle and critical hexagon in f combine to produce a temporary hexagon or a terminal hexagon with a hole.
(4) A temporary component leads to a terminal one (possibly with a hole).

This completes the induction. □

REMARK 8.2. By maximality of $\mathcal{F}$, the surface $F_n = \mathcal{F}(S \times \{n\})$ has no outermost bent arcs with outward orientation. A bent arc with inward orientation would violate the second induction hypothesis of Lemma 8.1. So F_n contains no bent arcs. F_n may contain facial curves, but the second induction hypotheses shows that all of these are innermost with transverse orientation pointing toward the bounded surgery disk.

If we replace S in Lemma 8.1 by a disjoint union of S with a collection of normal surfaces we obtain the following corollary.

COROLLARY 8.3. *If S' is any normal surface in $|T|$ which does not intersect S then $\mathcal{F} \cap S' = \emptyset$, perhaps after a normal isotopy of S' (rel S).* □

Suppose that τ is a model tetrahedron in the given triangulation T. We again abuse notation, writing $\tau - \mathcal{F}_k$ for the complement of the preimage (in τ) of the image of $\mathcal{F}_k$ (in $|T|$).

LEMMA 8.4. *For all $k \geq 1$, $\tau - \mathcal{F}_k$ is a disjoint collection of balls.*

PROOF. Our induction hypothesis is as follows: $\tau - \mathcal{F}_k$ is a disjoint collection of balls, unless $k = 0$ and τ contains the almost normal annulus of S. In that exceptional case $\tau - \mathcal{F}_0$ is a disjoint collection of balls and one solid torus.

The base case is trivial. Suppose B is a component of $\tau - \mathcal{F}_k$. There are now two cases to consider. Either B is cut by an exceptional tightening disk or it is not. Assume the latter. Then B is a three-ball by induction and after the $k + 1^{\text{th}}$ stage of the isotopy $B \cap \mathcal{F}_{k+1}$ is a regular neighborhood (in B) of a collection of disjoint arcs and disks in ∂B. Hence $B - \mathcal{F}_{k+1}$ is still a ball.

If B is adjacent to the almost normal piece of F_0 then let D_0 be the exceptional tightening disk. Set $B_\epsilon = B - \text{neigh}(D_0)$. Each component of B_ϵ is a ball, and the argument of the above paragraph shows that they persist in the complement of $\mathcal{F}_1$. □

Recall that $\partial\mathcal{F}_k = S \cup F_k$. As usual, for a model tetrahedron τ we write $\tau \cap F_k$ for the preimage of F_k in τ. A corollary of Lemma 8.4 is the following.

COROLLARY 8.5. *For all k, the connected components of $\tau \cap F_k$ are planar.* □

The connected components of $\tau \cap F_n$ warrant closer attention.

LEMMA 8.6. *If $n \geq 1$ then each component of $\tau \cap F_n$ has at most one normal curve boundary component. This normal curve must be short.*

PROOF. Let $\tau \in T$ be a tetrahedron. Let P be a connected component of $\tau \cap F_n$. By Lemma 8.1 the boundary ∂P is a collection of facial curves and normal curves in $\partial\tau$. Let α be any normal curve in ∂P. Let $\{\alpha_j\}$ be the normal arcs of α.

CLAIM. α has length three or four.

PROOF OF CLAIM. Call the collection of critical rectangles and hexagons in $\partial\tau \cap \mathcal{F}$ that meet α the *support* of α. To prove the claim we have two cases. First suppose that only critical rectangles support α. So α is normally isotopic to a normal curve $\beta \subset \partial\tau \cap S$. The first step of the tightening procedure prevents β from being a boundary of the almost normal piece of S. It follows that α must be short.

Otherwise α_1, a normal subarc of α, is on the boundary of a critical hexagon $h \subset f$. Let β be a normal curve of S meeting h and let $\beta_1 \subset \beta$ be one of the normal arcs in ∂h. Let e be the edge of f which α_1 does not meet. This edge is partitioned into three pieces; $e_h \subset h$, e', and e''. We may assume that β_1 separates e_h from e'. See Figure 8.

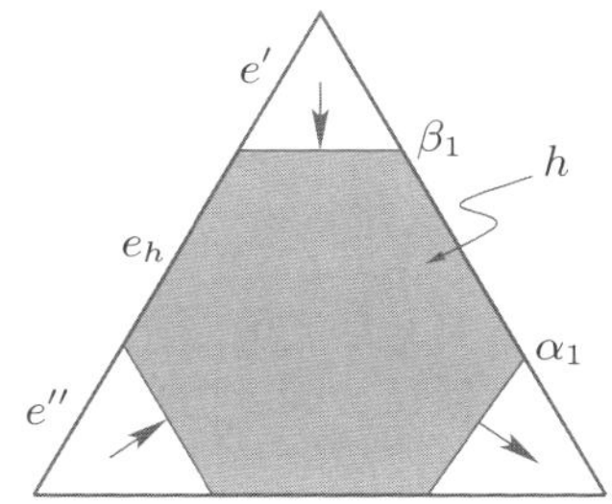

FIGURE 8. The normal arcs α_1 and β_1 are on the boundary of the critical hexagon h. Note that β does not meet e' or interior(h).

Note that a normal curve of length ≤ 8 has no parallel normal arcs in a single face. Thus β meets e' exactly once at an endpoint of e'. Since α and β do not cross it follows that β separates α from e' in $\partial\tau$.

Similarly, α is separated from e''. Thus α does not meet e at all. By Lemma 4.2 the normal curve α is short. This finishes the proof of the claim. □

CLAIM. The component $P \subset \tau \cap F_n$ has at most one boundary component which is a normal curve.

Proving this will complete the lemma. So suppose that ∂P contains two normal curves: α and β. Let A be the annulus cobounded by α and β in $\partial\tau$, the boundary of the model tetrahedron.

Suppose now that the transverse orientation, that α inherits from F_n, points away from A. Thus A and the support of α intersect. There are several cases to examine, depending on the length of α and the components of the support of α.

(1) Suppose α has length three:
  (a) If only critical rectangles support α then a normal triangle of S separates α and β.
  (b) If one critical hexagon and two critical rectangles support α then the almost normal octagon and the exceptional tightening disk together separate α and β. See left hand side of Figure 9.
  (c) If two critical hexagons and one critical rectangle support α then a normal triangle or a normal quad of S separates α from β. See right hand side of Figure 9.
  (d) If only critical hexagons support α then a normal triangle of S separates α and β.

(2) Suppose α has length four:
  (a) If only critical rectangles support α then a normal quad of S separates α and β.
  (b) If one critical hexagon and three critical rectangles support α then S could not have been an almost normal surface. See left hand side of Figure 10.
  (c) If two critical hexagons and two critical rectangles support α then a normal triangle of S separates α and β. See right hand side of Figure 10.

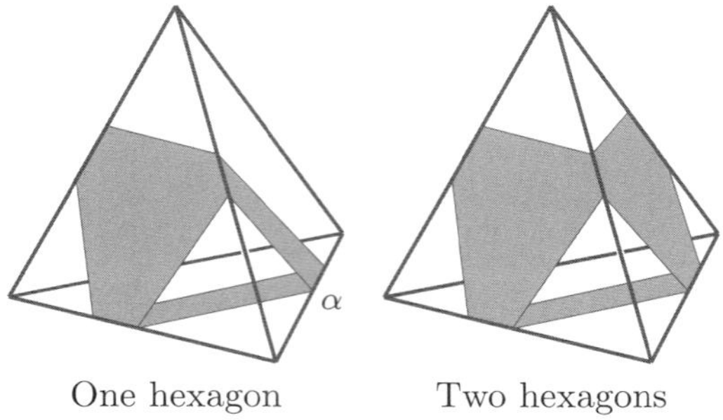

FIGURE 9. Diagrams for cases (1b) and (1c).

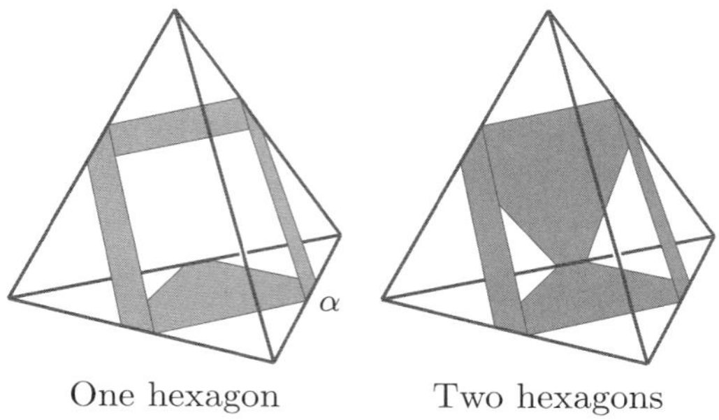

FIGURE 10. Diagrams for cases (2b) and (2c).

When α has length four it cannot be supported by more than two critical hexagons.

To recap: in all cases except 1(b) and 2(b), the support of α (possibly together with a terminal rectangle or hexagon) closes up, implying the existence of a normal disk of S with boundary a core curve of the annulus A. As this disk lies in S observe that $S \cap P \neq \emptyset$ and thus $S \cap F_n \neq \emptyset$. This contradicts the fact that $\mathcal{F}$ is an embedding (Lemma 8.1). Case 1(b) is similar, except that the support of α meets other critical or terminal components to form the octagon piece of S. So P must intersect either S or the exceptional tightening disk, again a contradiction of Lemma 8.1. Lastly, in case 2(b), S could not have been almost normal.

So deduce that the transverse orientation that α inherits from F_n must point toward A. Thus A and the support of α are disjoint. Let γ be an arc which runs along P from α to β. Let α' be a push-off of α along A, towards β. This push-off bounds a disk in one of the components of $\tau - \mathcal{F}$, by Lemma 8.4. This disk does not intersect $P \subset F_n \subset \mathcal{F}$ and hence fails to intersect γ. This is a contradiction. □

REMARK 8.7. By Lemma 8.1 all facial curves of F_i are innermost. It follows that the "tubes" analyzed in Lemma 8.6 do not run through each other.

9. Capping off

Here we construct our candidate for C_S, the compression body associated to S.

Let $\mathcal{F} \subset |T|$ be the image of the map constructed above. Recall that $\partial\mathcal{F} = S \cup F_n$ where S is the almost normal surface we started with and F_n is the surface obtained by "tightening" S. Note that, since $\mathcal{F}$ is the embedded image of $S \times [0, n]$, in fact F_n is isotopic to S in $|T|$. They are not normally isotopic as $\text{weight}(F_n) < \text{weight}(S)$.

DEFINITION 9.1. A two-sphere which is embedded in $|T|$ but disjoint from T^2 is called a *bubble.*

From Lemma 8.6, Corollary 8.5, and Remark 8.7 we deduce:

COROLLARY 9.2. *Let F'_n be the surface obtained by surgering all facial curves of F_n. Then F'_n is a disjoint collection of bubbles and normal surfaces. Each bubble bounds a ball with interior disjoint from $T^2 \cap F'_n$.* □

Construct C_S as follows: For every facial curve α of F_n attach a two-handle to $\mathcal{F}$ along α. Attach so that the core of the two-handle is the subdisk of T^2 cut out by α. Call this $\mathcal{F}'$. As noted in Remark 8.7 all facial curves of F_n are innermost. So $\mathcal{F}'$ is an embedded compression body. At this point there may be components of $\partial_-\mathcal{F}'$ which are not normal. By Corollary 9.2 all of these are bubbles bounding a ball disjoint from all of the other bubbles. Cap off each bubble to obtain C_S. Set $\text{norm}(S) = \partial_- C_S$. The next section proves that $v(\text{norm}(S))$ does not depend on the choices made in the construction of $\mathcal{F}$.

REMARK 9.3. Normal two-spheres may appear in the normalization procedure. In particular, if S is an almost normal two-sphere then, for one of the two possible transverse orientations, there will always be a normal two-sphere appearing in $\text{norm}(S)$. This is why two-spheres are permitted in $\partial_- C$ in Definition 6.1.

10. Proof of the normalization theorem

Suppose that S is almost normal and equipped with a transverse orientation. Recall from Definition 6.2 that C_S, a compression body in $|T|$, is *associated* to S if $\partial_+ C_S = S$, $\partial_- C_S$ is normal, the transverse orientation on S points into C_S, and any normal surface $S' \subset |T|$ disjoint from S may be normally isotoped to one disjoint from C_S.

We now have the following.

THEOREM 10.1. *Given a transversely oriented almost normal surface S there exists a compression body C_S associated to S and C_S is unique (up to normal isotopy). Furthermore there is a algorithm that, given the triangulation T and the surface vector $v(S)$, computes the surface vector of* $\text{norm}(S) = \partial_- C_S$.

PROOF. We proceed in several steps.

CLAIM. There exists a compression body C_S associated to S.

PROOF. There are two cases. Either the transverse orientation for S points at the exceptional surgery disk (implying that S contained an almost normal annulus) or the transverse orientation points at an exceptional tightening disk.

In the first case, C_S is obtained by thickening S slightly and adding a regular neighborhood of the exceptional surgery disk. It is clear that C_S is a compression body, $\partial_+ C_S = S$, and $\partial_- C_S$ is normal. Suppose that S' is any normal surface in T which is disjoint from S. Then, perhaps after a normal isotopy of S' (rel S), we have that S' is disjoint from the exceptional surgery disk for S. It follows that S' may be isotoped out of C_S.

In the second case the transverse orientation of S points at an exceptional tightening disk of S. As in Section 7 form $\mathcal{F}$ with $\partial\mathcal{F} = S \cup F_n$. As in Section 9 attach two-handles to $\mathcal{F}$ along the facial curves of F_n to obtain $\mathcal{F}'$. Cap off the bubbles with their three-balls to obtain C_S. Again, C_S is a compression body with $\partial_+ C_S = S$.

Suppose now that S' is some normal surface in T which is disjoint from S. Then, by Corollary 8.3, the surface S' is disjoint from $\mathcal{F}$, perhaps after a normal isotopy of S' (rel S). Since S' is normal it cannot meet any of the disks (in T^2) bounded by facial curves of F_n. So $S' \cap \mathcal{F}' = \emptyset$ as well. Finally, suppose that A is a bubble component of $\partial_-\mathcal{F}'$. Let B be the three-ball which A bounds (such that $B \cap T^2 = \emptyset$). Then no component of S' meets B as $S' \cap A = \emptyset$ and S' is normal. Deduce that $S' \cap C_S = \emptyset$. This finishes the claim. □

CLAIM. The associated compression body C_S is unique (up to normal isotopy).

PROOF. Suppose that C_S and C'_S are both associated to S. Let $A = \partial_- C_S$ and $A' = \partial_- C'_S$. Then A and A' are normal surfaces, both disjoint from S. It follows that there exists a normal isotopy $\mathcal{H}$ which moves A' out of C_S (rel S) and another normal isotopy $\mathcal{H}'$ which moves A out of C'_S (rel S).

Consider any face $f \in T^2$ and any normal arc $\alpha \subset f \cap S$. Let $X \subset f \cap C_S$ be the component containing α. Take X' to be the component of $f \cap C'_S$ which contains α. We must show that X and X' have the same combinatorial type. Suppose not. After possibly interchanging X and X' there are only six situations to consider:

(1) X is a critical rectangle and X' is a terminal rectangle.
(2) X is a critical rectangle and X' is a critical hexagon.
(3) X is a critical rectangle and X' is a terminal hexagon.
(4) X is a critical hexagon and X' is a terminal hexagon.

In any of these four cases let δ be the normal arc of $A = \partial_- C_S$ on the boundary of X. Note that $\partial X'$ contains α (as does ∂X) and also another normal arc $\beta \subset f \cap S$ which does not meet X (as $S = \partial_+ C_S$). Now note that it is impossible for $\mathcal{H}'$ to normally isotope δ out of X' while keeping S fixed pointwise (as δ would have to cross β).

(5) X is a terminal rectangle and X' is a critical hexagon.
(6) X is a terminal rectangle and X' is a terminal hexagon.

In either of these cases let β be the other normal arc of $S \cap \partial X$. Then β intersects the interior of X', a contradiction. This proves the claim. □

CLAIM. There is a algorithm that, given the triangulation T and the surface vector $v(S)$, computes the surface vector of $\partial_- C_S = \operatorname{norm}(S)$.

PROOF. We follow the proof of Lemma 8.1: We keep track of the intersection $f \cap \mathcal{F}_k$ for every face f of every model tetrahedron τ. The intersection is a union of components, with all allowable types shown (up to symmetry) in Figures 6 and 7. There is at most one hexagon in each face and perhaps many rectangles, arranged in three families, one for each vertex of f. At stage n there are no bent arcs remaining. Now delete all facial curves of F_n and all normal arcs of S. The normal arcs left completely determine $\operatorname{norm}(S)$ and from this we may find the surface vector $v(\operatorname{norm}(S))$. This proves the claim. □

Thus we are done with the proof of Theorem 10.1. □

The algorithm just given is inefficient. It depends polynomially on size(T) and weight(S). In the next section we improve this to a algorithm which depends polynomially on size(T) and log(weight(S)).

As a corollary of Theorem 10.1:

COROLLARY 10.2. *If $S \subset |T|$ is a transversely oriented almost normal two-sphere then C_S is a three-ball, possibly with some open three-balls removed from its interior. These have closures disjoint from each other and from S.* □

Now an orientable surface in an orientable three-manifold may be transversely oriented in exactly two ways. By Theorem 10.1, if S is an almost normal surface, for each transverse orientation there is a associated compression body. Call these C_S^+ and C_S^-.

From Corollary 10.2 deduce:

THEOREM 10.3. *If $S \subset |T|$ is an almost normal two-sphere and both $\partial C_S^+ - S$ and $\partial C_S^- - S$ are (possibly empty) collections of vertex-linking two-spheres, then $|T|$ is the three-sphere.*

PROOF. By hypothesis $\partial C_S^+ - S$ is a collection of vertex linking spheres. For each of these add to C_S^+ the corresponding vertex neighborhood. Let $\mathbb{B}^+$ be the resulting submanifold of $|T|$. By the Alexander trick $\mathbb{B}^+$ is a three-ball. Do the same to C_S^- to produce $\mathbb{B}^-$. The Alexander trick now implies that the manifold $|T| = \mathbb{B}^+ \cup_S \mathbb{B}^-$ is homeomorphic to the three-sphere. □

11. An example

Here we give a brief example of the normalization procedure. Let T be the one vertex triangulation shown in Figure 11.

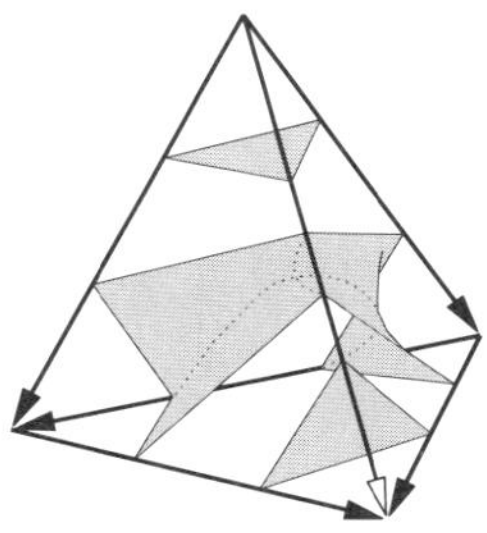

FIGURE 11. A one-tetrahedron triangulation of S^3. It is straightforward to list all normal and almost normal surfaces in T. It is a pleasant exercise to draw the graph T^1 as it actually sits in S^3. It is somewhat harder to draw the two-skeleton.

The front two faces (1 and 2) are glued to each other as are the back faces (0 and 3). The faces are glued to give the edge identifications shown. The surface S depicted in T is an almost normal two-sphere with two triangles and one almost normal octagon.

The sphere S has two exceptional tightening disks: D meeting the edge $(0,3)$ of the model tetrahedron and D' meeting edge $(1,2)$.

Tightening along D gives F_1 which is the vertex link. Tightening along D' and then along bent arcs to obtain an F–tightening sequence F_1', F_2', F_3'. As a note of caution: F_1' drawn in the model tetrahedron has four bent arcs – however $F_1' \cap T^2$ contains only two. These are independent of each other and doing these moves in some order gives F_2' and F_3'. Then F_3' is a weightless two-sphere in T with a single facial curve and no other intersection with the two-skeleton. Finally, surger the facial curve of F_3' and cap off the two resulting bubbles.

Thus: on the D side of S the normalization is the vertex link. On the D' side the normalization is the empty set. It follows from Theorem 10.3 that $|T|$ is the three-sphere.

12. Normalizing quickly

The normalization procedure can be accelerated. Suppose that T is a triangulation of a three-manifold and S is assumed to be a transversely oriented almost normal surface.

THEOREM 12.1. *There is a polynomial-time algorithm that, given any such T and surface vector $v(S)$, produces as output $v(\text{norm}(S))$, the normalization of S.*

As in Remark 6.3 we will restrict to the case where S is separating. Recall that N_S is the closure of the component of $|T| - S$ into which the transverse orientation points. Then $\widehat{N}_P$ is the union of all product blocks in N_S and $\widehat{N}_C$ is the union of all the core blocks. Also N_P is a regular neighborhood of $\widehat{N}_P$, taken in N_S. Finally $N_C = \overline{N_S - N_P}$. We will prove Theorem 12.1 by altering our original normalization procedure three times. First we will show that the order of the tightening moves is irrelevant. Then we will show that surgeries on facial curves and capping off of bubbles may happen during the normalization procedure, instead of being held until the end. Finally we show that tightening inside of N_P can be done quickly. These three modifications combine to give an efficient algorithm.

12.1. Changing the order of the tightening moves. As stated in Remark 7.1 the isotopy $\mathcal{F}\colon S \times [0,n] \to M$ need not be unique. However the first sentence of Theorem 10.1 implies the following.

LEMMA 12.2. *Any order for the tightening moves (performed in the construction of $\mathcal{F}$) gives the same surface* $\text{norm}(S)$ *once the facial curves of F_n have been surgered and bubbles have been capped off.* □

12.2. Surgery on facial curves and deleting bubbles. We now alter the tightening procedure in a more substantial fashion.

Recall that $S \subset |T|$ is a transversely orientable separating almost normal surface. Recall that D is the exceptional tightening disk for S. Transversely orient S to point into the component of $|T| - S$ which meets D. Here is the *G-tightening procedure*:

(1) Let $G_0 = S$. Let $D_0 = D$.
(2) Do a small normal isotopy of G_0 in the transverse direction while tightening G_0 along D_0. Call the surface so obtained G_0'. Surger all facial curves of $f \cap G_0'$ for every $f \subset T^2$ to obtain G_0''. Delete any bubble components

of G_0'' (two-sphere components which are contained in the interior of tetrahedra). Call the resulting surface G_1. Note that G_1 inherits a transverse orientation from G_0.

(3) Suppose, at step $k \geq 1$, that G_k has an outermost bent arc α with the transverse orientation of G_k pointing into the tightening disk D_k. So D_k is cut out of T^2 by α. Perform a small normal isotopy of G_k in the transverse direction while tightening G_k across D_k. Call the surface so obtained G_k'. Surger all facial curves of $f \cap G_k'$ for every $f \in T^2$ to obtain G_k''. Delete any bubble components of G_k''. Call the resulting surface G_{k+1}. Note that G_{k+1} inherits a transverse orientation from G_k.

(4) Suppose, at step $k \geq 1$, that there is no outermost bent arc $\alpha \subset G_k$. Set $n = k$ and halt.

LEMMA 12.3. *The surface G_n is normally isotopic to* norm(S), *the normalization of S.*

PROOF. Recall that Lemma 8.1 gives a complete classification of the possible components of intersection of image($\mathcal{F}_k$) with the faces of T^2. Again, see Figures 6 and 7. The only components containing a facial curve are the terminal rectangle with hole and terminal hexagon with a hole.

Since the terminal rectangles and hexagons with a hole do not contain normal or bent arcs of F_k they remain unchanged in the F-tightening procedure until F_n is reached. Then all facial curves are surgered and bubbles capped off. Thus it makes no difference to the resulting surface norm(S) if we surger facial curves and delete bubbles as soon as they appear. □

12.3. Tightening in I–bundle regions. We now give the final modification of the tightening procedure. Suppose that $v(S)$ is an almost normal surface vector. Suppose also that S has a transverse orientation pointing at an exceptional tightening disk.

Recall that N_S is the blocked submanifold cut from $|T|$ by the surface S (so that the transverse orientation points into N_S). Also, N_P is the I-bundle region of N_S while $N_C = \overline{N_S - N_P}$ is the core of N_S.

We now introduce the final data structures required in the proof — core(S), annuli(S), and product(S) — closely following Section 5 and Theorem 5.2.

Put a copy of the horizontal boundary of N_C in core(S). That is, record in core(S) all of the gluing information between edges of disks which are in the horizontal boundary of core blocks.

Next, place copies of all components of $\partial_v N_C$ into annuli(S) and record how they meet the surfaces contained in core(S).

Finally, define product(S) to be the list $\{2 \cdot v(N)\}$. Here N ranges over the components of N_P and $v(N)$ is the corresponding block vector found by Theorem 5.2. We record, using positions in stacks, how the components of $\partial_v N_P$ are identified with the components of annuli(S). We also record, for each component $N \subset N_P$, how each position partitions the numbers $v_k(N)$.

We now turn to constructing a sequence of surfaces H_k. Each H_k will be represented by core(H_k), annuli(H_k), and product(H_k). Here is the *H-tightening procedure.*

(1) Let core(H_0) = core(S), annuli(H_0) = annuli(S), and product(H_0) = product(S). Let $D_0 = D$. Suppose now that we are at step k of the procedure and there is a tightening disk D_k for H_k.
(2) If D_k is disjoint from annuli(H_k), and so does not meet N_P, then perform the tightening move as in the G-sequence. This effects only the pieces in core(H_k) and we use the tightening move to compute core(H_{k+1}). Set annuli(H_{k+1}) = annuli(H_k), product(H_{k+1}) = product(H_k) and go to stage $k+1$.
(3) Suppose D_k meets some component of annuli(H_k). Thus D_k meets a component of N_P; call this component N. Obtain product(H_{k+1}) by deleting the vector $2 \cdot v(N)$ from product(H_K). To get annuli(H_{k+1}) delete $\partial_v N$ from annuli(H_k). For each stack ℓ and positions p contained in ℓ: update the position of p using the partition of $v_\ell(N)$ determined by p.

    Set core$'(H_k)$ = core$(H_k) \cup \partial_v N$. Let $D'_k = D_k - N$; that is, remove a small neighborhood of T^1 from D_k. See Figure 12. Then D'_k is a surgery disk for core$'(H_k)$. So surger along D'_k, surger along all facial curves of core$'(H_k)$, and delete all bubbles in core$'(H_k)$. This finally yields core(H_{k+1}). Go on to stage $k+1$.
(4) If at stage k there is no tightening disk then set $n = k$. Sum the vectors in product(H_n) and add to this vector the number of normal disks of each type in core(H_n). Output the final sum $v(H_n)$.

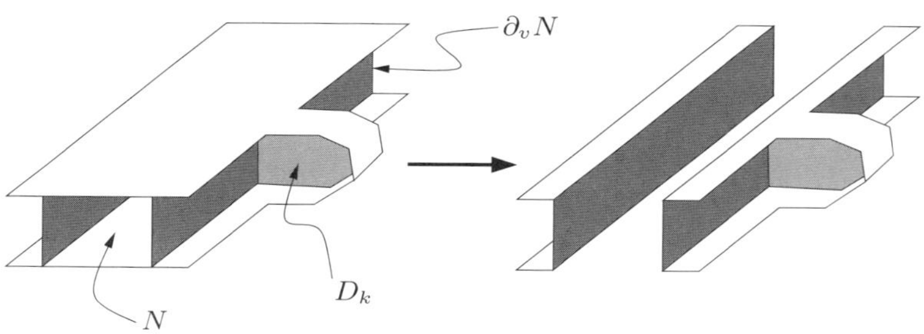

FIGURE 12. Removing the horizontal boundary of N and adding the vertical.

This completes the description of the H–tightening procedure.

12.4. Correctness and efficiency.

PROOF OF THEOREM 12.1. Note that if the transverse orientation on S points towards an exceptional surgery disk of S then the theorem is trivial. So suppose instead that a tightening disk is pointed at.

CLAIM. The H-tightening procedure outputs $v(\text{norm}(S))$.

PROOF. It suffices to show that H_n is normally isotopic to norm(S). Suppose that we are given k, ℓ so that $G_k = H_\ell$, perhaps after a normal isotopy. Let D be the given tightening disk. If D does not meet the product region then $G_{k+1} = H_{\ell+1}$ and we are done.

Suppose instead that D meets the product region. Recall that $\partial D = \alpha \cup \beta$ where $\beta \subset T^1$. The arc β is contained in $T^1 \cap \partial_v \widehat{N}_P$ while only a small neighborhood of $\partial\alpha$ (taken in α) is contained in N_P. Suppose that N is the component of N_P containing β.

We now show that we can reorder the tightening moves in the G-procedure so that there is a k' with $G_{k+k'} = H_{\ell+1}$. It then follows from Lemma 12.3 that the H procedure produces $v(\text{norm}(S))$.

Recall that $\widehat{N}$ and N are I-bundles. Let π be the bundle map crushing fibers to a point. Let $E = \pi(N)$. Let $\widehat{E} = \pi(\widehat{N})$. Note that $\widehat{E}$ is not necessarily a surface. However E is a surface with boundary, $\widehat{E}$ naturally embeds in E, and there is a small deformation retraction of E to $\widehat{E}$. Note that $\widehat{E}$ and E inherit cell structures from $\widehat{N}$ and N. Choose a spanning tree U for the one-skeleton $\widehat{E}^1$ of $\widehat{E}$ rooted at $b = \pi(\beta)$. Choose an ordering of the vertices of U, $\sigma\colon U^0 \to (\mathbb{N} \cap [1, k'])$, so that for any vertex d with parent c we have $\sigma(c) < \sigma(d)$. Here $k' = |U^0|$ is the number of vertices in U^0.

We now have a sequence of tightening moves to perform in the G procedure. At step one do the tightening move along the disk D, surger all facial curves, and delete bubbles. At step $i > 1$, examine the edge e between c and d (where $\sigma(d) = i$ and c is the parent of d). Then, by induction and the fact that $\sigma(c) < \sigma(d) = i$ there is a bent arc of G_{k+i-1} in the rectangle $\pi^{-1}(e)$ with endpoints on the segment $\pi^{-1}(d) \subset T^1$. Do this tightening move, surger facial curves, delete bubbles, and go to step $i + 1$.

After $k' = |U^0|$ steps we obtain the surface $G_{k+k'}$ which is normally isotopic to the following: $(G_k - \partial_h N) \cup \partial_v N$ surgered along the disk D', surgered along facial curves, with bubbles deleted. Here $D' = \overline{D - N}$. So $G_{k+k'}$ agrees with $H_{\ell+1}$ and the claim is proved. □

CLAIM. Precomputation for the H procedure takes time at most polynomial in size(T) and $\log(\text{weight}(S))$.

PROOF. This follows from Remark 5.1 and Theorem 5.2. □

CLAIM. The number of steps of the H procedure is at most linear in size(T).

PROOF. Each step reduces the weight of core(H_k) by two or removes a vector from product(H_k). Since the weight of core(H_k) is at most linear (Remark 5.1), and since there are at most a linear number of components of N_P (Remark 5.3), the claim follows. □

CLAIM. Performing each step of the H procedure takes time at most polynomial in size(T) and $\log(\text{weight}(S))$.

PROOF. If there is a tightening disk D_k contained in core(H_k) then it can be found in polynomial time.

If D_k is disjoint from N_P then we only have to alter core(H_k) in the tetrahedra adjacent to D_k. There are at most a linear number of such tetrahedra.

If the tightening disk meets a component $N \subset N_P$ then updating the positions in all stacks is at most a polynomial amount of arithmetic. As in the disjoint case, core$'(H_k)$ is at most linear in size. It follows that surgering facial curves and deleting bubbles can be done in polynomial time. □

Thus we can compute the desired result, $v(\text{norm}(S))$, in time which is at most a product of polynomials in size(T) and $\log(\text{weight}(S))$. This completes the proof of Theorem 12.1. □

13. Crushing: "New triangulations for old"

Crushing triangulations along normal surfaces is an important step in Casson's algorithm [**C97**]. As usual, we refer the reader to Jaco and Rubinstein's work [**JR03**]. The notion of crushing is also explained in detail in Chapter 3 of Barchechat's thesis [**B03**].

Let T be a triangulation of a closed three-manifold. Let τ_i be a model tetrahedron. Fix attention on the quad in τ partitioning the vertices $\{0,1,2,3\}$ into $\{0,a\}$ and $\{b,c\}$. We say that the quad is of *type* a.

Let θ be the permutation $(0a)(bc)$. Let $\{(i,j_s,\sigma_s)\}_{s=0}^3$ be the four face pairings with i as the first element. Here σ_s glues the sth face of τ_i to some face of τ_{j_s}. Note that $\{(j_s,i,\sigma_s^{-1})\}_{s=0}^3$ are also face pairings in T.

Define a new triangulation T' by *crushing* the tetrahedron τ_i along the ath quad, as follows: Delete τ_i from T. Delete all of the face pairings $\{(i,j_s,\sigma_s)\}_{s=0}^3$. Replace the face pairing (j_s,i,σ_s^{-1}) (if $i \neq j_s$) with

$$\left(j_s, j_{\theta(s)}, \sigma_{\theta(s)} \cdot R_{(s,\theta(s))} \cdot \sigma_s^{-1}\right),$$

for $s \in \{0,1,2,3\}$. Here $R_{(s,\theta(s))}$ is the rotation of the model tetrahedron, about the edge with vertices $\{0,1,2,3\} - \{s,\theta(s)\}$, which takes face s to face $\theta(s)$. Finally, no face of any model tetrahedron in T' is glued to itself – thus T' is a triangulation.

To keep track of this operation it may help to refer to the picture of a quad of type 3 shown on the right hand side of Figure 2.

Now suppose that p is a *polarization* of the triangulation T; that is, p is a map from the set of tetrahedra to the set $\{0,1,2,3\}$. Produce a new triangulation T' by *crushing* T along p: To begin with let T' be an exact copy of T. Now, for each $i = 1,2,\ldots,\text{size}(T)$ do one of two things; If $p(\tau_i) = 0$ simply go on to $i+1$. If $p(\tau_i) \neq 0$ then remove τ_i by crushing along the $p(\tau_i)$ quad, as above, and go on to $i+1$.

We now have:

THEOREM 13.1. *There is a polynomial-time algorithm that, given a triangulation T and a polarization p, produces T', the triangulation of T crushed along p.* □

Crushing T along the polarization determined by a non-vertex-linking normal surface S will be called *crushing T along S*.

THEOREM 13.2. *Suppose T is a triangulation so that the connect sum $\#|T|$ is a homology three-sphere. Suppose S is a non-vertex-linking normal two-sphere. Then the triangulation T', obtained by crushing T along S, satisfies $\#|T'| \cong \#|T|$.*

PROOF. Theorem 5.9 of Jaco and Rubinstein's paper [**JR03**] essentially claims this result for any closed, orientable three-manifold $|T|$ with the caveat that some connect summands of $|T|$ homeomorphic to lens spaces may by omitted from the crushed $|T'|$. See also [**B03**, Theorem 3.1].

However, by Lemma 3.6 no non-trivial lens space appears as a connect summand of the homology three-sphere $|T|$. Finally, omitting S^3 summands does not change the connect sum. The result follows. □

14. Rubinstein and Thompson's theorem

We use Casson's version [**C97**] of the proof of Theorem 1.1. Chapter 6 of [**B03**] gives a more detailed exposition of Casson's algorithm.

THEOREM 14.1 (Casson [**C97**]). *The three-sphere recognition problem lies in* **PSPACE**; *there is a polynomial-space algorithm that, given a triangulation T, decides whether or not $|T|$ is homeomorphic to the three-sphere.*

PROOF. We give only a sketch of Casson's version of the Rubinstein-Thompson algorithm. Begin with a triangulation $T_0 = T$. Check, using Theorems 3.4 and 3.5, that T_0 is a homology three-sphere. Inductively we have a triangulation T_i.

If T_i is not zero-efficient then apply Lemma 4.13 to find $S_i \subset |T_i|$, a fundamental non-vertex-linking normal two-sphere. Let T_{i+1} be the triangulation obtained by crushing along S_i. This requires Theorem 13.1.

If T_i is zero-efficient use Lemma 4.13 to search for almost normal two-spheres. If some component of T_i does not contain an almost normal two-sphere then by Theorem 13.2 and Theorem 4.12 the manifold $|T|$ was not the three-sphere. If S_i is an almost normal two-sphere inside a component T' of T_i then let $T_{i+1} = T_i - T'$.

This completes the description of Casson's algorithm. If T_n is non-empty, then $|T|$ was not the three-sphere. If T_n is empty then $|T|$ was homeomorphic to the three-sphere. Both of these again use Theorem 13.2.

Note that $\text{size}(T_i) + i \leq \text{size}(T)$ as either crushing along a polarization or deleting a component always reduces the number of tetrahedra by at least one. This completes the sketch. □

15. Showing the problem lies in NP

We are now in a position to prove:

THEOREM 15.1. *The three-sphere recognition problem lies in* **NP**.

PROOF. Suppose that T is a triangulation of the three-sphere. The certificate is a sequence of pairs $(T_i, v(S_i))$ with the following properties.

- $T = T_0$.
- S_i is a normal or almost normal two-sphere, contained in $|T_i|$, with
$$\text{weight}(S_i) \leq \exp(\text{size}(T_i)).$$
- If S_i is normal then S_i is not vertex linking and T_{i+1} is obtained from T_i by crushing along S_i.
- if S_i is almost normal then S_i normalizes to vertex linking two-spheres (or the empty set), in both directions. Also, T_{i+1} is obtained from T_i by deleting the component T' of T_i which contains S_i.
- Finally, the last triangulation T_n is empty, as is S_n.

Note that existence of the certificate is given by our proof of Theorem 14.1. The only task remaining is to check the certificate. There are two subtle points – we do not verify that the S_i are fundamental nor do we check that the T_i containing almost normal two-spheres are zero-efficient.

Instead, since the S_i are fundamental, they obey the weight bounds given in Lemma 4.8; that is, $\text{weight}(S_i) \leq \exp(\text{size}(T_i))$.

Suppose a certificate $(T_i, v(S_i))$ is given as above, for the triangulation T. Check, using Theorems 3.4 and 3.5, that T is a triangulation of a homology three-sphere.

By Lemma 3.1 check that $T = T_0$. Using Theorem 4.6 verify that S_i is a connected normal or almost normal surface. Using Lemma 4.5 compute the Euler characteristic of S_i. (Here we are using the fact that $\text{weight}(S_i) \leq \exp(\text{size}(T_i))$ in order to compute Euler characteristic in time polynomial in size(T_i).) This verifies that S_i is a two-sphere.

If S_i is normal, by Theorem 13.1, crush T_i along S_i in time at most polynomial in size(T_i). Then check, using Lemma 3.1, that T_{i+1} agrees with the triangulation obtained by crushing T_i.

If S_i is almost normal, we need to check that T', the component of T_i containing S_i, has $|T'| \cong S^3$. Using Theorem 12.1 normalize S_i in both directions in time at most polynomial in size(T_i). If all components of the two normalizations norm(S_i^+) and norm(S_i^-) are vertex linking two-spheres then T' is a triangulation of the three-sphere, by Theorem 10.3. Finally, use Lemma 3.1 to check that the triangulation $T_i - T'$ is identical to T_{i+1}. □

References

[A] I. Agol, Personal communication.

[AHT02] I. Agol, J. Hass, and W. Thurston, *3-manifold knot genus is NP-complete*, in *Proceedings of the Thirty-Fourth Annual ACM Symposium on Theory of Computing*, pages 761–766, ACM, New York, 2002. arXiv:math/0205057.

[B03] A. Barchechat, *Minimal triangulations of 3-manifolds.* PhD thesis, U.C. Davis, 2003. arXiv:math/0307302.

[Bu99] B. A. Burton, *Regina: Normal surface and 3-manifold topology software.* `http://regina.sourceforge.net/`, 1999–2009.

[C97] A. J. Casson, *The three-sphere recognition algorithm*, Lecture at MSRI, USA, 1997.

[CG87] A. J. Casson and C. McA. Gordon, *Reducing Heegaard splittings.* Topology Appl., **27** (1987), no. 3, 275–283.

[DC91] B. R. Donald and D. R. Chang, *On the complexity of computing the homology type of a triangulation.* In *32nd Annual Symposium on Foundations of Computer Science (San Juan, PR, 1991)*, pages 650–661. IEEE Comput. Soc. Press, Los Alamitos, CA, 1991. http://citeseer.ist.psu.edu/chang94computing.html.

[G87] D. Gabai, *Foliations and the topology of* 3*-manifolds. III.* J. Differential Geom., **26** (1987), no. 3, 479–536.

[GJ79] M. R. Garey and D. S. Johnson, *Computers and intractability: A guide to the theory of NP-completeness.* A Series of Books in the Mathematical Sciences, W. H. Freeman and Co., San Francisco, Calif., 1979.

[HLP99] J. Hass, J.C. Lagarias, and N. Pippenger, *The computational complexity of knot and link problems.* J. ACM, **46** (1999), no. 2, 185–211. arXiv:math/9807016.

[H01] A. Hatcher, *Notes on basic 3-manifold topology.* 2001. http://www.math.cornell.edu/~hatcher/3M/3Mdownloads.html.

[Il89] C. S. Iliopoulos, *Worst-case complexity bounds on algorithms for computing the canonical structure of finite abelian groups and the Hermite and Smith normal forms of an integer matrix.* SIAM J. Comput., **18** (1989), no. 4, 658–669.

[Iv01] S. V. Ivanov, *Recognizing the 3-sphere.* Illinois J. Math. **45** (2001), no. 4, 1073–1117. http://www.math.uiuc.edu/~hildebr/ijm/winter01/final/ivanov.html.

[Iv08] S. V. Ivanov, *The computational complexity of basic decision problems in* 3*–dimensional topology.* Geom. Dedicata, **131** (2008), 1–26.

[JO84] W. Jaco and U. Oertel, *An algorithm to decide if a* 3*-manifold is a Haken manifold*, Topology, **23** (1984), no. 2, 195–209.

[JR03] W. Jaco and J. H. Rubinstein, 0*-efficient triangulations of 3-manifolds.* J. Differential Geom., **65** (2003), no. 1, 61–168. arXiv:math/0207158.

[JT95] W. Jaco and J. L. Tollefson, *Algorithms for the complete decomposition of a closed* 3*-manifold.* Illinois J. Math., **39** (1995), no. 3, 358–406.

[K01] S. A. King, *The size of triangulations supporting a given link.* Geom. Topol., **5** (2001), 369–398. arXiv:math/0007032.

[M95] S. V. Matveev, *Algorithms for the recognition of the three-dimensional sphere (after A. Thompson).* Mat. Sb., **186** (1995), no. 5, 69–84.

[M03] S. V. Matveev, *Algorithmic topology and classification of 3-manifolds*, volume 9 of *Algorithms and Computation in Mathematics*, Springer-Verlag, Berlin, 2003.

[P94] C. H. Papadimitriou, *Computational complexity.* Addison-Wesley Publishing Company, Reading, MA, 1994.

[R97] J. H. Rubinstein, *Polyhedral minimal surfaces, Heegaard splittings and decision problems for* 3*-dimensional manifolds.* In *Geometric topology (Athens, GA, 1993)*, pages 1–20. Amer. Math. Soc., Providence, RI, 1997.

[R92] J. H. Rubinstein, *The solution to the recognition problem for* S^3. Lectures in Haifa, Israel, 1992.

[R95] J. H. Rubinstein, *An algorithm to recognize the* 3*-sphere.* In *Proceedings of the International Congress of Mathematicians, Vol. 1, 2 (Zürich, 1994)*, pages 601–611, Birkhäuser, Basel, 1995.

[S01] Saul Schleimer, *Almost normal Heegaard splittings.* PhD thesis, U.C. Berkeley, 2001. http://warwick.ac.uk/~masgar/Maths/thesis.pdf.

[T94] A. Thompson, *Thin position and the recognition problem for* S^3. Math. Res. Lett., **1** (1994), no. 5, 613–630.

Department of Mathematics, University of Warwick, Coventry, CV4 7AL, UK
E-mail address: s.schleimer@warwick.ac.uk

Proceedings of Symposia in Pure Mathematics
Volume **82**, 2011

Open problems in geometric topology

ABSTRACT. This is a report on the problem session that was held near the end of the conference on May 28, 2009, based on notes taken by Michael Usher and Dylan Thurston. The problem session was moderated by John Etnyre, Peter Kronheimer, Peter Ozsváth, and Saul Schleimer.

CONTENTS

1. Knot theory

PROBLEM 1.1 (K. Baker). *When do homotopic knots K_1 and K_2 in a given 3-manifold Y have identical-coefficient surgeries which are homeomorphic? When does it additionally hold that the dual knots K_1^* and K_2^* are homotopic?*

PROBLEM 1.2 (K. Baker). *Given a rational number p/q, does there exist an infinite family $\{K_i\}_{i=1}^{\infty}$ with the property that the p/q-surgeries $S^3_{p/q}(Y_i)$ are mutually homeomorphic, independently of i?*

Osoinach [**Os**] produced examples with $p/q = 0/1$, and Teragaito [**Te**] modified Osoinach's construction to give examples with $p/q = 4/1$ and $S^3_4(K_i)$ a Seifert fibered space; however the problem for other coefficients remains open. [**Ki**, 3.6(D)] asks whether $S^3_{p/q}(K_i)$ can be arranged to be a homology sphere.

PROBLEM 1.3 (J. Bloom). *Does Khovanov homology detect Conway mutation* [**Con**] *of knots?*

2010 *Mathematics Subject Classification.* 57Mxx, 57Q45, 57R17, 20F65.

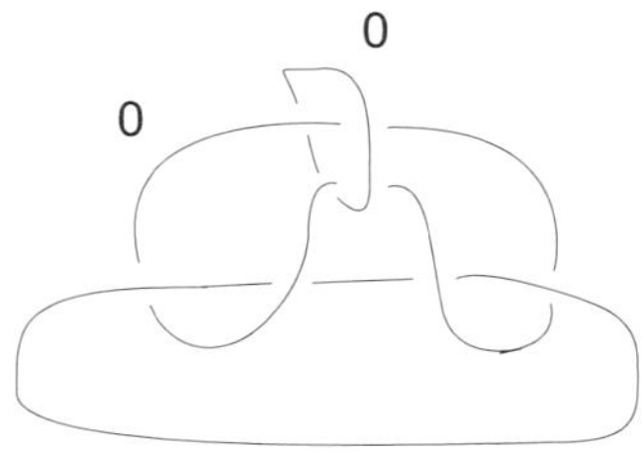

FIGURE 1. Does HFK detect this knot?

The Jones polynomial (of which Khovanov homology is a categorification) is invariant under mutation, as are the colored Jones and HOMFLYPT polynomials. Wehrli [**We1**] gave an example in which the Khovanov homology of a *link* changes under mutation. However, Bloom [**Bl**] has shown that, for knots, odd Khovanov homology is mutation-invariant, which in particular shows that (as was also proven by Wehrli [**We2**]) the Khovanov homology with $\mathbb{F}_2$-coefficients is unchanged under mutation.

PROBLEM 1.4 (D. Ruberman, following J. Cha). *Do there exist parts of classical knot theory which cannot be seen by Heegard Floer theory? Possible candidates include the Alexander* module *and higher-order signatures.*

Of course, the Alexander *polynomial* manifests itself as the graded Euler characteristic of HFK; the classical signature is closely related to the τ invariant [**OzSz1**].

PROBLEM 1.5 (P. Ozsváth). *It's known* [**OzSz3**] *that knot Floer homology detects the unknot. Does it also detect the knot in $\#^{2n} S^1 \times S^2$ given as the n-fold connect sum of the "Borromean" knot given by the surgery diagram in Figure 1? This knot is distinguished as the only fibered knot of genus n in any manifold with the fundamental group of $\#^{2n} S^1 \times S^2$, generalizing the unknot which corresponds to the case $n = 0$.*

1.1. Knot concordance.

PROBLEM 1.6 (C. Leidy–S. Harvey). *In the Cochran–Orr–Teichner filtration* [**COT**] *of the smooth knot concordance group, what is the structure of the groups $\mathcal{F}_{n.5}/\mathcal{F}_{n+1}$ ($n \in \mathbb{N}$)?*

These groups have not yet even been shown to be nontrivial. By contrast, as was discussed in S. Harvey's talk at the conference, for each n the group $\mathcal{F}_n/\mathcal{F}_{n.5}$ has been shown to contain many different subgroups isomorphic to $\mathbb{Z}^\infty$ [**CHL1**] and to $\mathbb{Z}_2^\infty$ [**CHL2**]. One would also like to know the status of certain particular types of knots in the filtration; for instance:

PROBLEM 1.7 (C. Leidy–S. Harvey). *Can 2-torsion be found in the groups $\mathcal{F}_n/\mathcal{F}_{n.5}$ by infecting (see* [**CHL2**, Section 2]*) ribbon knots by negative amphichiral knots.*

All of the torsion that has so far been found in $\mathcal{F}_n/\mathcal{F}_{n.5}$ (for $n \geq 1$) is 2-torsion arising from constructions involving infection by negative amphichiral knots. In light of this, one might ask:

PROBLEM 1.8 (J. Cha). *Is it possible to detect the fact that a knot is "infected" by the fact that it represents 2-torsion in $\mathcal{F}_n/\mathcal{F}_{n.5}$?*

PROBLEM 1.9 (C. Leidy–S. Harvey). *Is there any two-torsion in $\mathcal{F}_n/\mathcal{F}_{n.5}$ that doesn't arise from infection by negative amphichiral knots?*

PROBLEM 1.10 (C. Leidy–S. Harvey). *For $n \geq 1$, is there any k-torsion in $\mathcal{F}_n/\mathcal{F}_{n.5}$ with $k \neq 2$? In particular, bearing in mind that a result of Levine* [**Le**] *implies that $\mathcal{F}_0/\mathcal{F}_{0.5} \cong \mathbb{Z}^\infty \oplus \mathbb{Z}_2^\infty \oplus \mathbb{Z}_4^\infty$, is there 4-torsion for $n \geq 1$?*

The *rational* knot concordance group is by definition that group generated by knots in rational homology spheres under the connect sum operation, with two such considered equivalent if they are related in the obvious way by a rational homology cobordism, see [**Cha**].

PROBLEM 1.11 (J. Cha). *Understand in detail the map from the standard knot concordance group to the rational concordance group induced by inclusion.*

Not much is known about this, though [**Cha**, Theorem 1.4] finds infinite subgroups of both the kernel and the cokernel.

1.2. Higher-dimensional knot theory.

PROBLEM 1.12 (D. Ruberman). *Let $K \subset S^4$ be a 2-knot, and suppose that $\pi_1(S^4 \setminus K)$ has finitely generated commutator subgroup. Is K fibered? As a special case, if $\pi_1(S^4 \setminus K) \cong \mathbb{Z}$, is K the trivial 2-knot?*

A result of Stallings [**St**] shows that this holds for knots in S^3.

PROBLEM 1.13 (D. Ruberman). *Is every link in S^4 (or more generally in S^{2n}) slice?*

Equivalently (in light of results from [**Ke**] showing that every even-dimensional knot is slice), is the link concordant to a "boundary link" (one whose components each bound disjoint "Seifert surfaces")? In all odd dimensions, Cochran–Orr [**CoO**] (and, later and by a different method, Gilmer–Livingston [**GiL**]) found infinitely many concordance classes of links not containing any boundary links.

2. The mapping class group and other problems about groups in geometric topology

PROBLEM 2.1 (S. Schleimer). *Can Heegaard Floer homology be used to obtain information about the conjugacy problem in the mapping class group of a Riemann surface?*

Here "the conjugacy problem" refers to the problem of, given two elements in the group, determining if they are conjugate. Hemion [**He**] gave a combinatorial algorithm to solve the problem, though without any reasonable complexity bound. As discussed in her lecture at the conference, Hamenstadt [**Ham**] has shown that mapping class groups admit biautomatic structures, which implies an exponential bound on the length of a conjugating element. For pseudo-Anosov elements Masur-Minsky [**MaM2**] show that the problem can be solved in linear time.

PROBLEM 2.2 (S. Schleimer). *How much geometry does Heegaard Floer homology see? For instance, for a pseudo-Anosov mapping class ϕ, consider the mapping tori $M(\phi^n)$ of iterates of ϕ. Is there a relationship between the growth of $rk(\widehat{HF}(M(\phi^n)))$ and the growth of geometric quantities associated to $M(\phi^n)$? Geometrically, $M(\phi^n)$ converges as $n \to \infty$ to a doubly-degenerate manifold, but what happens to $HF^\bullet(M(\phi^n))$?*

PROBLEM 2.3 (T. Mrowka). *Where again $M(\phi)$ denotes the mapping torus of a mapping class ϕ, does $HF^{\bullet}(M(\phi))$ measure a geometric notion of the complexity of ϕ?*

PROBLEM 2.4 (T. Hall). *Prove or disprove the Andrews–Curtis conjecture* [**AnCu**]. *Could any new invariants help distinguish whether a given presentation yields the trivial group?*

This conjecture asserts that if $\langle x_1, \ldots, x_n | r_1, \ldots, r_n \rangle$ is a presentation of the trivial group, then this presentation can be reduced to the trivial presentation $\langle x_1, \ldots, x_n | x_1, \ldots, x_n \rangle$ by a sequence of the following four types of moves: inverting r_i; interchanging r_i with r_j; conjugating r_i by some word; and replacing r_i by $r_i r_j$. Of course, presentations correspond to 2-handlebodies, and the last move listed naturally corresponds to a handleslide. The consensus guess is that the conjecture is probably false. Certain proposed counterexamples would give interesting candidates for exotic S^4's [**GoS**, Remark 5.1.11].

3. Three-manifolds

PROBLEM 3.1 (P. Kronheimer). *Let Y be any closed 3-manifold which is not diffeomorphic to S^3. Does there always exist a nontrivial representation*

$$\rho\colon \pi_1(Y) \to SU(2)?$$

Note that Kronheimer-Mrowka's celebrated proof of Property P for knots [**KrM1**] rested on showing that the answer is affirmative when Y is obtained by +1-surgery on a knot other than the unknot (strictly speaking, that paper replaced $SU(2)$ by $SO(3)$), and indeed any Y obtained by Dehn surgery on a nontrivial knot with rational coefficient $r \in [-2, 2]$ has fundamental group admitting a representation to $SU(2)$ with non-cyclic image [**KrM2**].

PROBLEM 3.2 (P. Ozsváth). *If Y is an integer homology 3-sphere and $\widehat{HF}(Y) = \mathbb{Z}$, must Y be a connected sum of Poincaré homology spheres?*

PROBLEM 3.3 (Y. Ni). *Can we use Heegaard Floer homology to study specific Heegaard splittings of a given 3-manifold? Or, more broadly, to obtain invariants of a closed surface in a 3- or 4-manifold?*

Analogously to a construction in Khovanov homology, one could probably obtain invariants of surfaces in $\mathbb{R}^4$; however, the relevant invariants in Khovanov homology have been shown to depend only on the genus of the surface [**Ca**].

PROBLEM 3.4 (S. Schleimer). *Given a hyperbolic 3-manifold, classify its Heegaard splittings.*

This appears to be a rather hard problem, as so far a classification exists only for the exteriors of two-bridge knots [**Ko**] (and hence also for their large surgeries in light of a result of [**MoR**]). Even estimating the Heegaard genera of hyperbolic manifolds tends to be somewhat difficult, but see the survey [**So**] for some results in this direction.

PROBLEM 3.5 (S. Schleimer). *How does the Heegaard genus of a manifold with torus boundary behave under Dehn filling?*

If the manifold is hyperbolic and the surgery coefficient is large then the Heegaard genus does not change [**MoR**]. Additional results for more general 3-manifolds appear in [**RiS1**],[**RiS2**].

PROBLEM 3.6 (S. Schleimer). *Give a practical method for computing the Hempel distance associated to a given Heegaard surface.*

If $Y = H_i \cup_\Sigma H_2$ is a Heegaard splitting, the Hempel distance $d(\Sigma)$ is the minimal distance in the curve complex from a compressing disk for Σ in H_1 to a compressing disc for Σ in H_2. One can obtain bounds on the Hempel distance based on the genera of certain other surfaces in Y (*e.g.*, [**Har**], [**ScT**]), but specific computations tend to be difficult.

PROBLEM 3.7 (S. Schleimer). *Consider the "sphere complex" $\mathbb{S}_n$, whose simplices given by disjoint systems of certain spheres in $\#n(S^1 \times S^2)$ (see* [**Hat**]*). Is $\mathbb{S}_n$ δ-hyperbolic?*

$\mathbb{S}_n$ is the splitting complex of the free group F_n, and has been useful in studying the automorphism group of F_n, see [**Hat**],[**HaV**]. Analogously, the curve complex (with simplices given by disjoint systems of curves in $\#n(S^1 \times S^1)$) is δ-hyperbolic by a famous result of Masur-Minsky [**MaM1**].

PROBLEM 3.8. *Is there a categorification of the Reshitikhin–Turaev invariants of 3-manifolds?*

Such a categorification could be viewed as a version of Khovanov homology [**Kh**] for 3-manifolds. Note that Cautis–Kamnitzer [**CaKa**] have categorified the Reshitikhin-Turaev tangle invariants associated to the standard representation of $\mathfrak{sl}(m)$.

PROBLEM 3.9 (P. Ozsváth–T. Mrowka). *Find a categorification for (any or all versions* [**Fl**],[**KMOS**, Theorem 2.4],[**OzSz2**, Theorem 1.7] *of) Floer's exact triangle, or prove that no such theory can exist.*

This appears challenging in part because the appropriate cobordism maps generally commute only up to homotopy.

PROBLEM 3.10 (P. Ozsváth). *Develop methods for computing various flavors of Floer homology.*

While the Sarkar–Wang algorithm [**SaW**] computes the version $\widehat{HF}$ of Heegaard Floer homology, other variants (including HF^+, which is needed in the construction of four–manifold invariants) did not admit known algorithmic descriptions at the time of the problem session. A few months later, the preprint [**MOT**] appeared, giving algorithms for the computation of the $\mathbb{Z}/2$-versions of all the Heegaard Floer groups and the four–manifold invariants; however these algorithms are still rather inefficient. Naturally, one would also like to have additional effective methods for computing monopole or instanton Floer homologies.

PROBLEM 3.11 (R. Lipshitz). *To what extent are Floer-theoretic invariants continuous with respect to appropriate notions of convergence of spaces (*e.g.*, Gromov–Hausdorff limits of hyperbolic 3-manifolds, larger-and-larger-coefficient surgeries on a given knot in a 3-manifold, higher-and-higher order cabling...)?*

4. Four-manifolds

PROBLEM 4.1 (P. Kronheimer). *In the Barlow surface, can the Poincaré dual of the canonical class be represented by a smoothly embedded, genus two surface? More generally, in any of the other symplectic 4-manifolds that have been constructed more recently which are homeomorphic but not diffeomorphic to $\mathbb{C}P^2 \# k\overline{\mathbb{C}P^2}$* (e.g., [**AkPa**],[**PPS**]*), is the Poincaré dual of the canonical class represented by a smoothly embedded surface of genus $10-k$?*

For instance, a connected symplectic representative of the Poincaré dual of the canonical class would necessarily have the desired genus. Taubes' $SW = Gr$ equivalence [**Ta1**] provides a smoothly embedded (though not always connected) symplectic representative of the Poincaré dual of the canonical class of a symplectic four-manifold with $b^+ > 1$; however the manifolds in question have $b^+ = 1$ so the story is more complicated for them. For those small exotic manifolds which admit a complex structure (such as the Barlow surface and that in [**PPS**]), the fact that $b^+ = 1$ implies that one has $p_g = 0$, so there is no holomorphic representative of the Poincaré dual of the canonical class. Meanwhile, [**LL**, Corollary 2] shows that these manifolds admit symplectic representatives of *twice* the Poincaré dual of the canonical class in all cases.

PROBLEM 4.2 (P. Kronheimer). *Let X be, say, the $K3$ surface, and let X' be some fake (homotopy equivalent but not diffeomorphic) copy of X, with $\phi\colon X \to X'$ a homotopy equivalence. Compare $Diff_0(X)$ to $Diff_0(X')$.*

For instance, in the diagram

$$\begin{array}{ccc} Diff_0(X) & & \\ & \searrow & \\ & & Map(X,X), \\ & \nearrow_{\phi^*} & \\ Diff_0(X') & & \end{array}$$

do $Diff_0(X)$ and $Diff_0(X')$ have the same image on π_n for all n?

In a somewhat different vein, the behavior of finite subgroups of the diffeomorphism group of a homotopy $K3$ surface is quite sensitive to the smooth structure [**ChKw**].

PROBLEM 4.3 (J. Etnyre). *Given a smooth 4-manifold X and a class $A \in H_2(X;\mathbb{Z})$, let $g(A,X)$ denote the minimal genus of any smoothly embedded surface representing A. Under what circumstances can one find a smooth manifold X' homeomorphic to X and with $g(A,X') < g(A,X)$? In particular, can this ever be done with X equal to the $K3$ surface?*

Using the adjunction inequality (see, *e.g.*, [**GoS**, Theorem 2.4.8]) and standard surgery operations, it's not difficult to find examples where an exotic $K3$ surface has *larger* minimal genus function than does the $K3$ surface, but the adjunction inequality suggests that it would be difficult to decrease the minimal genus function without some new tools.

PROBLEM 4.4 (T. Mrowka). *Let X_K denote the result of knot surgery* [**FiS2**] *on the $K3$ surface using a knot K which has Alexander polynomial $\Delta_K = 1$. Is the minimal genus function $g(\cdot, X_K)$ the same as that for the $K3$ surface?*

The assumption that $\Delta_K = 1$ ensures that the Seiberg–Witten invariant of X_K is the same as that of the $K3$ surface, so the adjunction inequality cannot shed any light on this question. Relatedly, consider:

PROBLEM 4.5 (D. Auckly). *Suppose that X and X' are homeomorphic and that the minimal genus functions $g(\cdot, X)$ and $g(\cdot, X')$ coincide. Are X and X' diffeomorphic?*

PROBLEM 4.6 (D. Auckly). *Does there exist an exotic smooth structure on the 4-torus T^4?*

Part of what causes this to be a challenging problem given current techniques is that many of the surgery operations that are often used to produce exotic 4-manifolds (*e.g.*, [**FiS2**]) would, when applied to T^4, result in a change in the fundamental group. Note that for all $n \geq 5$ exotic T^n's do exist ([**HsS**],[**HsW**],[**Wa**]).

PROBLEM 4.7 (P. Kronheimer). *Given a natural number $p \geq 2$, let B_p denote the rational ball arising in Fintushel-Stern's rational blowdown construction* [**FiS1**]. *For which p does B_p embed into the quintic surface?*

PROBLEM 4.8 (J. Etnyre, following R. Fintushel-R. Stern). *If X_1 and X_2 are two homeomorphic smooth closed four-manifolds, can one be obtained from the other by a sequence of surgeries on nullhomologous tori?*

For instance, the knot surgery operation [**FiS2**] can be described as a sequence of such surgeries. Also, for every $2 \leq k \leq 8$ an there is an infinite collection of exotic $\mathbb{C}P^2\#k\overline{\mathbb{C}P^2}$'s that can be obtained by surgery on a single nullhomologous torus in a certain homotopy $\mathbb{C}P^2\#k\overline{\mathbb{C}P^2}$ (for $5 \leq k \leq 8$ this was shown in [**FiS3**], and a few months after the conference a different construction for $2 \leq k \leq 7$ was presented in [**FiS4**]).

PROBLEM 4.9 (J. Etnyre). *Suppose that X_1 and X_2 are a pair of homeomorphic smooth four-manifolds which are related by a sequence of surgeries on nullhomologous tori. Since X_1 and X_2 are homeomorphic, there is a (by definition contractible)* Akbulut cork *$W \subset X_1$ and an involution $\Phi\colon \partial W \to \partial W$ so that $X_2 = (X_1 \setminus W) \cup_\Phi W$* [**Mat**],[**CHMS**]. *Is it possible to explicitly identify W?*

5. Manifold topology in general dimensions

PROBLEM 5.1 (M. Hogancamp, following M. Hill). *Give an explicit construction of a 62-manifold with Kervaire invariant one. Then generalize this to construct a 126-manifold with Kervaire invariant one.*

An old result of Browder [**Br**] showed that the Kervaire invariant vanishes for all manifolds of dimension not of the form $2^k - 2$. There are explicit examples of Kervaire-invariant-one manifolds in dimensions 2, 6, 14, and 30 [**Jo**], while in dimension 62 the behavior of the Adams spectral sequence implies [**BJM**] that a Kervaire-invariant-one manifold must exist, but no such manifold has yet been constructed. As M. Hill discussed in his talk at the conference, recent landmark work of Hill–Hopkins–Ravenel [**HHR**] proves that the Kervaire invariant vanishes in all dimensions larger than 126, leaving 126 as the only dimension for which the problem is unresolved. Hill suggests that the constructions would likely be related to the Lie groups E_7 and E_8.

PROBLEM 5.2 (Y. Rudyak). *Let $f\colon M^n \to N^n$ be a degree-one map from one closed oriented manifold to another. Must it hold that*

$$cd(\pi_1(M)) \geq cd(\pi_1(N))?$$

Here cd denotes cohomological dimension. In the case where $cd(\pi_1(M)) = 1$ (which is to say that $\pi_1(M)$ is free) the answer is affirmative by Theorem 5.2 of [**DrRu**].

PROBLEM 5.3 (Y. Rudyak). *For a closed manifold M let $crit(M)$ denote the minimal number of critical points of a smooth function on M, and let $cat(M)$ denote the Lusternik-Schnirelmann category of M. If $crit(M) \geq crit(N)$, does it follow that $cat(M) \geq cat(N)$?*

By definition, $cat(M)$ is one less than the minimal possible size of a cover of M by contractible open subsets. Note that $crit(M) \geq cat(M) + 1$; however there are many examples where the inequality is strict.

6. Symplectic topology

PROBLEM 6.1 (Y. Rudyak). *What groups arise as the fundamental groups of closed symplectically aspherical manifolds? In particular, does there exist a group Γ with the property that, for every $n \in \mathbb{Z}_{>0}$, there is a closed symplectically aspherical manifold M^{2n} of dimension $2n$ with $\pi_1(M^{2n}) = \Gamma$?*

Recall that a symplectic manifold (M, ω) is called symplectically aspherical provided that, for every $A \in \pi_2(M)$, one has $\int_A \omega = 0$ (some conventions additionally require that $\langle c_1(TM), A\rangle = 0$ for all $A \in \pi_2(M)$).

In particular if M is closed, it can't be simply connected, since if it were the Hurewicz theorem would force ω to be exact and then Stokes' theorem would prevent ω from being nondegenerate. By passing to covers, one sees additionally that $\pi_1(M)$ cannot be finite. A variety of results and examples relating to the problem can be found in [**IKRT**] and [**KRT**]. Among finitely generated abelian groups G, [**KRT**, Theorem 1.2] shows that G is the fundamental group of a symplectically aspherical manifold iff $G = \mathbb{Z}^2$ or $rk(G) \geq 4$.

PROBLEM 6.2 (P. Kronheimer). *Does every simply-connected, non-spin symplectic 4-manifold contain a Lagrangian $\mathbb{R}P^2$?*

Of course, the fact that the normal bundle of a Lagrangian submanifold is isomorphic to its tangent bundle shows that a Lagrangian $\mathbb{R}P^2$ necessarily has $\mathbb{Z}_2$-intersection number 1, and in particular $\mathbb{R}P^2$ cannot arise as a Lagrangian submanifold of $\mathbb{R}^4$, or of any simply-connected spin manifold.

PROBLEM 6.3 (K. Wehrheim, following L. Polterovich). *Let T denote the following monotone Lagrangian torus, considered as a submanifold of $S^2 \times S^2 \subset \mathbb{R}^3 \times \mathbb{R}^3$:*

$$T = \{(\vec{v}, \vec{w}) \in S^2 \times S^2 | \vec{v} \cdot \vec{w} = -1/2,\ v_3 + w_3 = 0\}.$$

*Is T displaceable (*i.e., *is there a Hamiltonian diffeomorphism $\phi\colon S^2 \times S^2 \to S^2 \times S^2$ such that $\phi(T) \cap T = \varnothing$)?*

In the months following the conference, this question was answered **negatively** by Fukaya, Oh, Ohta, and Ono [**FOOO**, Remark 3.1]. Note that, where Δ is the diagonal, $S^2 \times S^2 \setminus \Delta$ can be identified with T^*S^2, and under this identification T corresponds to a Lagrangian submanifold of T^*S^2 which had earlier been shown [**AlFr**] to nondisplaceable. Recent work of Chekanov and Schlenk [**CheS**] constructs nondisplaceable Lagrangian "twist tori" in $(S^2)^n$, and in the case that $n = 2$ it seems likely that T is equivalent to such a twist torus, which would give another proof of its nondisplaceability. Yet another proof of the nondisplaceability of T is outlined in the recent preprint [**ElP**].

PROBLEM 6.4 (K. Wehrheim, following L. Polterovich). *Moving up a dimension from the previous question, is the monotone Lagrangian submanifold*

$$L = \{(\vec{u}, \vec{v}, \vec{w}) \in S^2 \times S^2 \times S^2 | \vec{u} + \vec{v} + \vec{w} = 0, \vec{u} \cdot \vec{v} = \vec{v} \cdot \vec{w} = -1/2\}$$

displaceable in $S^2 \times S^2 \times S^2$.

In light of recent developments, note that if L is equivalent to a twist torus [**CheS**], then it would be nondisplaceable.

PROBLEM 6.5 (K. Wehrheim). *(When) can Lagrangian submanifolds in symplectic quotients be lifted? In other words, given a Hamiltonian action of a Lie group G on a symplectic manifold (M, ω) with moment map $\mu\colon M \to \mathfrak{g}^*$, and given a Lagrangian submanifold ℓ of the symplectic reduction $\mu^{-1}(0)/G$, is there a Lagrangian submanifold $L \subset M$ which meets $\mu^{-1}(0)$ transversely and so that projection $\mu^{-1}(0) \to \mu^{-1}(0)/G$ restricts as a diffeomorphism $L \cap \mu^{-1}(0) \to \ell$? If ℓ is monotone, can L also be taken to be monotone?*

There are simple examples where the answer is no; for instance the reduction of the standard rotation action of S^1 on S^2 is a point, and setting ℓ equal to this point we note that no Lagrangian $L \subset S^2$ meets the equator $\mu^{-1}(0)$ transversely in just one point. For the standard S^1 action on $\mathbb{C}P^2$ (with quotient $\mathbb{C}P^1$) one has $\mu^{-1}(0) = S^3$ with projection given by the Hopf map, and the answer is again no. However, it's conceivable that the construction could work for the action of S^1 on a blowup of $\mathbb{C}P^2$.

More broadly, one would like to have a better general understanding of Lagrangian submanifolds of symplectic manifolds with Hamiltonian group actions.

7. Contact topology

PROBLEM 7.1 (J. Etnyre). *Given a Legendrian knot K in a tight contact 3-manifold Y, is the contact manifold resulting from Legendrian surgery on K necessarily tight?*

If Y is allowed to have boundary, a tight contact structure on the genus-four handlebody shows that the answer is no [**Ho**, Theorem 4.1]. However, the closed case remains unresolved. Note that a number of important contact topological properties of closed 3-manifolds are preserved by Legendrian surgery, such as weak [**EtH2**], strong [**Wei**], and Stein [**El2**] fillability, and nonvanishing of the Ozsváth-Szabó contact invariant [**LS1**].

PROBLEM 7.2 (J. Etnyre). *If L_1 and L_2 are two Legendrian knots in S^3 which are* not *Legendrian isotopic, can the respective Legendrian surgeries on them be contactomorphic?*

For any n, there is a tight contact manifold (M_n, ξ_n) containing distinct Legendrian knots $L_1, \ldots, L_n$ so that Legendrian surgery on the L_i produce the same contact manifold [**Et**, Corollary 2], but as yet there are no examples with $M_n = S^3$.

PROBLEM 7.3 (J. Etnyre). *Which closed 3-manifolds admit tight contact structures? In particular, do all hyperbolic 3-manifolds admit tight contact structures?*

Etnyre-Honda [**EtH1**] showed some years ago that the Poincaré homology sphere Σ admits no tight contact structures compatible with its nonstandard orientation (and hence that the connect sum $\Sigma \# \overline{\Sigma}$ admits no tight contact structures at all). More recently Lisca-Stipsicz [**LS2**] determined precisely which Seifert fibered spaces admit tight contact structures. In the class of hyperbolic manifolds, not much is known beyond some isolated examples (for instance the Weeks manifold admits a tight contact structure).

PROBLEM 7.4 (J. Etnyre). *Which odd-dimensional manifolds admit contact strucutures?*

The fact that every 3-manifold admits a contact structure goes back to Martinet [**Mar**]. In higher dimensions, at least if the contact structure is to be cooriented, there is a topological obstruction arising from the fact that, if the manifold has dimension $2n+1$, the structure group needs to reduce to $U(n)$ (such a reduction is called an *almost contact structure*). In dimensions 5 and 7, this translates to the requirement that the second Stiefel-Whitney class should admit an integral lift. Geiges (see [**Ge**, Chapter 8]) has shown that any almost contact structure on an oriented simply connected 5-manifold arises from a contact structure, thus reducing the existence question on simply-connected 5-manifolds to characteristic classes. For more results in dimension 5 and 7 see [**GeTh**],[**GeSt**]. In dimensions above 7 very little is known; the existence of a contact structure on T^{2n+1} for every n was only established in 2002 [**Bo1**].

PROBLEM 7.5 (J. Etnyre). *Understand the space of contact structures on a given manifold.*

Eliashberg [**El3**, Theorem 2.4.2] showed that the space of tight contact structures on S^3 which are fixed at a given point is contractible. On the other hand, Geiges-Gonzalo [**GeGo**] found, for each member of the standard sequence ξ_n of tight contact structures on T^3, an element of infinite order in the fundamental group of the space of contact structures based at ξ_n. Infinite subgroups of some other homotopy groups of spaces of contact structures were subsequently found by Bourgeois [**Bo2**]. Ding-Geiges [**DiGe**] have recently shown that the fundamental group of the space of contact structures on $S^1 \times S^2$ (based at the standard tight one) is $\mathbb{Z}$.

With respect to overtwisted contact structures, Eliashberg [**El1**] showed that, given an overtwisted disk Δ in a contact 3-manifold (M, ξ), the space of overtwisted contact structures on M coinciding with ξ near Δ is homotopy equivalent to the space of 2-plane fields coinciding with ξ near Δ. Thus, up to homotopy, understanding the space of overtwisted contact structures on a given manifold is essentially a classical (albeit nontrivial) matter.

PROBLEM 7.6 (L. Ng). *Formulate "embedded sutured contact homology" for Legendrian knots.*

Recall here that the complement of a Legendrian knot has a standard description as a sutured manifold, and so has an associated sutured (Heegaard) Floer homology which is isomorphic to its knot Floer homology [**Ju**]. Meanwhile, embedded contact homology has recently been proven to be isomorphic to monopole Floer homology [**Ta2**]. Thus the putative embedded sutured contact homology should be isomorphic to knot Floer homology (or at any rate the monopole version thereof [**KrM3**]) and may lead to some interesting links between the contact homology world and the Heegaard Floer world.

Legendrian knots do have a Legendrian contact homology [**Che**] constructed in the spirit of symplectic field theory; however, this invariant vanishes for stabilized Legendrian knots, in contrast to the Legendrian knot invariants constructed from Heegaard Floer theory as in [**OzSzT**].

Since the conference, a version of sutured embedded contact homology has been defined [**CGHH**], though certain foundational questions, such as independence of the choice of auxiliary data, remain unresolved.

References

[AkPa] A. Akhmedov and B. D. Park, *Exotic smooth structures on small 4-manifolds*. Invent. Math. **173** (2008), no. 1, 209–223.

[AlFr] P. Albers and U. Frauenfelder, *A nondisplaceable Lagrangian torus in T^*S^2*. Comm. Pure Appl. Math. **61** (2007), no. 8, 1046–1051.

[AnCu] J. Andrews and M. Curtis, *Free groups and handlebodies*. Proc. Amer. Math. Soc. **16** (1965), no. 2, 192–195.

[BJM] M. Barratt, J. Jones, and M. Mahowald, *Relations amongst Toda brackets and the Kervaire invariant in dimension 62*. J. London Math. Soc. **30** (1984), no. 3, 533–550.

[Bl] J. Bloom. *Odd Khovanov homology is mutation invariant*, Math. Res. Lett. **17** (2010), no. 1, 1–10.

[Bo1] F. Bourgeois. *Odd dimensional tori are contact manifolds*, Int. Math. Res. Notices **2002** (2002), 1571–1574.

[Bo2] F. Bourgeois. *Contact homology and homotopy groups of the space of contact structures*, Math. Res. Lett. **13** (2006), no. 1, 71-85

[Br] W. Browder. *The Kervaire invariant of framed manifolds and its generalization*, Ann. Math. (2) **90** 157-186.

[Ca] C. Caprau. *The universal $\mathfrak{sl}(2,\mathbb{C})$ cohomology via webs and foams*, Topology Appl. **156** (2009), no. 9, 1684–1702.

[CaKa] S. Cautis and J. Kamnitzer, *Knot homology via derived categories of coherent sheaves. II. $\mathfrak{sl}(m)$ case*. Invent. Math. **174** (2008), no. 1, 165-232.

[Cha] J. Cha, *The structure of the rational concordance group of knots* Mem. Amer. Math. Soc. **189**, AMS, Providence, 2007.

[Che] Yu. Chekanov, *Differential algebra of Legendrian links*. Invent. Math. **150** (2002), no. 3, 441483.

[CheS] Yu. Chekanov and F. Schlenk, *Notes on monotone Lagrangian twist tori*. arXiv:1003.5960.

[ChKw] W. Chen and S. Kwasik, *Symmetries and exotic smooth structures on a K3 surface*. J. Topology. **1** (2008) 923962.

[CHL1] T. Cochran, S. Harvey, and C. Leidy, *Knot concordance and higher-order Blanchfield duality*. Geom. Topol. **13** (2009), 1419–1482.

[CHL2] T. Cochran, S. Harvey, and C. Leidy, *2-torsion in the n-solvable filtration of the knot concordance group*. arXiv:0907.4789.

[CoO] T. Cochran and K. Orr, *Not all links are concordant to boundary links*. Bull. Amer. Math. Soc. (N.S.) **23** (1990), no. 1, 99–106.

[COT] T. Cochran, K. Orr, and P. Teichner, *Knot concordance, Whitney towers, and L^2-signatures*. Ann. Math. (2) **157** (2003), no. 2, 433–519.

[CGHH] V. Colin, P. Ghiggini, K. Honda, and M. Hutchings, *Sutures and contact homology I*. arXiv:1004.2942.

[Con] J. Conway, *An enumeration of knots and links, and some of their algebraic properties*. In *Computational Problems in Abstract Algebra (Proc. Conf., Oxford, 1967)*. Pergamon, Oxford, 1970, 329-358.

[CHMS] C. Curtis, M. Freedman. W.-C. Hsiang, and R. Stong, *A decomposition theorem for h-cobordant smooth simply-connected compact 4-manifolds*. Invent. Math. **123** (1996), no. 2, 343348

[DiGe] F. Ding and H. Geiges, *The diffeotopy group of* $S^1 \times S^2$ *via contact topology*. arXiv:0903.1488.

[DrRu] A. Dranishnikov and Y. Rudyak, *On the BersteinSvarc theorem in dimension 2*. Math. Proc. Camb. Phil. Soc. **146** (2009), 407–413.

[El1] Y. Eliashberg, *Classification of overtwisted contact structures on 3-manifolds*. Invent. Math. **98** (1989), no. 3, 623–637.

[El2] Y. Eliashberg, *Topological characterization of Stein manifolds of dimension* > 2. Internat. J. Math. **1** (1990), 2946

[El3] Y. Eliashberg, *Contact 3-manifolds twenty years since J. Martinets work*. Ann. Inst. Fourier. **42** (1992), 165–192.

[ElP] Y. Eliashberg and L. Polterovich, *Symplectic quasi-states on the quadric surface and Lagrangian submanifolds*, arXiv:1006.2501.

[EnP] M. Entov and L. Polterovich, *Rigid subsets of symplectic manifolds*. Compositio Math. **145** (2009), 773–826.

[Et] J. Etnyre, *On contact surgery*. Proc. Amer. Math. Soc. **136** (2008), no. 9, 3355–3362.

[EtH1] J. Etnyre and K. Honda, *On the nonexistence of tight contact structures*. Ann. Math. (2) **153** (2001), 749–766.

[EtH2] J. Etnyre and K. Honda, *Tight contact structures with no symplectic fillings*. Invent. Math. **148** (2002), 609–626.

[FiS1] R. Fintushel and R. Stern, *Rational blowdowns of smooth 4-manifolds*. J. Differential Geom. **46** (1997), no. 2, 181–235.

[FiS2] R. Fintushel and R. Stern, *Knots, links, and 4-manifolds*. Invent. Math. *134* (1998), no. 2, 363–400.

[FiS3] R. Fintushel and R. Stern, *Surgery on nullhomologous tori and simply connected 4-manifolds with* $b^+ = 1$. J. Topology. **1** (2008), 1–15.

[FiS4] R. Fintushel and R. Stern, *Pinwheels and nullhomologous surgery on 4-manifolds with* $b^+ = 1$. arXiv:1004.3049.

[Fl] A. Floer, *Instanton homology, surgery, and knots*. In *Geometry of low-dimensional manifolds, 1 (Durham, 1989)*. London Math. Soc. Lecture Note Ser., **150**, Cambridge Univ. Press, Cambridge, 1990, 97–114.

[FOOO] K. Fukaya, Y.-G. Oh, H. Ohta, and K. Ono, *Toric degeneration and non-displaceable Lagrangian tori in* $S^2 \times S^2$. arXiv:1002.1660.

[Ge] H. Geiges, *An Introduction to Contact Topology*. Cambridge Univ. Press, Cambridge, 2008.

[GeGo] H. Geiges and J. Gonzalo, *On the topology of the space of contact structures on torus bundles*. Bull. London Math. Soc **36** (2004), 640–646.

[GeSt] H. Geiges and A. Stipsicz, *Contact structures on product five-manifolds and fibre sums along circles*. Math. Ann. 348 (2010), no. 1, 195–210.

[GeTh] H. Geiges and C. Thomas, *Contact structures on 7-manifolds*. In *Geometry, topology, and dynamics*. CRM Proc. Lecture Notes **15**, AMS, Providence, 1998, 53–67.

[GiL] P. Gilmer and C. Livingston, *The Casson-Gordon invariant and link concordance*. Topology **31** (1992), no. 3, 475-492.

[GoS] R. Gompf and A. Stipsicz, *4-manifolds and Kirby calculus*. AMS, Providence, 1999.

[Ham] U. Hamenstädt, *Geometry of the mapping class group, II: a biautomatic structure*. arXiv:0912.0137.

[Har] K. Hartshorn, *Heegaard splittings of Haken manifolds have bounded distance*. Pacific J. Math. **204** (2002), no. 1, 61–75

[Hat] A. Hatcher, *Homological stability for automorphism groups of free groups*. Comment. Math. Helv. **70** (1995), 39-62.

[HaV] A. Hatcher and K. Vogtmann, *Isoperimetric inequalities for automorphism groups of free groups*. Pacific J. Math. **173** (1996), 425-441.

[He] G. Hemion, *On the classification of homeomorphisms of 2-manifolds and the classification of 3-manifolds*. Acta Math. **142** (1979), 123-155.

[HHR] M. Hill, M. Hopkins, and D. Ravenel, *On the non-existence of elements of Kervaire invariant one.* arXiv:0908.3724.

[Ho] K. Honda, *Gluing tight contact structures.* Duke Math. J. **115** (2002), no. 3, 435–478.

[HsS] W.-C. Hsiang and J. Shaneson, *Fake tori.* In *Topology of Manifolds. (Proc. Inst., Univ. of Georgia, Athens, Ga., 1969)*, edited by J. Cantrell, 18–51.

[HsW] W.-C. Hsiang and C.T.C. Wall, *On homotopy tori. II.* Bull. London. Math. Soc. **1** (1969), 341–342.

[IKRT] R. Ibañez, J. Kędra, Yu. Rudyak, and A. Tralle, *On fundamental groups of symplectically aspherical manifolds, II.* Math. Z. **256** (2007), no. 4, 825–835.

[Jo] J. Jones, *The Kervaire invariant of extended power manifolds.* Topology **17** (1978), no. 3, 249–266.

[Ju] A. Juhasz, *Holomorphic discs and sutured manifolds.* Algebr. Geom. Topol. **6** (2006), 1429–1457.

[KRT] J. Kędra, Yu. Rudyak, and A. Tralle, *On fundamental groups of symplectically aspherical manifolds.* Math. Z. **248** (2004), no. 4, 805–826.

[Ke] M. Kervaire, *Les noeuds de dimensions supérieures.* Bull. Soc. Math. France **93** (1965), 225-271

[Kh] M. Khovanov, *A categorification of the Jones polynomial.* Duke Math. J. **101** (2000), no. 3, 359-426.

[Ki] *Problems in low-dimensional topology*, Edited by Rob Kirby. AMS/IP Stud. Adv. Math., 2.2, Geometric topology (Athens, GA, 1993), 35–473, AMS, Providence, 1997.

[Ko] T. Kobayashi, *Heegaard splittings of exteriors of two bridge knots.* Geom. Topol. **5** (2001), 609–650.

[KrM1] P. Kronheimer and T. Mrokwa, *Witten's conjecture and property P.* Geom. Topol. **8**, (2004), 295-310.

[KrM2] P. Kronheimer and T. Mrowka, *Dehn surgery, the fundamental group and* $SU(2)$. Math. Res. Lett. **11** (2004), 741–754.

[KrM3] P. Kronheimer and T. Mrowka, *Knots, sutures and excision.* J. Differential Geom. **84**, no. 2 (2010), 301–364.

[KMOS] P. Kronheimer, T. Mrowka, P. Ozsváth, and Z. Szabó, *Monopoles and lens space surgeries.* Ann. Math. (2) **165** (2007), no. 2, 457–546.

[Le] J. Levine, *Invariants of knot cobordism.* Invent. Math. **8** (1969), 98–110; addendum, ibid., **8** (1969), 355.

[LL] T. J. Li and A. K. Liu, *Uniqueness of symplectic canonical class, surface cone and symplectic cone of* 4*-manifolds with* $b^+ = 1$. J. Differential Geom. **58** (2001), 331–370.

[LS1] P. Lisca and A. Stipsicz, *Seifert fibered contact threemanifolds via surgery.* Algebr. Geom. Top. **4** (2004), 199-217.

[LS2] P. Lisca and A. Stipsicz, *On the existence of tight contact structures on Seifert fibered 3-manifolds.* Duke Math. J. **148** (2009), no. 2, 175–209.

[MOT] C. Manolescu, P. Ozsváth, and D. Thurston, *Grid diagrams and Heegaard Floer invariants*, arXiv:0910.0078.

[Mar] J. Martinet, *Formes de contact sur les variétés de dimension 3.* Lect. Notes. Math. **209** (1971), 142–163.

[MaM1] H. Masur and Y. Minsky, *Geometry of the complex of curves I: hyperbolicity.* Invent. Math. **138** (1999), 103–149.

[MaM2] H. Masur and Y. Minsky, *Geometry of the complex of curves II: hierarchical structure.* GAFA **10** (2000), 902–974.

[Mat] R. Matveyev, *A decomposition of smooth simply-connected h-cobordant 4-manifolds.* J. Differential Geom. 44 (1996), no. 3, 571-582.

[MoR] Y. Moriah and H. Rubinstein, *Heegaard structures of negatively curved 3-manifolds.* Comm. Anal. Geom. **5** (1997), no. 3, 375–412.

[Os] J. Osoinach, Jr., *Manifolds obtained by surgery on an infinite number of knots in* S^3. Topology **45** (2006), no. 4, 725–733.

[OzSz1] P. Ozsváth and Z. Szabó, *Knot Floer homology and the four-ball genus.* Geom. Topol. **7** (2003) 615–639.

[OzSz2] P. Ozsváth and Z. Szabó, *Holomorphic disks and three-manifold invariants: Properties and applications.* Ann. Math. (2) **159** (2004), 1159-1245.

[OzSz3] P. Ozsváth and Z. Szabó, *Holomorphic disks and genus bounds.* Geom. Topol. **8** (2004) 311–334.

[OzSzT] P. Ozsváth, Z. Szabó, and D. Thurston, *Legendrian knots, transverse knots and combinatorial Floer homology.* Geom. Topol. **12** (2008), no. 2, 941–980.

[PPS] H. Park, J. Park, and D. Shin, *A simply connected surface of general type with* $p_g = 0$ *and* $K^2 = 4$. Geom. Topol. **13** (2009), no. 3, 1483–1494.

[RiS1] Y. Rieck and E. Sedgwick, *Persistence of Heegaard structures under Dehn filling.* Topology Appl. **109** (2001), no. 1, 41–53.

[RiS2] Y. Rieck and E. Sedgwick, *Finiteness results for Heegaard surfaces in surgered manifolds.* Comm. Anal. Geom. **9** (2001), no. 2, 351–367.

[SaW] S. Sarkar and J. Wang, *An algorithm for computing some Heegaard Floer homologies.* Ann. Math. (2) **171** (2010), no. 2, 1213–1236.

[ScT] M. Scharlemann and M. Tomova, *Alternate Heegaard genus bounds distance.* Geom. Topol. **10** (2006), 593–617.

[So] J. Souto, *Geometry, Heegaard splittings and rank of the fundamental group of hyperbolic 3-manifolds.* In *Workshop on Heegaard splittings (Technion, Summer 2005).* Geom. Topol. Monographs **12** (2007), 351–399.

[St] J. Stallings, *On fibering certain* 3*-manifolds.* In *Topology of* 3*-manifolds and related topics (Proc. The Univ. of Georgia Institute, 1961).* Prentice-Hall, Englewood Cliffs, N.J., 95–100.

[Ta1] C. Taubes, *The Seiberg–Witten and Gromov invariants.* Math. Res Lett. **2** (1995), 221–238.

[Ta2] C. Taubes, *Embedded contact homology and Seiberg-Witten Floer cohomology,* I,II,III,IV,V. Geom. Topol. **14** (2010), 2497–3000.

[Te] M. Teragaito, *A Seifert Fibered Manifold with Infinitely Many Knot-Surgery Descriptions.* Int. Math. Res. Notices. **2007** (2007), Article ID rnm028, 1–16.

[Wa] C. T. C. Wall, *On homotopy tori and the annulus theorem.* Bull. London Math. Soc. **1** (1969), 95-97.

[We1] S. Wehrli, *Khovanov homology and Conway mutation.* arXiv:math/0301312.

[We2] S. Wehrli, *Mutation invariance of Khovanov homology over* $\mathbb{F}_2$. arXiv:0904.3401

[Wei] A. Weinstein, *Contact surgery and symplectic handlebodies.* Hokkaido Math. J. **20** (1991), 241-251.

Titles in This Series

83 **Hisham Sati and Urs Schreiber, Editors,** Mathematical Foundations of Quantum Field Theory and Perturbative String Theory

82 **Michael Usher, Editor,** Low-Dimensional and Symplectic Topology (University of Georgia, Athens, Georgia, 2009)

81 **Robert S. Doran, Greg Friedman, and Jonathan M. Rosenberg, Editors,** Superstrings, geometry, topology, and C*-algebras

80.2 **D. Abramovich, A. Bertram, L. Katzarkov, R. Pandharipande, and M. Thaddeus, Editors,** Algebraic Geometry (Seattle, 2005)

80.1 **D. Abramovich, A. Bertram, L. Katzarkov, R. Pandharipande, and M. Thaddeus, Editors,** Algebraic Geometry

79 **Dorina Mitrea and Marius Mitrea, Editors,** Perspectives in Partial Differential Equations, Harmonic Analysis and Applications: A Volume in Honor of Vladimir G. Maz'ya's 70th Birthday

78 **Ron Y. Donagi and Katrin Wendland, Editors,** From Hodge Theory to Integrability and TQFT

77 **Pavel Exner, Jonathan P. Keating, Peter Kuchment, Toshikazu Sunada, and Alexander Teplyaev, Editors,** Analysis on graphs and its applications

76 **Fritz Gesztesy (Managing editor), Percy Deift, Cherie Galvez, Peter Perry, and Wilhelm Schlag, Editors,** Spectral theory and mathematical physics: A Festschrift in honor of Barry Simon's 60th birthday, Parts 1 and 2 (California Institue of Technology, Pasadena, CA, March 27–31, 2006)

75 **Solomon Friedberg (Managing editor), Daniel Bump, Dorian Goldfeld, and Jeffrey Hoffstein, Editors,** Multiple Dirichlet series, automorphic forms, and analytic number theory (Bretton Woods, New Hampshire, July 11–14, 2005)

74 **Benson Farb, Editor,** Problems on mapping class groups and related topics, 2006

73 **Mikhail Lyubich and Leon Takhtajan, Editors,** Graphs and patterns in mathematics and theoretical physics (Stony Brook University, Stony Brook, NY, June 14–21, 2001)

72 **Michel L. Lapidus and Machiel van Frankenhuijsen, Editors,** Fractal geometry and applications: A jubilee of Benoît Mandelbrot, Parts 1 and 2 (San Diego, California, 2002 and École Normale Supérieure de Lyon, 2001)

71 **Gordana Matić and Clint McCrory, Editors,** Topology and Geometry of Manifolds (University of Georgia, Athens, Georgia, 2001)

70 **Michael D. Fried and Yasutaka Ihara, Editors,** Arithmetic fundamental groups and noncommutative algebra (Mathematical Sciences Research Institute, Berkeley, California, 1999)

69 **Anatole Katok, Rafael de la Llave, Yakov Pesin, and Howard Weiss, Editors,** Smooth ergodic theory and its applications (University of Washington, Seattle, 1999)

68 **Robert S. Doran and V. S. Varadarajan, Editors,** The mathematical legacy of Harish-Chandra: A celebration of representation theory and harmonic analysis (Baltimore, Maryland, 1998)

67 **Wayne Raskind and Charles Weibel, Editors,** Algebraic K-theory (University of Washington, Seattle, 1997)

66 **Robert S. Doran, Ze-Li Dou, and George T. Gilbert, Editors,** Automorphic forms, automorphic representations, and arithmetic (Texas Christian University, Fort Worth, 1996)

65 **M. Giaquinta, J. Shatah, and S. R. S. Varadhan, Editors,** Differential equations: La Pietra 1996 (Villa La Pietra, Florence, Italy, 1996)

64 **G. Ferreyra, R. Gardner, H. Hermes, and H. Sussmann, Editors,** Differential geometry and control (University of Colorado, Boulder, 1997)

63 **Alejandro Adem, Jon Carlson, Stewart Priddy, and Peter Webb, Editors,** Group representations: Cohomology, group actions and topology (University of Washington, Seattle, 1996)

62 **János Kollár, Robert Lazarsfeld, and David R. Morrison, Editors,** Algebraic geometry—Santa Cruz 1995 (University of California, Santa Cruz, July 1995)

ISBN 978-0-8218-5235-4
9 780821 852354
PSPUM/82